Knowledge Intensive Business Services and Regional Competitiveness

Research interest in the service sector has boomed in recent years as deindustrialisation became entrenched. Instead of being regarded as merely supplementary to traditional industry and manufacturing, services have generated progressively rising levels of growth in developed economies while at the same time coming to be recognised as major drivers of innovation.

Among the factors that have helped service companies notch up swifter growth rates than all other sectors are the outsourcing of such services by other sectors, including the development of information and communication technologies, and changes to the regulatory, legal and market frameworks as well as globalisation and internationalisation. The result is a cluster of highly innovative firms that can loosely be grouped under the heading of Knowledge Intensive Business Services (KIBS). *Knowledge Intensive Business Services and Regional Competitiveness* charts the development of these firms and explores their success through four mutually linked parts: KIBS and industrial dynamics; KIBS and their context; KIBS and their contribution to regional competitiveness and economic development; and finally, KIBS and public policy.

This book is suitable for researchers and policy makers interested in the rise of these influential actors and their influence on regional competitiveness.

João J. M. Ferreira is Associate Professor in Management at the University of Beira Interior (UBI), Portugal.

Mário L. Raposo is Professor of Business and Economics, University of Beira Interior (UBI), Portugal.

Cristina I. Fernandes is a Professor at the Polytechnic Institute of Castelo Branco, Portugal.

Marcus Dejardin is Professor of Economics at the University of Namur, Belgium.

Routledge advances in regional economics, science and policy

1 Territorial Patterns of Innovation
An inquiry on the knowledge economy in European regions
Edited by Roberta Capello and Camilla Lenzi

2 The Political Economy of City Branding
Ari-Veikko Anttiroiko

3 Structural Change, Competitiveness and Industrial Policy
Painful lessons from the European periphery
Edited By Aurora A.C. Teixeira, Ester Silva and Ricardo Mamede

4 Airports, Cities and Regions
Edited by Sven Conventz and Alain Thierstein

5 Rural and Regional Futures
Edited by Anthony Hogan and Michelle Young

6 Untamed Urbanisms
Edited by Adriana Allen, Andrea Lampis and Mark Swilling

7 Explorations in Urban and Regional Dynamics
A case study in complexity science
Joel Dearden and Alan Wilson

8 Quality of Life in Cities
Equity, sustainable development and happiness from a policy perspective
Edited by Alessandra Michelangeli

9 Spatial Dynamics in the Experience Economy
Edited by Anne Lorentzen, Karin Topsø Larsen and Lise Schrøder

10 Chicago
An economic history
John F. McDonald

11 Broadband Telecommunications and Regional Development
Tony H. Grubesic and Elizabeth A. Mack

12 Knowledge Economy in the Megalopolis
Interactions of innovations in transport, information, production and organizations
T.R. Lakshmanan, William P. Anderson and Yena Song

13 Regional Disparities in the Enlarged European Union
Geography, innovation and structural change
Valentina Meliciani

14 Tourist Clusters, Destinations and Competitiveness
Theoretical issues and empirical evidences
Edited by Francesco Capone

15 Regional Development in Australia
Being regional
Robyn Eversole

16 Knowledge Intensive Business Services and Regional Competitiveness
Edited by João J. M. Ferreira, Mário L. Raposo, Cristina I. Fernandes and Marcus Dejardin

Knowledge Intensive Business Services and Regional Competitiveness

Edited by João J. M. Ferreira,
Mário L. Raposo,
Cristina I. Fernandes and
Marcus Dejardin

Routledge
Taylor & Francis Group
LONDON AND NEW YORK

First published 2016
by Routledge
2 Park Square, Milton Park, Abingdon, Oxon OX14 4RN

and by Routledge
711 Third Avenue, New York, NY 10017

Routledge is an imprint of the Taylor & Francis Group, an informa business

British Library Cataloguing in Publication Data
A catalogue record for this book is available from the British Library

Library of Congress Cataloging in Publication Data
A catalog record for this book has been requested

ISBN: 978-1-138-85936-4 (hbk)
ISBN: 978-1-315-71730-2 (ebk)

Typeset in Times New Roman
by diacriTech, Chennai

Contents

Figures

Tables

Contributors

Joaquín Alcazar, Polytechnic University of Valencia, Spain.

Anna-Leena Asikainen, Luxembourg Institute of Science and Technology.

Jadwiga Berbeka, Faculty of Management, Cracow University of Economics, Poland.

Krzysztof Borodako, Faculty of Management, Cracow University of Economics, Poland.

Alexandra M. Braga, CIICESI Research Unit, School of Technology and Management of Felgueiras – Polytechnic Institute of Porto, Felgueiras, Portugal, and CETRAD Research Unit, University of Tras-os-Montes e Alto Douro (UTAD), Vila Real, Portugal.

John R. Bryson, City-Region Economic Development Institute (City-REDI), Birmingham Business School, University of Birmingham, UK.

Antonella Caru, University of Bocconi, Milan, Italy.

Peter W. Daniels, School of Geography, Earth and Environmental Sciences, University of Birmingham, UK.

Marcus Dejardin, University of Namur and Université catholique de Louvain, Belgium.

Benoit Desmarchelier, Xi'an Jiaotong-Liverpool University, China.

Faridadh Djellal, Lille 1 University, France.

David Doloreux, HEC Montréal, Department of International Business, Montréal, Québec, Canada.

Cristina I. Fernandes, Polytechnic Institute of Castelo Branco and NECE Research Unit-UBI, Portugal.

João J. M. Ferreira, Department of Business and Economics and NECE Research Unit, University of Beira Interior, Portugal.

Daniel Feser, Faculty of Economic Sciences, Chair of Economic Policy and SME Research, University of Goettingen, Germany.

Faiz Gallouj, Lille 1 University, France.

Hung-Nien Hsieh, Department of Architecture and Urban Planning, Chung-Hua University, Hsinchu, Taiwan.

Tai-Shan Hu, Department of Architecture and Urban Planning, Chung-Hua University, Hsinchu, Taiwan.

Tatiana A. Iakovleva, Associate Professor, University of Stavanger, Norway.

Viroj Jienwatcharamongkhol, Department of Economics, Lund University, Sweden.

Chien-Yuan Lin, Graduate Institute of Building and Planning, National Taiwan University, Taipei, Taiwan.

Giovanni Mangiarotti, Luxembourg Institute of Science and Technology.

Carla S. Marques, CETRAD Research Unit, University of Tras-os-Montes e Alto Douro (UTAD), Vila Real, Portugal.

Alicia Mas-Tur, University of Valencia, Spain.

Kjersti Vikse Meland, Senior Scientist, Stiftelsen Polytec, Regional Research Institute, Norway.

Till Proeger, Faculty of Economic Sciences, Chair of Economic Policy and SME Research, University of Goettingen, Germany.

Mário L. Raposo, Department of Business and Economics and NECE Research Unit, University of Beira Interior, Portugal.

Vanessa Ratten, Associate Professor, Department of Management and Marketing, La Trobe Business School, La Trobe University, Australia.

Belén Ribeiro-Navarrete, ‘Entrepreneurship: Student to entrepreneur’ Chair, Grupo Maicerías Españolas-Arroz DACSA, University of Valencia, Spain.

Norat Roig-Tierno, Polytechnic University of Valencia, Spain.

Michał Rudnicki, Faculty of Management, Cracow University of Economics, Poland.

Richard Shearmur, School of Urban Planning, University of McGill, Montreal, Canada.

Sam Tavassoli, Centre for Innovation, Research and Competence in the Learning Economy (CIRCLE), Lund University, Lund, Sweden, and Department of Industrial Economics, Blekinge Institute of Technology, Karlskrona, Sweden.

Marja Toivonen, VTT Technical Research Centre, Finland.

Knowledge Intensive Business Services and regional competitiveness

An overview

João J. M. Ferreira, Mário L. Raposo, Cristina I. Fernandes and Marcus Dejardin

KIBS and regional competitiveness: an overview

Research interest in the service sector has been rising ever since 1980, when regional development studies in Europe and North American began to focus on deindustrialisation related issues (Kline and Rosenberg, 1986; Von Hippel, 1988; Johne and Storey, 1998; Miles, 2000). Thus far, services had been perceived as merely subsidiary to transformative and manufacturing activities. In the last two decades, attention has mounted especially on service activities given that they have not only generated progressively rising levels of development but have also been identified as major drivers of innovation (Wood, 2005).

Despite the growing awareness that innovation is not simply confined to technical processes and products, some recent innovation research has centred only on the observation of technical innovation and especially in the transformative industrial sector (Becker and Dietz, 2004; Huergo and Jaumandreu, 2004; Lynskey, 2004). Indeed, greater importance has only more recently been given to the service sector (Gallouj and Weinstein, 1997; Tether, 2003).

Services were seen at the time as moderately contributing to technical progress and productivity increases, in contrast with transformative industrial companies. Nevertheless, rapid growth in the Knowledge Intensive Business Service (KIBS) sector has more recently demonstrated the crucial role played by innovation processes within the services industry (Muller, 2001; Howells and Tether, 2004; Koch and Stahlecker, 2006; Fernandes *et al.*, 2015). Above all, this role proves of such relevance given that the sector enables bridges of knowledge and innovation between companies and scientific output (Miles *et al.*, 1995; Czarnitzki and Spielkamp, 2003). Furthermore, the origins of a third industrial revolution might be traced to the importance of KIBS companies (Tether and Hipp, 2002). Although the debate over the growth of KIBS revolves around their new specialised fields of competence and the growth in the tertiary sector in general, there is a rising level of acknowledgement that not only do new manufacturing processes but also new services and innovations in more general terms trace their origins to the KIBS sector (KaraÃmerlioglu and Carisson, 1999; Freel, 2006).

The importance of studying these new services is clearly emphasised in the literature. Compared to all the other services to companies, KIBS have notched up swifter growth rates than all other sectors. This achievement is due to a range of issues, in particular, the outsourcing of such services by other sectors, the development of information and communication technologies, changes to the regulatory, legal and market frameworks as well as globalisation and internationalisation (Bengtsson and Dabhilkar 2009).

Taking into consideration that these companies are knowledge intensive, there is clearly a need to highlight how such knowledge is currently perceived as a crucial resource for companies and taking on an ever higher profile in recognising and leveraging entrepreneurial opportunities. The growing body of literature on the importance of entrepreneurship at the regional level, as well as location characteristics tend to concur that knowledge is at the core of the founding of new companies and hence opting to emphasise the knowledge spread mechanisms (spillovers) operating in connection with universities and other R&D institutions. Correspondingly, the knowledge generated emerges out of means of cooperation between companies and state funded research institutions (Audretsch and Lehmann, 2005). According to Acs *et al.* (2006), entrepreneurial activities may be expected to steadily grow in both size and effectiveness given that investment in new knowledge proves relatively high but within the framework of which, companies, especially KIBS, make recourse to genuine sources of knowledge (universities and R&D institutions).

Despite the plethora of voices arguing the importance and the role that KIBS play in the economy and in regional dynamics in particular (Marshall *et al.*, 1987; Hansen, 1993; Muller and Zenker, 2001; Czarnitzki and Spielkamp, 2003; Miles, 2003), there have, however, been very few studies on the interplay between KIBs and regional competitiveness. Nevertheless, there has been a steady increase in the numbers of these service sector companies. In particular, small KIBS have been recognised as playing a dynamic and central role in the "new" knowledge based economies. This recognition has fostered new and growing levels of research on this service sector (Wong and He, 2005).

As suppliers of knowledge intensive services, their provision in any specific location is frequently deemed an important leverage of the competitiveness of regional industries and economies. Hence, associated with the role actually played by KIBS, theories have inevitably emerged on locations, serving as the foundation for the identification of the factors that may be behind the decisions of entrepreneurs in choosing one site over another for their business (Autant-Bernard *et al.*, 2006; Van Praag and Versloot, 2007; Ferreira *et al.*, 2010; Lafuente *et al.*, 2010).

The location of KIBS within urban areas, emphasizing agglomeration economies, has collected much of attention. The rural/urban dichotomy contrasting the location of entrepreneurial activities in rural areas is however a point of particular importance and interest (Fernandes *et al.*, 2015). In 2009, the OECD established a typology taking into account the different geographies of regions (OECD, 2009). Nowadays, the European Union and many other OECD member states have introduced policies designed to foster entrepreneurship as a key tool for rural development. In Europe, diversification in the rural base of production was stipulated as an objective of rural development policies (European Commission, 1997).

Similarly, there is growing demand and interest in founding and nurturing new business that are perceived as key factor in the development and revitalisation of some specific European areas. Given the paucity of studies on this type of service sector (KIBS) and given they play an increasingly active role in regional innovation and competitiveness, there is correspondingly an imperative to research the contribution actually made by such companies from evidence collected internationally.

The ambition of this book is to discuss the main issues, trends, challenges, and opportunities, related to the dynamics of Knowledge Intensive Business Services and regional competitiveness, focusing on innovation, entrepreneurial activities and their outcome. It aims to be a significant step towards new insights, to contribute for improved analyses of the dynamics of Knowledge Intensive Business Services in regional context. Its content will allow international comparison and benchmarking.

Overview of the book contents

This book collects among the most recent developments about the KIBS, their determinants and effects in interrelationships with regional competitiveness. It is composed of four mutually linked parts, namely KIBS and industrial dynamics; KIBS and their context; KIBS and their contribution to regional competitiveness and economic development; and finally, KIBS and public policy.

Part I: KIBS and industrial dynamics

The first part of the book, KIBS and industrial dynamics, consists of three initiating contributions. They reflect the importance of the subject in literature and in the assessment of the phenomenon in connection with industrial dynamics. The first chapter, *Knowledge Intensive Business Services research: a bibliometric study of leading international journals (1994–2014)*, by A. Braga and C. Marques, presents a bibliometric analysis of the scientific production within the field of Knowledge Intensive Business Services (KIBS), over the past 20 years. Resulting from a systematic review of 140 scientific articles published in refereed journals, the core concepts of KIBS are identified, as much as its intellectual structure, the scientific journals with a significant impact on this research field and the affiliation and collaboration networks associated to this strand of literature. This contribution also provides co-citations networks of authors, journals and their respective clusters, and aimed to identify the most relevant topics related to KIBS in order to support future research.

The second chapter undertaken by B. Desmarchelier, F. Djellal and F. Gallouj, *KIBS and the dynamics of industrial clusters: a Complex Adaptive Systems approach*, argues that KIBS represent an important – and often omitted – factor for explaining these dynamics, and that taking KIBS into account requires considering an alternative and integrative approach: the Complex Adaptive Systems. The authors study an existing cluster – Skywin (aeronautics in Wallonia region, Belgium) – within this framework. They show that KIBS are responsible for the emergence of a small-world effect within this cluster's innovation networks. Such an effect represents an interesting proxy for assessing clusters' attractiveness and potential for further growth.

The last chapter in Part I, *Bad news travels fast: the role of informal networks for SME-KIBS cooperation*, by D. Feser and T. Proeger, aims to analyse the role of informal networks for innovative cooperation with Knowledge Intensive Business Services (KIBS) in a peripheral regional innovation system in Germany, based upon interviews with SMEs. The authors focus on informal networks disseminating information, which compensate for SMEs' lack of information about KIBS, but also foster the dissemination of worst-case examples. The resulting skepticism within their informal networks reduces SMEs' inclination to conduct innovative cooperation. They argue that this reduces a region's overall innovative capabilities and that, consequently, policy-makers should build trust between SMEs and KIBS by establishing formal networks maintained by institutions considered to be impartial by SMEs.

Part II: KIBS and their context

Within the second part, KIBS and their context, the reader will be informed of the interrelations between KIBS and their environment with a focus on regional determinants.

The fourth chapter by D. Doloreux and R. Shearmur, *Does the geographic distribution of Knowledge Intensive Business Services affect the use of services for innovation? Empirical evidence from Quebec KIBS manufacturers*, propose to examine the extent to which recourse to KIBS is associated with the local presence of KIBS providers. Based upon a sample of 804 manufacturing establishments in the province of Quebec, this question is approach in three stages. The first stage analyses the association between KIBS use and location for all manufacturing establishments included in the sample. The second stage examines whether this association is different between innovative and non-innovative manufacturing firms. In the final stage, they investigate whether the relationship differs according to innovation type – product, process, and both. The authors found that the general propensity to use KIBS is not affected by location in (or close to) a KIBS cluster, but that certain types of innovator use more KIS when in non-metropolitan KIS-rich environments.

In the fifth chapter, *Institutions and spin-offs: determining factors for establishment and early market entry success of innovation based spin-offs from KIBS-firms*, by K. V. Meland and T. A. Iakovleva, explores the role of institutional factors on the establishment and early market-entry success of corporate innovation based spin-offs from KIBS firms. A corporate spin-off is defined as a firm established by a parent company to implement a new activity or a new product/service. Based on empirical data from six cases in three industries in Norway, the authors argue that the effectiveness of establishment and early market-entry success of spin-offs is conditioned by sectoral, regional and managerial institutions. Furthermore, a strong patent regime was found as a condition in the establishment of spin-offs in both oil and gas and maritime sectors, but not in the ICT sector.

The sixth chapter, *Survival of Knowledge Intensive Business Services firms: the role of agglomeration externalities*, by S. Tavassoli and V. Jienwatcharamongkhol, analyses the role of various types of agglomeration externalities on the survival rate of newly established KIBS firms. In particular, the authors trace the population of newly established Swedish KIBS firms from 1997 to 2010 and investigate the role of Marshallian, Jacobian, and Urbanisation externalities on the survival of these firms. They found that only Jacobian externalities (diversity) positively affect the survival of KIBS firms. Not all Jacobian externalities matter though. Only the higher the "related variety" of the region in which a KIBS firm is founded, the higher will be the survival chance of the firm, while "unrelated variety" does not have any significant effect. The result is robust after controlling for extensive firm characteristics and individual characteristics of the founders.

Part III: KIBS and their contribution to regional competitiveness and economic development

The third part of the book, entitled KIBS and their contribution to regional competitiveness and economic development, collects five contributions examining this typical and important question, at the core of this volume.

In chapter seven, *Entrepreneurship and KIBS: key factors in the growth of territories*, the authors, J. Alcazar, N. Roig-Tierno, A. Mas-Tur and B. Ribeiro-Navarrete, analyse KIBS as tools that enhance entrepreneurship, and how they can encourage the growth of territories. According to the authors, policies linked to innovation and driving knowledge intensive entrepreneurship have become increasingly prominent among both academics and the political agenda of policy makers at all administrative levels (local, regional, national, and supranational). Innovation, entrepreneurship, and knowledge, especially when properly combined, play a fundamental role in economic well-being. Knowledge Intensive Business Services (KIBS) act as sources of external knowledge, while contributing to innovation and hence the growth of territories.

Chapter eight, *Contribution of knowledge intensive activities to regional competitiveness: production function approach*, by A. L. Asikainen and G. Mangiarotti, empirically examines the relationship between high-tech knowledge intensive services and regional competitiveness across the productivity distribution. Almost all economic indicators show large differences between regions in EU27. The current production model in most economies emphasises the role of knowledge creation, absorption and exploitation. New knowledge can be produced by public and private actors. The results indicate that the quality of labour force drives development in the most competitive regions, whereas in the regions of lower competitiveness investments in capital have the highest impact. These results call for tailored policy instruments to narrow the gap between regions.

Chapter nine, *KIBS as a factor in meetings industry competitiveness creation in Krakow, Poland*, by K. Borodako, J. Berbeka and M. Rudnicki, aims to investigate if event-oriented KIBS can be included in the standard classification of KIBS and how they support the increased competitiveness and innovation of Krakow's meetings industry. The flow of knowledge and experience between the organiser, professional conference organisers (PCOs) and other partners in the project confirms that the key motivation for cooperating with KIBS (all kinds of services) are their high level of knowledge, the quality of the final product, reduction of costs and finally increased competiveness.

Chapter 10, *Interactive relationships, development effects and knowledge intermediaries among KIBS firms and their clients: a comparison of the Hsinchu and Tainan regions, Taiwan*, by T. S. Hu, C. Y. Lin and H. N. Hsieh, first examines the roles and functions of KIBS in the evolution of an area innovation system and then analyses interactions between KIBS firms and their clients. Moreover, this chapter focuses on interpreting the role of KIBS as knowledge intermediary mediating and transmitting knowledge among actors, and explores the mechanism of knowledge exchanges in different types of geographic innovation systems. The examined local innovation systems are in the Hsinchu and Tainan Science-Based Industrial Parks and surrounding areas in Taiwan. The results indicate that the cumulative interactions increased demand for and reliance on KIBS, increased the need for specialised support in numerous different functions, and resulted in KIBS firms gradually playing more important roles in the industrial interface.

Chapter eleven, *Regional competitiveness and localised Knowledge Intensive Business Services: the case of the Gold Coast, Australia* by V. Ratten, has as purpose to analyse Knowledge Intensive Business Services and regional innovation in Australia using a case study methodology from the Gold Coast region of Australia. A conceptual framework is developed that leads to a set of research propositions focusing on how institutions, social and industry context influence entrepreneurial innovation and regional competitiveness. The main findings suggest that innovation and entrepreneurship help regional development when coupled with economic growth. The results highlight how business investment in innovation and government policy can advance regions and make them more competitive in the global marketplace. In addition, the research has policy implications that support the role of innovation in society and theoretical implications, which stress the importance of Knowledge Intensive Business Services.

Part IV: KIBS and public policy

The last part is devoted to KIBS and public policy. It brings together two contributions able to give the reader some indication or guideline about what kind of public resolution would be in accordance with a dedicated KIBS policy.

The twelfth chapter undertaken by J. Bryson and P. W. Daniels, *Skills, competitiveness and regional policy: Knowledge Intensive Business Services in the West Midlands, UK*, states that KIBS firms sell advice, expertise and knowledge and this implies that their core asset rests on the quality of their people. Skills, capabilities and competencies are critical for the competitiveness of KIBS firms. This chapter explores KIBS firms and skills in the West Midlands region of the UK. The focus is on understanding skills, hard-to-fill vacancies, technical skills versus soft skills, commercial skills and the relationship between KIBS, skills and competitiveness. Part of the analysis aims to identify skills and capabilities as core elements of the competitiveness of KIBS firms.

Thirteenth and last chapter of this book, *Prospects and policies in the development of Knowledge Intensive Business Services in Europe*, is developed by M. Toivonen and A. Caru and it examines the current and future opportunities and challenges of KIBS. Four scenarios are presented based on the work of the High Level Expert Group on Business Services – an initiative of European Commission 2013–14. Two scenarios are driven by developments in the operational environment (digitalisation and systemic changes) and the other two by developments in the business of KIBS and their customers. The chapter identifies policy options that regional, national and international stakeholders could apply to strengthen the innovative impact of KIBS.

We are confident that the attentive reader will highly appreciate the important contributions gathered here. It is also more modestly our sincere wish to the reader for an interesting reading. Enjoy discovering!

References

Acs, Z. Audretsch, D. Braunerhjelm, P., Carlsson, B. (2006). The knowledge spillover theory of entrepreneurship, CESIS Electronic Working Paper Series.

Audretsch, D., Lehmann, E. (2005). Does the knowledge spillover theory of entrepreneurship hold for regions? *Research Policy*, 34(8), pp. 1191–202.

Autant-Bernard, C., Mangematin, V., Massard, N. (2006). Creation of Biotech SMEs in France. *Small Business Economics*, 26, pp. 173–87.

Becker, W., Dietz, J. (2004). R&D Cooperation and innovation activities of firms – evidence for the German IPO data. *Economics of Innovation and New Technology*, 15(1), pp. 71–81.

Bengtsson, L., Dabhilkar, M. (2009). Manufacturing outsourcing and its effect on plant performance – lessons for KIBS outsourcing. *Journal of Evolutionary Economics*, 19(2), pp. 231–57.

Czarnitzki, D., Spielkamp, A. (2003). Business services in German: bridges for innovation. *The Services Industries Journal*, 23(2), pp. 1–30.

European Commission (1997). *Rural developments*. CAP 2000 Working Document, V/1117/97.

Fernandes, C., Ferreira, J., Marques, C. (2015). Innovation management capabilities in rural and urban knowledge intensive business services: empirical evidence. *Service Business*, 9, pp. 233–56.

Ferreira, J., Marques, C., Fernandes, C. (2010). Decision–making for location of new knowledge intensive businesses on ICT sector: Portuguese evidences. *International Journal of E-Entrepreneurship and Innovation*, 1(1), pp. 60–82.

Freel, M. (2006). Patterns of technological innovation in knowledge-intenive business services. *Industry and Innovation,* 13(3), pp. 335–58.

Gallouj, F., Weinstein, O. (1997). Innovation in services. *Research Policy*, 26, pp. 537–56.

Hansen, N. (1993). Producer services, productivity and urban income. *Review of Regional Studies*, 3, pp. 255–64.

Howells, J., Tether, B., Gallouj, F., Djellal, F., Gallouj, C., Blind, K., Elder, J., Hipp, C., Montobbio, F., Corrocher, N., Macpherson, A., Banach, D. (2004). Innovation in services: issues at stake and trends. Brussels, Luxembourg: European Commission.

Huergo, E., Jaumendreu, J. (2004). How does probability of innovation change with firm age? *Small Business Economics*, 22 (3–4), pp. 193–207.

Johne, A., Storey, C. (1998). New service development: a review of the literature and annotated bibliography, *European Journal of Marketing*, 32 (3/4), pp. 184–252.
KaraÃmerlioglu, D. K., Carisson, B. (1999). Manufacturing in decline? A matter of definition. *Economy, Innovation, New Technology*, 8, pp. 175–96.
Kline, S.J., Rosenberg, N. (1986). An overview of innovation||, in: Laudau, R., Rosenberg, N. (Eds), *The positive sum strategy: harnessing technology for economic growth*. Washington: National Academy Press, pp. 275–306.
Koch, A., Stahlecker, T. (2006). Regional innovation systems and foundation of knowledge intensive business services. *European Planning Studies*, 14 (2), pp. 123–46.
Lafuente, E., Vaillant, Y., Serarols, C. (2010), Location decisions of knowledge-based entrepreneurs: Why some Catalan KISAs choose to be rural? *Technovation*, 30 (11–12), pp. 590–600. doi:10.1016/j.technovation.2010.07.004
Lynskey, M. (2004). Determinants of innovative activity in Japanese technology-based start-up firms. *International Small Business Journal*, 22 (2), pp. 159–96.
Marshall, J., Damesick, P., Wood, O. (1987). Understanding the location and role of producer services in the United Kingdom. *Environment and Planning A*, 19 (5), pp. 575–95.
Miles, I. (2000). Services innovation: coming of age in the knowledge-based economy. *International Journal of Innovation Management,* 43(4), pp. 371–89.
Miles, I. (2003). Services and the knowledge-based economy, in: Tidd, J., Hull, F. M. (Eds), *Service innovation, organizational responses to technological opportunities and market imperatives*, London: Imperial College Press, pp. 81–112.
Miles, I., Kastrinon, N., Flanagan, K., Bilderbeek, R., den Hertog, P., Huntink, W., Bouman, M. (1995). Knowledge intensive business services. Users and sources of innovation. Brussels: European Commission.
Muller, E. (2001). *Innovation interactions between knowledge intensive business and small and medium sized enterprises*. Heidelberg, New York: Physica-Velarg.
Muller, E., Zenker, A. (2001). Business services as actors of knowledge transformation: the role of KIBS in regional and national innovation systems. *Research Policy*, 30 (9), pp. 1501–16.
OECD, (2009). *Rural Policy Review, Spain*. OECD, Paris.
Tether B., Hipp, C. (2002). Knowledge intensive technical and other services: patterns of competitiveness and innovation compared. *Technology Analysis & Strategic Management* 14(2), pp. 163–82.
Tether, B. S. (2003). The sources and aims of innovation in services: variety between and within sectors. *Economics of Innovation and New Technology*, 12 (6), pp. 481–505.
Van Praag, M., Versloot, P. (2007). What is the value of entrepreneurship? A review of recent research. *Small Business Economics,* 29, pp. 351–82.
Von Hippel, E. (1988). *The sources of innovation*. New York: Oxford University Press.
Wong, P., He, Z. (2005). A comparative study of innovative behaviour in Singapore's KIBS e manufacturing firms. *Services Industries Journal,* 25, pp. 23–42.
Wood, P. (2005). A service-informed approach to regional innovation – or adaptation? – *The Services Industries Journal*, 25 (4), pp. 429–45.

Part I

KIBS and industrial dynamics

1 Knowledge Intensive Business Services research

Bibliometric study of leading international journals (1994–2014)

Alexandra M. Braga and Carla S. Marques

Introduction

Over the last 20 years, Knowledge Intensive Business Services (KIBS) have grown considerably in many European and Asian countries and they have a significant influence on innovation activities across the whole economy (e.g. Shi *et al.*, 2014; Hu *et al.*, 2013; Abreu *et al.*, 2010; Viljamaa *et al.*, 2010; Wood, 2005; Miles *et al.*, 2000; Mas-Verdú *et al.*, 2011). For this reason, KIBS have recently become an important field of both theoretical (e.g. Murray *et al.*, 2009; Bettiol *et al.*, 2012; Chae, 2012; Gimzauskiene and Staliuniene, 2010) and empirical study (e.g. Miozzo and Grimshaw, 2005; Yam *et al.*, 2011; Palacios-Marques *et al.*, 2011; Santos-Vijande *et al.*, 2013a; Carmona-Lavado *et al.*, 2013).

Regional innovation research still echoes national studies by assuming the primacy for regional competitiveness of process-orientated, technologically driven innovation. It has nevertheless recognised the growing importance for such innovation of regional institutional interaction and flexibility and of key service expertise, especially through KIBS (Wood, 2005). Santos-Vijande *et al.*, (2013a) argue that as the dynamism of the KIBS sector has an impact on the whole economy, it is also necessary to understand the most advisable management practices in KIBS to foster innovation and improved performance, although relatively few studies have approached this issue.

In order to assess the KIBS structure in a certain field, an important method – bibliometric analysis – can be used to analyse the trends in the published research. Bibliometric studies have been used in several areas of business and economics (Dragos *et al.*, 2014), entrepreneurship (Ávilla *et al.*, 2014), technology entrepreneurship (Ferreira *et al.*, 2015), innovation (Toivanen, 2014), social innovation and social entrepreneurship (Philips *et al.*, 2015), service innovation (Zhu and Guan, 2013), etc. However, no prior evidence of a systematic literature review in the leading international journals in this area has been found. In light of this consideration, this study aims to map and analyse the scientific production within the field of KIBS, using the publications database ISI Web of Science – WoS, for the period between 1994 and 2014.

Specifically, our objectives are: (1) to identify how the topic is defined in the international literature and the progress achieved in the research field; (2) to evaluate and measure the research productivity, key authors and scientific journals with the highest impact on this research field and the networks of association between the respective institutions and countries of origin; (3) and to analyse and map citations, co-citations and research themes to identify which topics and dimensions are related to KIBS in order to support future research.

This paper is organised as follows. In Section 2, the emergence of the field of study on KIBS and an overview of the literature on its concept are discussed. Section 3 presents and discusses the methodological features of the research, the sample and introduces the bibliometric analysis method. The subsequent section presents the results in terms of the KIBS' core areas and presents visual maps of the KIBS network research. The last section concludes the paper, presenting observations and suggesting opportunities for future research.

Knowledge Intensive Business Services

Although the term "Knowledge-intensive business services" has been used since the early nineties, only recently it has become a major theme of investigation and empirical research (Mas-Verdú *et al.*, 2011). Despite this relatively recent concern of the academia in studying KIBS, the literature has already provided many definitions of KIBS firms that, in many cases, do not differ significantly, but rather display different nuances. The different definitions of KIBS found in the literature can be explained by the purpose of the studies, in which a definition serves a particular purpose.

Bettencourt *et al.*, (2002, p.100), describe KIBS firms as those aiming to generate value-added service activities, and that these activities consist in "the accumulation, creation, or dissemination of knowledge for the purpose of developing a customised service or product solution to satisfy the client's needs." The knowledge that serves as the basis for their business can, according to Miozzo and Grimshaw (2005), be social and institutional knowledge (e.g. accountancy; management consultancy) or technical knowledge (computer R&D; engineering services). Many authors (e.g. Borodako *et al.*, 2014; Muller and Zenker, 2001; Fernandes and Ferreira, 2013; Huang and Ji, 2013; Hakanen, 2014) refer to the concept presented by Miles *et al.*, (1995), who have distinguished KIBS as traditional professional KIBS (P-KIBS) and new technology-based services (T-KIBS). P-KIBS help their clients to navigate or negotiate complex systems such as social, physical, psychological, and biological systems (for example, marketing or consultancy services). T-KIBS are services that rely heavily on professional knowledge (e.g. IT services, communication, and computer services), thus, their employment structures are heavily weighted towards engineers and scientists.

In light with this consideration, Wong and He (2005) include three major KIBS sectors in their study: IT and related services, business and management consulting, and engineering and technical services. Based on Borodako *et al.*, (2014), the third type of division is made according to the relationship of the KIBS to the (client) company and the market. Here, three groups of KIBS are identified: market KIBS (key services: market research; advertising; and research and experimental development in social sciences and humanities); enterprise KIBS (IT and programming services; legal services; accounting and tax advisory services; management advisory and PR services; temporary employment agencies; and other recruitment services); and technical KIBS (multilevel KIBS – connecting both the above groups of market and enterprise services: architectural activities; technical testing and analysis; research and experimental development in natural sciences and engineering; engineering activities).

According to Borodako *et al.*, (2014), most definitions in the literature stress the following key aspects of KIBS: they are offered by private business to other business (e.g. Hertog, 2000); they are based on knowledge or expertise – mostly highly advanced and related to a specific field; and the consumption of the service usually improves the client company's intellectual capital. When focusing on the role of KIBS services in client innovation, three different aspects can be perceived: KIBS act as (1) facilitators (if it supports a client firm in its innovation process); (2) carriers (if it plays a role in transferring existing innovations from one firm or industry to the client firm or industry); or (3) sources of innovation (if it plays a major role in initiating and developing innovations in client firms, mostly in close interaction with the client firm) (Hauknes, 1998).

A strong characteristic of KIBS firms, given the nature of their business and the importance of knowledge on the society, is the impact they have on the economic tissue. Wong and He (2005), with this respect, refer that KIBS firms are "group of services which are very actively integrated into innovation systems by joint knowledge development with their clients, and which consequently create considerable positive externalities and possibly accelerate knowledge intensification across the economy".

In the academia, KIBS literature has addressed the concept from several different perspectives. The topic of KIBS can be interpreted in different ways and types of study. Table 1.1 provides some examples of how the literature has dealt with KIBS concept.

Table 1.1 KIBS concepts from the literature

Reference	*Definitions of KIBS*
Miles et al. (1995)	KIBS are services involving economic activities which are intended to result in the creation, accumulation or dissemination of knowledge.
Muller and Zenker (2001)	KIBS do not only "transmit" knowledge, in fact they play a crucial role in terms of "knowledge re-engineering". KIBS has potentially as receptors, interfaces and "catalysators" in terms of knowledge-creation and diffusion. KIBS can be described as services offered by firms, usually to other firms, incorporating 'a high intellectual value-added'.

(continued)

Table 1.1 KIBS concepts from the literature (*continued*)

Reference	*Definitions of KIBS*
Wong and He (2005, p. 27)	"KIBS firms' innovation efforts extend far beyond their internal organisation to the service relationship and directly into the domain of service clients by providing competence enhancing knowledge services to their clients".
Bettiol et al. (2011)	The KIBS sector constitutes a service subsector that includes establishments whose primary activities are mainly concerned with providing knowledge intensive inputs to the business processes of other organisations, including private and public sector clients
Santos-Vijande et al. (2013)	KIBS are private companies or organizations which have a high degree of professional knowledge
Corrocher and Cusmano (2014)	KIBS are key players in innovation systems, particularly in advanced regions where manufacturing competitiveness largely depends on knowledge contents provided by highly specialised suppliers.
Shi et al. (2014)	KIBS are becoming a major force in promoting innovation and that effect is highly related to the average level of human capital.
Doloreux and Laperriere (2014)	The KIBS firm has developed a core portfolio of services, methods or solutions and achieves growth through the penetration of new markets and/or client groups that demonstrate similar needs.

Many studies analyse the relevance of KIBS to innovation (e.g. He and Wong, 2009; Santos-Vijande *et al.*, 2013b; Mas-Tur and Soriano, 2014; Alvarez-Gonzalez and Gonzalez-Morales, 2014; Shi *et al.*, 2014; Doloreux and Laperriere, 2014; Santos-Vijande *et al.*, 2013b; He and Wong, 2009) and it is increasingly recognised that KIBS are key to innovation systems (e.g. Mas-Verdú *et al.*, 2011, Corrocher and Cusmano, 2014; Hu et al., 2013) and are vectors of knowledge transmission (e.g. Skjolsvik *et al.*, 2007; Larsen, 2001; Muller and Zenker, 2001).

According to Di Maria *et al.*, (2012), the literature so far pointed out that the spatial proximity is necessary for sustaining the interaction between KIBS and the client. Nevertheless, there are few theoretical or empirical analysis focusing on the role of the relationship with the local context (Koch and Strotmann, 2006; Doloreux and Shearmur, 2012; Huggins and Johnston, 2012; Peiker *et al.*, 2012; Aslesen and Isaksen, 2007), which may be vital for KIBS development (Koch and Strotmann, 2006).

Recent papers also analyse the relevance of KIBS with regards to the penetration in new external markets (e.g. Doloreux and Laperriere, 2014; Di Maria *et al.*, 2012, Abecassis-Moedas *et al.*, 2012, Peiker *et al.*, 2012).

Bibliometric analysis of the KIBS literature

Selection of the articles

Considering the growth of academic interest in KIBS, this study attempts to provide a comprehensive review of the existing studies, through a systematic review of the literature. Bibliometrics is the mathematical and statistical analysis of communication in the form of documents aiming to providing a relatively robust and less subjective method to analyse the foundations of a scholarly discipline (Wallin, 2012). Bibliometric studies may be used to examine, for instance, the most cited works, the co-citation networks and, to understand the intellectual structure of literature (Ramos-Rodríguez and Ruíz-Navarro, 2004). The analysis of co-citations is often used to identify papers with higher impact (Zitt and Bassecoulard, 1994). According to Smith (1981), two documents are considered co-cited when they are cited together in other documents. Previous research has applied bibliometric analysis to e.g. measuring publication in leading management journals as a measure of institutional research performance (Stahl *et al.*, 1988).

In this study, the clusters and respective networks of references were obtained following the methodological guidelines proposed by van Eck and Waltman (2010). The simple graphical representations were provided by software packages such as SPSS and Pajek. For the analyses, we used the software VOSviewer (www.vosviewer.com)[1] and CitNetExplorer[2] (www.citnetexplorer.nl) which supported the construction of the bibliometric maps, and TreeCloud.org (http://treecloud.univ-mlv.fr) to generate "tree of words."

This research was based on a sample of international and national scientific papers collected from the Social Science Citation Index (SSCI), compiled by the Thomson Reuters online database, which contains, in addition to the publications, bibliographic information about authors, affiliations and citations.

The data collection was conducted through the indexed databases ISI Web of Sciences,[3] over the last two decades (between 1994 and 2014) and according to the following criteria (Table 1.2). First, we searched for publications using the research terms in the topic: "KIBS" or "Knowledge Intensive Business Services" or "Knowledge-Intensive Business Services," and we found 267 articles (we found only one difference of 2 articles for the period 1900–2014, which were related to patents). Then, we refined the results for the following criteria: (a) document types: articles (excluding proceeding papers, review, and editorial materials) (and we reduced the results to 181 articles); (b) data bases: web of science core collection (resulting in 167 articles); (c) research domain: social sciences (158 articles); (d) research area: business economics (there was no difference if we included operations research management science) and, finally, we found 140 articles. The papers were selected on the basis of their title, abstract and keywords. The citations identified were reviewed according to the inclusion and exclusion criteria (Table 1.3).

The search performed resulted in 140 scientific articles with publications dates between 1994 (one article and the first being published since 1900, according to the criteria in our study) and 2014 (17). We considered articles published between January 1994 and December 2014.[4] The unit of analysis in this research is the publication, and the variables correspond to authors and respective affiliations, journals, number of citations and cited references. The process of literature collection took place during December 2014 until May 2015.

Table 1.2 Settings of the research

Basic search	*Timespan*	*Databases*	*Research domain*	*Research areas*	*Document type*
TOPIC: "KIBS" ***OR:*** "Knowledge Intensive Business Services" ***OR:*** "Knowledge-Intensive Business Services"	From 1994 to 2014	Web of Science™ core collection	Social sciences	Business economics	Article

Table 1.3 Inclusion and exclusion criteria

Criteria	*Reasons for inclusion*	*Reasons for exclusion*
Pre-1994		Contributions toward Knowledge Intensive Business Services have developed in the past 20 years
All countries	To ensure a cross-cultural view of KIBS	
Editorial, patent, clinical trial, meeting, review, other		Focus on high-quality peer-reviewed research
Theoretical and empirical articles	To capture all existing studies	
Science technology, arts humanities		To focus in the social sciences area – limited to one research domain

The emergence and evolution of KIBS

The literature on the Knowledge Intensive Business Services is a relatively new field of research that has spread remarkably in the past 20 years. Knowledge Intensive Business Services research has flourished in 1994, mainly in Europe and USA. The earlier published paper found in WoS was written by Simone Strambach (1994), from University of Stuttgart, Germany, and it was published in Tijdschrift Voor Economische en Sociale Geografie, a journal published by Wiley-Blackwell (USA), which web of science categories are economics and geography (2013 impact factor: 1,012). The article entitled "Knowledge Intensive Business Services in the Rhine-Neckar Area" emphasises the importance of network relationships for knowledge intensive service firms and shows that network relationships play a key role in the interaction between suppliers and clients. Later, other authors have explored this link (e.g. Plaza *et al.*, 2011; Hakanen, 2014; Najafi-Tavani *et al.*, 2014).

The data in Figure 1.1 shows an increase in the number of articles on KIBS published, per year, with particular emphasis on the last decade (about 94 per cent of the total publications). It is also important to mention that half of the papers (70) were published over the last three years (Figure 1.2). Since 2008, this number has been greater than (or equal to) 10 every year. In 2012 and 2013 the highest number of publications in the field, was achieved, with 30 and 23 articles published, respectively).

The 140 articles considered in our sample display an average citation rate of 12.2 per cent, with 31 of the articles never being cited and 55 have been cited between one (17) and five times (5). Table 4 reveals the top-40 ranking of papers in terms of highest number of citations.

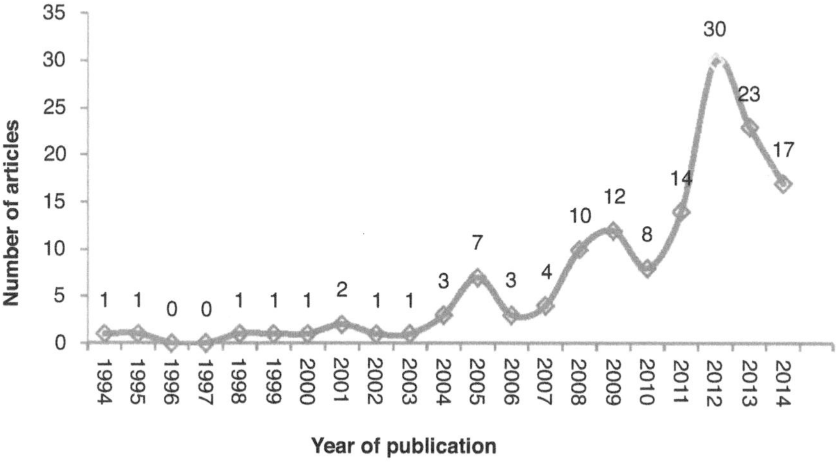

Figure 1.1 Number of articles by year of publication

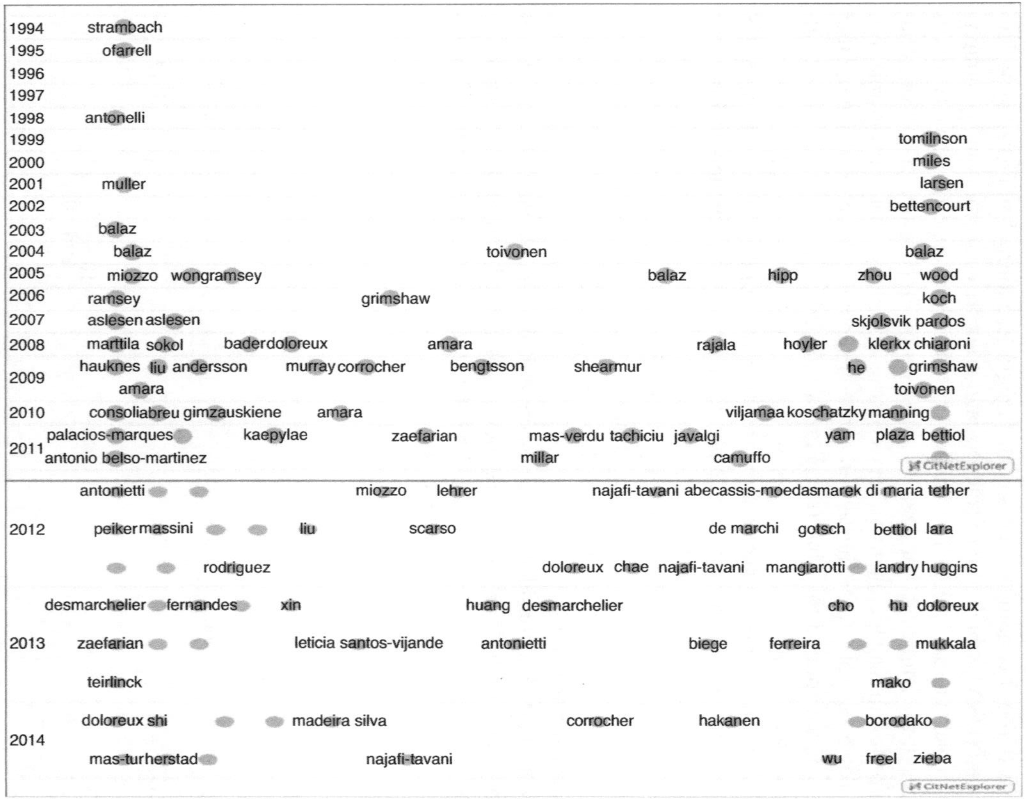

Figure 1.2 Articles published by year

Table 1.4 Most-cited articles in the field of KIBS

		Total citations			*Total citations*
1	**(Muller and Zenker, 2001)**	**216**	21	(Murray et al., 2009)	23
2	**(Hipp and Grupp, 2005)**	**172**	22	(Hauknes and Knell, 2009)	22
3	**(Bettencourt et al., 2002)**	**165**	23	(Shearmur and Doloreux, 2009)	20
4	**(Toivonen and Tuominen, 2009)**	**63**	24	(Skjolsvik et al., 2007)	19
5	**(Miozzo and Grimshaw, 2005)**	**62**	25	(Grimshaw and Miozzo, 2006)	19
6	**(Antonelli, 1998)**	**54**	26	(Doloreux and Shearmur, 2012)	18
7	(Amara et al., 2009)	42	27	(Tseng et al., 2011)	17
8	(Amara et al., 2008)	39	28	(Consoli and Elche-Hortelano, 2010)	17
9	(Yam et al., 2011)	34	29	(Hoyler et al., 2008)	16
10	(Klerkx and Leeuwis, 2008)	33	30	(Tomlinson, 1999)	16
11	(Larsen, 2001)	32	31	(Ofarrell and Moffat, 1995)	16
12	(Abreu et al., 2010)	30	32	(Bader, 2008)	14
13	(Wood, 2005)	30	33	(Mas-Verdú et al., 2011)	13
14	(Aslesen and Isaksen, 2007)	29	34	(Grimshaw and Miozzo, 2009)	11
15	(Aarikka-Stenroos and Jaakkola, 2012)	28	35	(Doloreux and Mattsson, 2008)	11
16	(Miles et al., 2000)	28	36	(Koch and Strotmann, 2006)	11
17	(Andersson and Hellerstedt, 2009)	27	37	(Ramsey et al., 2005)	10
18	(Wong and He, 2005)	27	38	(Manning et al., 2010)	9
19	(De Marchi, 2012)	26	39	(Koschatzky and Stahlecker, 2010)	9
20	(Corrocher et al., 2009)	25	40	(Bengtsson and Dabhilkar, 2009)	9

The top six studies with the highest number of citations (more than 50 citations) are:

1. Muller, E. and Zenker, A. (2001). Business services as actors of knowledge transformation: the role of KIBS in regional and national innovation systems. *Research Policy*, 30(9), Special Issue: SI, 1501–16. (215 citations)
2. Hipp, C. and Grupp, H. (2005). Innovation in the service sector: The demand for service-specific innovation measurement concepts and typologies. *Research Policy*, 34(4), 517–35. (172 citations)
3. Bettencourt, L. A., Ostrom, A. L., Brown, S. W., and Roundtree, R. I. (2002). Client co-production in knowledge-intensive business services. *California Management Review*, 44(4), 100–28 (165 citations)
4. Toivonen, M. and Tuominen, T. (2009). Emergence of innovations in services. *Service Industries Journal*, 29 (7), 887–902. (63 citations)
5. Miozzo, M. and Grimshaw, D. (2005). Modularity and innovation in knowledge-intensive business services: IT outsourcing in Germany and the UK. *Research Policy*, 34(9), 1419–39 (62 citations)
6. Antonelli, C. (1998). Localized technological change, new information technology and the knowledge-based economy: The European evidence. *Journal of Evolutionary Economics*, 8(2), 177–98 (54 citations)

The most cited paper (Muller and Zenker, 2001) provides an overview of the role and function of KIBS in innovation systems and their knowledge production, transformation and diffusion activities. This study focuses on innovation interactions between manufacturing small and medium sized enterprises (SMEs) and KIBS and concludes that innovation activities link SMEs and KIBS through the process of knowledge generation and diffusion. The investigation follows a methodology based on the examination of firm samples located in five different regions in France and Germany. Hipp and Grupp (2005) focused in the concept of innovation in the service sector, suggesting that the notion of innovation, well established in the manufacturing sector, cannot simply be transposed to the service sector. The authors analysed selected results of the German innovation survey and introduced a new typology aiming to obtain a better understanding of innovation in services. They draw special attention to the inclusion of Knowledge Intensive Business Services because of their particular importance for innovation processes. Bettencourt *et al.*, (2002) argued that a common characteristic of Knowledge Intensive Business Service (KIBS) firms is that clients routinely play a critical role in co-producing the service solution along with the service provider, which can have a strong effect on both the quality of the service delivered and on customers' satisfaction with the knowledge-based service solution. In the authors' perspective, by strategically managing client co-production, service providers can improve operational efficiency, develop more optimal solutions, and generate a sustainable competitive advantage. This was based on research conducted with an IT consulting firm and work done with other Knowledge Intensive Business Service providers. Toivonen and Tuominen (2009) provided analytical and detailed discussion on the nature of service innovations and their emergence. The theories examined are multi-disciplinary including general service theories, general innovation theories and theories associated to new service development and innovation management. This was based on two empirical case studies in Finland in the fields of real estate and construction services and of Knowledge Intensive Business Services. Drawing on an empirical study of IT outsourcing in the UK and Germany, Miozzo and Grimshaw, (2005) explored the lessons for modularity that can be drawn from the outsourcing of KIBS. In their perspective, given of the inseparability of information and production technologies, IT outsourcing is habitually accompanied by wider transformations in clients' production technologies, which results in the need for knowledge and organisational coordination in the form of the transfer of staff from the client and the retained IT organisation. According to this approach, modularity is often presented as a design strategy that stimulates innovation; however the intangibility of services exacerbates the conflicts between clients and suppliers, which may present obstacles to innovation. Antonelli (1998) focuses in the co-evolution of new information and communication technologies and the knowledge intensive business industry to show that new information technology affects the actual conditions of information, its basic characteristics of appropriation and tradability, favouring the role of business services as forces of interaction between knowledge components in the generation of new technology. Using input/output statistics of the European economy in the second half of 1980s, the author found the existence of a correlation between the use of business and communication services and confirmed their high output elasticity. The respective citation network is presented in Figure 1.3.

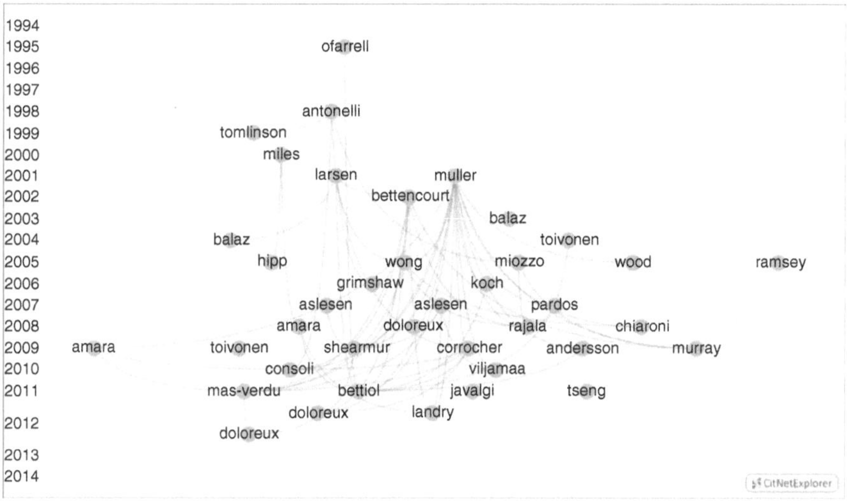

Figure 1.3 Citation network

Evolution and co-citation networks

The initial sample of 140 scientific papers was reduced to papers with at least 10 citations, resulting in a reduced sample of 37 articles quoted 1,435 times. Based on these 37 articles, we performed a co-citation analysis in order to build the respective network, and the size of the sample was reduced to 23 papers (see Figure 1.4) grouped into four clusters (see Table 1.5), which supports the main dimensions related to KIBS, namely: cluster 1 points for innovation: concepts and process, cluster 2 addresses the relation between Knowledge and KIBS, cluster 3 identify articles related to innovation networks and cooperation, and cluster 4 stands for Location and Relationship with Clients.

Concerning to the sources, the 140 papers included in the sample were published in 44 academic journals (with 1.707 citations) and as one can see in Table 1.6, 19 journals display, at least, 10 citations.

The journals with the highest citation number are *Research Policy* (592 citations), *Service Industries Journal* (329 citations), *California Management Review* (184 citations), *Journal of Evolutionary Economics* (104 citations), *Industry and Innovation* (62 citatitons), *Technovation* (46 citations), and *International Journal of Technology Management* (41 citations). With regards to the number of papers published, special emphasis should be given to *Service Industries Journal* (with 27 articles), followed by the *Research Policy* (with 9 articles), the *International Journal of Technology Management* (with 8 articles) and the *Industry and Innovation* (with 7 articles). Some of these papers are also those that have the greatest impact factor,[5] such as *International Journal of Technology Management* (2,704), followed by *Service Industries Journal* (2,617) and *Research Policy* (2,598). The respective network is presented in Figure 1.5 and as can be seen, the co-citation analysis reveals five clusters (Table 1.7).

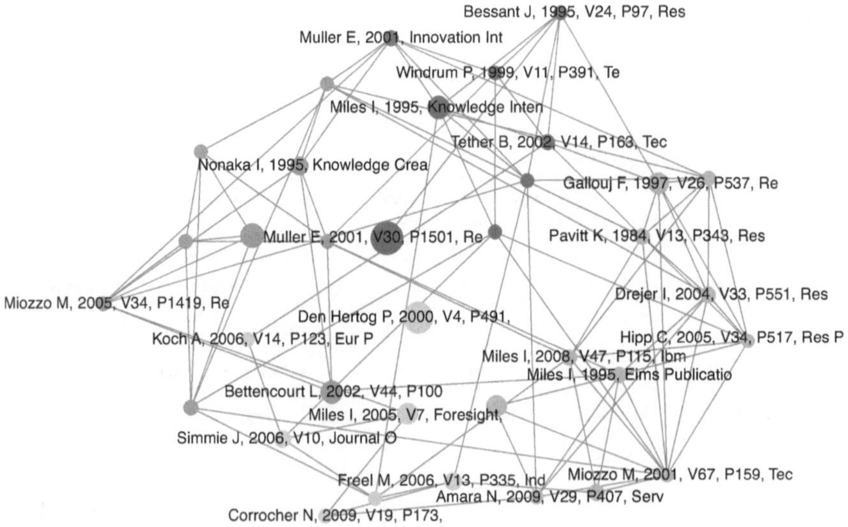

Figure 1.4 Co-citation network

Table 1.5 Resulting clusters from the co-citation analysis performed on the 23 most cited articles

Cluster 1: Innovation: concepts and process

Article	*Focus of the study*	*Method/ Sample*	*Main insights*
(Amara et al., 2009)	To develop indicators to capture forms or types of innovation in KIBS; To propose a conceptual framework inspired by the knowledge-based theory using different categories of knowledge assets as explanatory variables.	Multivariate probit regression models 1124 small and medium KIBS operating in the province of Québec, in Canada	Process, strategic, managerial and marketing innovations are complementary; and the different forms of innovation are explained by different explanatory variables
(Drejer, 2004)	To apply innovation concepts developed especially for services, thereby contributing to the existing divide between manufacturing and services.	–	Reference to Schumpeter, in particular innovation, as a contrast to activities based on routine systems, in service oriented studies would add a needed theoretical and conceptual strengthening to service innovation studies
(Gallouj and Weinstein, 1997)	To lay the foundations of a theory that can be used to interpret innovation processes in the service sector.	–	Various modes of innovation are highlighted and interpreted in terms of a characteristic dynamic.
(Hipp and Grupp, 2005)	To support the conceptual findings and to identify potential improvements (on innovation).	German innovation survey	Introduces a new typology with a view to obtain a better understanding of innovation in services. Special attention is directed towards the inclusion of KIBS that are of particular importance for innovation processes.

(continued)

Table 1.5 Resulting clusters from the co-citation analysis performed on the 23 most cited articles (*continued*)

Cluster 1: Innovation: concepts and process

Article	*Focus of the study*	*Method/ Sample*	*Main insights*
(Miles et al., 1995)	To highlight the contributions of KIBS to innovation; provide the agenda for coherent analyses of KIBS innovation processes; and, draw recommendations for a consideration of KIBS in policy-making.	Case studies of innovative KIBS	The knowledge intensity of all sectors of the economy is increasing. R&D becomes increasingly the basis of new techniques, and networks of innovators become increasingly the basis of accumulation of the knowledge that results in innovation.
(Miles, 2008)	To depict how service industries vary in such areas as products, markets, work organization, and technological characteristics - most being very distinctive from primary and secondary industries	It was used input/output and other data	Innovation survey data indicates that some service organizations behave very much like high-technology manufacturing. This is especially true of technology-based, knowledge-intensive business services (T-KIBS). Distinctive innovation patterns are displayed by KIBS based more on professional knowledge and by large network-based service firms, while many smaller service firms conform to a supplier-driven pattern.

Cluster 1: Innovation: concepts and process

Article	*Focus of the study*	*Method/ Sample*	*Main insights*
(Miozzo and Soete, 2001)	To outline a taxonomy of services based on their technological linkages with manufacturing and other service sectors. The effect of recent technological changes on the transformations in business organisation, industry structure, internationalization, and the role of transnational corporations in these technology-intensive service sectors is explored.	–	The taxonomy identifies a number of technology-intensive service sectors closely related to the use of information that are essential to growth.
(Pavitt, 1984)	To describe and explain sectoral patterns of technical change.	2000 significant innovations in Britain since 1945	Innovating firms principally in electronics and chemicals are relatively big, and they develop innovations over a wide range of specific product groups within their principal sector, but relatively few outside. Firms principally in mechanical and instrument engineering are relatively small and specialised, and they exist in symbiosis with large firms, in scale intensive sectors like metal manufacture and vehicles, who make a significant contribution to their own process technology. In textile firms, on the other hand most process innovations come from suppliers.

(*continued*)

Table 1.5 Resulting clusters from the co-citation analysis performed on the 23 most cited articles (*continued*)

Cluster 2: Knowledge: creation and sharing, co-production and transfer			
Article	*Focus of the study*	*Method/ Sample*	*Main insights*
(Bettencourt et al., 2002)	To develop co-production management model	25 in-depth interviews were conducted with twelve TechCo associates and thirteen clients.	The co-production model illustrates the importance of considering clients as "partial employees" of the service provider firms and applying traditional employee management practices to developing effective client partnerships.
(Miozzo and Grimshaw, 2005)	To explore the lessons for modularity that can be drawn from the outsourcing of KIBS.	Drawing on an empirical study of IT outsourcing in the UK and Germany.	This results in the need for knowledge and organisational coordination in the form of the transfer of staff from the client and the retained IT organisation. Modularity is often presented as a design strategy that stimulates innovation.
(Nonaka and Takeuchi, 1995)	The main contribution of the book "The Knowledge-Creating Company" is an outline of knowledge creation, use and forms of knowledge.	It includes a novel theory from two authors supported by their case studies from Japanese industry and an extensive philosophical introduction into Western and Eastern epistemology.	The types of implicit knowledge should add a third dimension which may also be important for knowledge-creating, innovative organizations.

Cluster 3: Innovation networks and cooperation

Article	*Focus of the study*	*Method/Sample*	*Main insights*
(Bessant and Rush, 1995)	This paper examines the implications of technology transfer within such models, identifying the components of managerial capabilities required to absorb and assimilate new inputs of technology required for successful transfer.	–	Recent models of the innovation activity depict the process as non-linear, and characterised by multiple interactions, systems integration and complex networks. Particular attention is paid to the intermediary roles which can be played by consultants in bridging the 'managerial gap', the changing nature and scope of services offered by consultants and the contributions they can make within technology policy.
(Miles et al., 1995)	This report aims to highlight the contributions of KIBS to innovation; provide the agenda for coherent analyses of KIBS innovation processes; and, draw recommendations for a consideration of KIBS in policy-making.	Case studies of innovative KIBS	There is much evidence that the knowledge intensity of all sectors of the economy is increasing. R&D becomes increasingly the basis of new techniques, and networks of innovators become increasingly the basis of accumulation of the knowledge that results in innovation.

(*continued*)

Table 1.5 Resulting clusters from the co-citation analysis performed on the 23 most cited articles (*continued*)

Cluster 3: Innovation networks and cooperation

Article	*Focus of the study*	*Method/Sample*	*Main insights*
(Muller and Zenker, 2001)	Focusing on innovation interactions between manufacturing small- and medium-sized enterprises (SMEs) and KIBS, the empirical analyses grasps KIBS position in five regional contexts.	–	The paper gives an overview of the role and function of KIBS in innovation systems and their knowledge production, transformation and diffusion activities. The analysis leads to the conclusion that innovation activities link SMEs and KIBS through the process of knowledge generation and diffusion.
(Tether and Hipp, 2002)	To examine patterns of innovation and sources of competitiveness, the purpose is to investigate how these patterns differ across services, and in particular how knowledge intensive and technical service firms differ from services more generally.	German service firms	The analysis finds a high degree of customization in the output of service firms, especially amongst the knowledge intensive and technical service firms, the innovation activities of which are also relatively more oriented to product innovation. Knowledge intensive and technical service firms also invest more heavily in information communication technologies, whilst other services invest heavily in non-ICTs. Thus significant diversity is found between the groups of firms examined, but much diversity also exists within the groups.

Cluster 3: Innovation networks and cooperation

Article	*Focus of the study*	*Method/Sample*	*Main insights*
(Windrun and Tomlinson, 1999)	The paper draws an important distinction between the quantity of services in a domestic economy and the degree of connectivity between services and other economic activities. Particular attention is paid to the role and impact of knowledge intensive service sectors to international competitiveness.	In addition to the UK and Germany, data is drawn from the Netherlands and Japan.	Using these four comparative cases it explores the distinction between a high representation of services in the domestic economy, and the innovation spill-overs facilitated by a high degree of connectivity between services and other economic sectors within a domestic.
(Corrocher et al., 2009)	To investigate the sectoral variety and common patterns across different typologies of KIBS	Original survey-based firm-level dataset: The case of Lombardy – a highly developed manufacturing area	When examining in more depth the variables that are associated with cluster membership, one finds that firm strategy is the most significant determinant, with size, customer location, and training also playing a role in defining cluster specificities.

(*continued*)

Table 1.5 Resulting clusters from the co-citation analysis performed on the 23 most cited articles (*continued*)

Cluster 4: Location and relationship with clients			
Article	*Focus of the study*	*Method/Sample*	*Main insights*
(Hertog, 2000)	To make an analysis of the role played by Knowledge Intensive Business Services (KIBS) in innovation. It presents a four-dimensional model of (services) innovation that point to the significance of such non-technological factors in innovation as new service concepts, client interfaces and service delivery system. The various roles of service firms in innovation processes are mapped out by identifying five basic service innovation patterns.	–	KIBS are seen to function as facilitator, carrier or source of innovation, and through their almost symbiotic relationship with client firms, some KIBS function as co-producers of innovation. In addition to discrete and tangible forms of knowledge exchange, process-oriented and intangible forms of knowledge flows are crucial in such relationships.
(Freel, 2006)	To draw broad comparisons between patterns of innovation expenditure and output, innovation networking, knowledge intensity and competition within KIBS and manufacturing firms. The principal interest of the paper is in identifying the factors associated with higher levels of innovativeness, within each sector, and the extent to which such "success" factors vary across sectors.	Estimation of the production functions takes the form of three ordered logit equations 'Survey of Enterprise in Northern Britain': 1,161 small firms (KIBS; N5563 and manufacturing firms; N5598). KIBS disaggregated as technology based KIBS (t-KIBS; N5264) and professional KIBS (p-KIBS; N5299).	The results of the analysis appear to offer support for some widely held beliefs about the relative roles of "softer" and "harder" sources of knowledge and technology within services and manufacturing

Cluster 4: Location and relationship with clients

Article	*Focus of the study*	*Method/Sample*	*Main insights*
(Koch and Stahlecker, 2006)	To analyse interrelationships between KIBS foundations and their respective innovation and production systems by performing qualitative and conceptual in-depth studies of three German metropolitan regions. The present contribution has mapped out some of the interrelationships between regional innovation systems and KIBS foundations in a qualitative and explorative way.	Qualitative and conceptual in-depth studies of three German metropolitan regions.	The analysis has shown that, and how, the regional techno-economic and institutional structures influence the early development of the KIBS sector. The main reasons for the observed different foundation patterns in the regions examined lie in the different endowment with (potential) incubator organizations providing knowledge, human capital, and opportunities for the foundation of KIBS as well as for their sustained development. Thus, especially in the early stages of the development of newly founded KIBS, geographical proximity to their suppliers and clients seems to play a crucial role. This fact can also be attributed to the prominent role of (tacit) knowledge in the examined sector.

(*continued*)

Table 1.5 Resulting clusters from the co-citation analysis performed on the 23 most cited articles (*continued*)

Cluster 4: Location and relationship with clients			
Article	*Focus of the study*	*Method/Sample*	*Main insights*
(Miles, 2005)	To examine KIBS in the European Union, highlighting key similarities and differences in their development across Member States. KIBS are one of the fastest growing areas of the European economy, and are increasingly important contributors to the performance of the sectors who are their clients.	Statistics on KIBS in the European Union are examined. Scenario analysis is used to examine policy issues concerning KIBS. These are based on deskwork: group discussion would be a valuable complement to this approach.	KIBS are continuing to grow at rapid rates, and are experiencing qualitative change. The growth is associated with outsourcing, the internationalization of services, and the growth in demand for certain forms of knowledge. Many KIBS sectors are becoming more concentrated (though most KIBS sectors feature a higher share of small firms than does the economy as a whole).
(Simmie and Strambach, 2006)	To develop a theoretical position for understanding the role of services in innovation in post-industrial societies. The paper suggests a systematic theoretical approach to understanding the currently under-theorised role of services in general and KIBS in particular in innovation. It also points to the importance of the geography of specialised services.	This study develops an evolutionary and institutional approach to understanding the role of certain specialist services in innovation and illustrates how significant they are for the economies of large metropolitan areas in England and Germany.	The paper argues that the role of KIBS in innovation may be understood theoretically in terms of evolutionary and institutional economics. Urban economies are path dependent interactive learning systems that develop individually through time. They are increasingly characterised by networked production systems in which KIBS play a key role in the transfer of bespoke knowledge between actors both within and from outside individual cities. As a result, KIBS make a significant and place specific contribution to innovation in the cities where they are located.

Table 1.6 Top sources of citations in the field of KIBS

	Total citations	*Total articles*	*2013 impact factor*
Research Policy	**592**	**9**	**2,598**
Service Industries Journal	**329**	**27**	**2,617**
California Management Review	184	2	1,944
Journal of Evolutionary Economics	104	5	,675
Industry and Innovation	62	7	1,116
Technovation	46	2	**2,704**
International Journal of Technology Management	41	8	,492
Regional Studies	39	6	1,756
Industrial Marketing Management	36	4	1,897
Journal of Economic Geography	26	3	**2,821**
Journal of Knowledge Management	25	4	1,257
Journal of International Marketing	23	1	2,000
Service Business	20	6	,878
Organizational Studies	19	1	**2,504**
Knowledge Management Research and Practice	16	3	,683
Tijdschrift Voor Economische En Sociale Geografies	14	2	1,012
Journal of Business and Industrial Marketing	13	3	,907
Human Relations	11	1	1,867
International Small Business Journal	11	1	1,397
Economia Política	7	5	,533

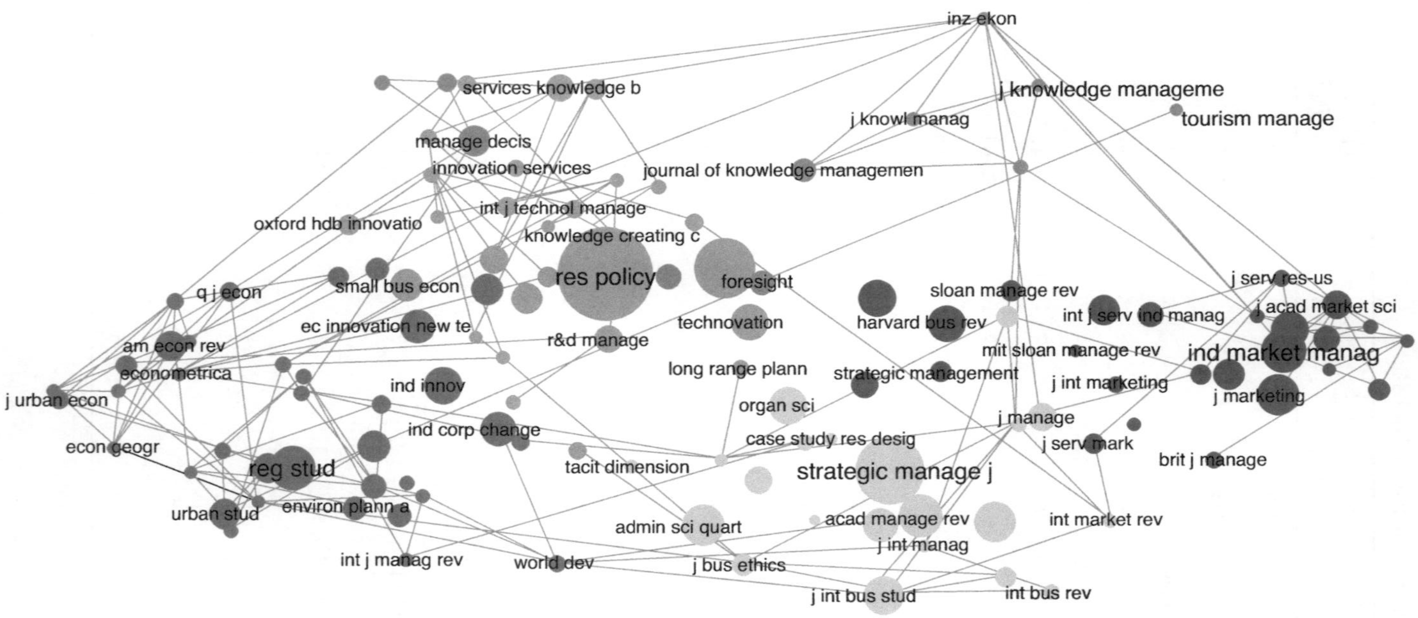

Figure 1.5 Network of co-cited sources in the 140 articles and respective clusters

Table 1.7 Clusters resulting from the most-cited sources (number of citations in brackets)

Cluster 1 – Economics, Geography and Environmental Studies	*Cluster 2 – Engineering, Operations Research and Management Studies*
Regional Studies (112)	Research Policy (450)
Industry and Innovation (72)	Service Industries Journal (187)
Economics of Innovation and New Technology (62)	Technovation (72)
Industrial and Corporate Change (62)	R&D Management (39)
European Planning Studies (56)	International Journal of Technology
Urban Studies (51)	Management (22)
Journal of Economic Geography (49)	
American Economic Review (47)	
Cluster 3 – Business	*Cluster 4 – Management*
Industrial Marketing Management (109)	Strategic Management Journal (259)
Journal of Marketing (92)	Academic of Management Review (112)
California Management Review (66)	Academic of Management Journal (93)
Harvard Business Review (73)	Organization Studies (40)
Journal of Business Research (70)	Management International Review (24)
Journal of Business & Ind. Marketing (27)	Journal of International Management (11)
Journal of International Marketing (17)	
Cluster 5 – Strategy, Management, Operations, Information and Library Science	
Management Decisions (54) Foresight (30) Journal of Knowledge Management (50)	

Regarding authorship, the results show that 275 authors[6] are responsible for the 140 articles included in the sample. It is interesting to note that the authors with more publications are: Doloreux, D. (8 publications) and Miozzo, M. (6 publications), followed by Santos-Vijande ML; Landry, R.; Amara, N.; Grimshaw, D.; Shearmur, R. and Balaz, V. (all with four publications, each one). It's also important to highlight that one can find 112 different first authors in the sample, from 92 different institutions and 30 different countries.

Table 1.8 shows the 50 most frequently cited authors, as well as the number of citations per author and the number of articles published by author.[7] As can be seen, 38 of this authors have at least 10 citations and the most cited authors are Muller, E. (215 citations), Hipp, C. (172 citations) and Bettencourt, LA (165 citations). The authors with higher numbers of articles published are Doloreux, D. (5 articles), Bader, MA (4 articles) and Santos-Vijande, L. (4 articles)

Following the overall analysis of the 140 articles, Figure 1.6 shows the co-citations of authors considering the 38 authors who were cited at least 10 times. These 38 authors were grouped into clusters as shown in Table 1.9.

Table 1.8 Top-cited authors in the field of KIBS

Authors	*Total citations*	*Total articles*	*Authors*	*Total citations*	*Total articles*
Muller, E	**215**	1	Consoli, Davide	17	1
Hipp, C	172	1	Tseng, Chun-Yao	17	1
Bettencourt, LA	165	1	Hoyler, Michael	6	1
Amara, Nabil	84	3	Ofarrell, PN	16	1
Miozzo, Marcela	64	2	Tomlinson, M	16	1
Toivonen, Marja	63	1	Bader, Martin A.	14	1
Antonelli, C	54	1	Balaz, V	13	4
Aslesen, Heidi Wiig	37	2	Mas-Verdu, Francisco	13	1
Yam, Richard C. M	34	1	Ramsey, Elaine	13	2
Doloreux, David	33	**5**	Santos-Vijande, ML	12	4
Andersson, Martin	32	2	Bettiol, Marco	11	2
Klerkx, Laurens	32	1	Koch, Andreas	11	1
Larsen, JN	32	1	Najafi-Tavani, Zhale	10	3
Abreu, Maria	30	1	Bengtsson, Lars	9	1
Grimshaw, Damian	30	2	Koschatzky, Knut	9	1
Wood, P	30	1	Manning, Stephan	9	1
Aarikka-Stenroos, Le	28	1	Toivonen, M	9	1
Miles, I	28	1	Javalgi, Rajshekhar	8	1

Authors	*Total citations*	*Total articles*	*Authors*	*Total citations*	*Total articles*
Wong, PK	27	1	Musolesi, Antonio	8	1
Corrocher, Nicoletta	26	3	Pardos, Eva	8	1
De Marchi, Valentina	26	2	Zaefarian, Ghasem	8	2
Murray, Janet Y	23	1	Camuffo, Arnaldo	7	1
Hauknes, Johan	22	1	Kaepylae, Jonna	7	1
Shearmur, Richard	20	1	Viljamaa, Anmari	7	1
Skjolsvik, Tale	19	1	Chiaroni, Davide	6	1

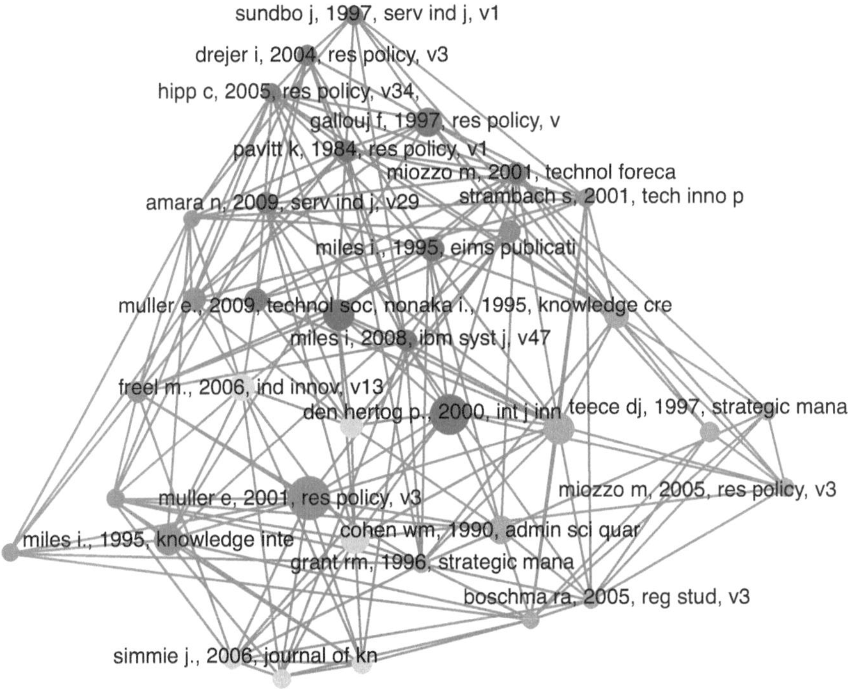

Figure 1.6 Network of co-cited authors in the 140 articles and respective clusters

Affiliation and collaboration networks

Concerning to affiliation and collaboration networks it's possible to find 156 institutions, from 34 countries that underlie the 140 articles included in the sample of this research. The institutions with more researchers publishing in this field

Table 1.9 Clusters of most-cited authors (number of citations in brackets)

Cluster 1	*Cluster 2*	*Cluster 3*	*Cluster 4*
Den Hertog, P (49)	Bettencourt, LA (29)	Muller, E (74)	Miles, I (28)
Miles, I (29)	Cohen, WM (28)	Miles, I (29)	Freel, M. (20)
Muller, E (28)	Nonaka, I (19)	Tether, BS (16)	Simmie, J (17)
Amara, N (13)	Barney, J (13)	Windrum, P (12)	Corrocher (15)
Drejer, I (13)	Boschma, RA (11)	Czarnitzdi, D (11)	Koch, A (12)
Pavitt, K (13)	Miozzo, M (11)	Wong, PK (11)	Aslesen, HW (10)
Hipp, C (11)	Zahra, SA (11)	Bessant, J (10)	
Sundbo, J (11)	Grant, RM (10)	Strambach, S (10)	
Miozzo, M (10)	Teece, DJ (10)		

are located in Europe (mainly England, Italy and Spain) or Canada, although authors almost from all continents (excluding Africa) were included in the sample. The institutions top five ranking includes University of Manchester (England), University of Padua (Italy), University of Ottawa (Canada), University of Laval (Canada) and University of Oviedo (Spain). Some of these institutions present the greatest number of co-authorships (Table 1.10).

As it can be seen, most of the paired of institutions term of co-authorship are geographically near, for instance, University of Leeds and University of Manchester, with 4 co-authored publications. Notwithstanding, one should also mention the international co-autorship: University of Manchester (England), Suffolk University (USA) and Bocconi University (Italy) with 2 co-authored papers.

Word networks

Aiming to increase our understanding of the subjects discussed in the publications of KIBS field, a lexical analysis of the words that can be more frequently found in the bibliographic database was conducted, considering the title and abstracts of the 140 papers included in the sample, which allowed to generate a "cloud words" (Figure 1.7) formed by the words that occurred more frequently in those texts (Table 1.11). Tittle and abstracts of all papers were exported to the French site TreeCloud.org that generates a "tree of words," where the words are grouped as clouds concerning their semantic proximity within the text. The result show three main groups of words, one of them related to studies and activities of the firms, manufacturing, services and KIBS, highlighting innovation and knowledge. A second group refers to management

Table 1.10 Top institutions with co-authored publications in the field of KIBS

Institution 1	*Number of articles*	*Institution 2*
University of Leeds (England)	4	University of Manchester (England)
Bocconi University (Italy)	2	Insubria University (Italy)
University of Oviedo (Spain)	2	University Autonoma of Madrid (Spain)
University of Oviedo (Spain)	2	University of Extremadura (Spain)
University of Laval (Canada)	2	University of Quebec (Canada)
University of Ottawa (Canada)	2	University of Quebec (Canada)
Seinäjoki Univ. of Applied Sciences (Finland)	2	Lappeenranta Univ. of Technology (Finland)
Institution 1	*Institution 2*	*Institution 3*
Suffolk University (USA)	Bocconi University (Italy)	University of Manchester (England) (2 articles)

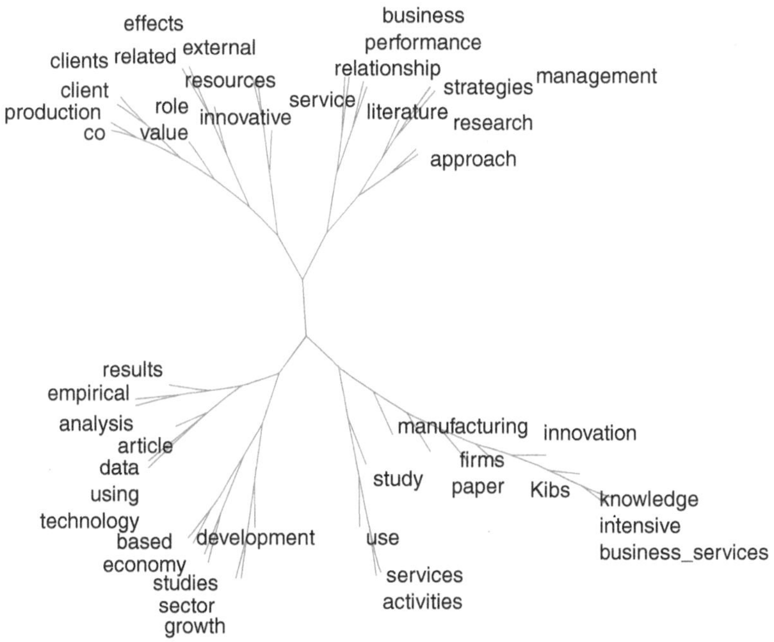

Figure 1.7 Word network

Table 1.11 Count higher frequencies of words

Word	*Word count*	*Word*	*Word count*
Knowledge	443	External	44
KIBS	343	Empirical	44
Innovation	290	Use/using	43/39
Service/services	116/108	Growth	42
Firms	212	Data	39
Intensive	197	Innovative	38
Business_services	187	Clients	37
Paper	113	Approach	37
Based	85	Literature	36
Study	75	Relationship	35
Performance	72	Value	35
Results	70	Resources	34
Research	67	Studies	34
Business	66	Effects	34
Development,	55	Client	33
Activities	55	Technology	33
Management	50	Production	33
Manufacturing	48	Sector	32
Analysis	46	Economy	32
Strategies	46	Article	32
Role	45	Related	32

strategies and business performance, with particular emphasis in external activities as the relationships and clients. The last group focuses on the results and technology uses and also in the growth and development of the sector and the economy.

Based on the bibliometric study presented so that, there seems to be evidence that KIBS research is extremely relevant, since the number of papers and researchers is not, yet, very high. In addition, in Portugal this field of research displays a very limited representation, with only there are just four papers published in Web of Science (the first one was published in 2012).

Conclusions and perspectives for research

Increasingly researches attached importance to the field of KIBS, which was an emerging research field. This paper used the complex network analysis of bibliometric analysis to study the KIBS field, in order to depict the intellectual structure of KIBS, highlighting the maturation of the field. The study also provides information about scientific journals, authors, affiliations and countries of the existing literature, in a coherence effort.

The paper used the Web of Science database, for the period between 1994 and 2014. We used the query terms "KIBS, "Knowledge Intensive Business Services" and "Knowledge-Intensive Business Services" in the bibliographic field "Topic" to search related publications, and we found 140 papers, after redefine document types (using only articles), research domain (Social Sciences) and research areas (business economics).

The study considered keywords, authors, sources and other subject categories of an article as actors to establish the keyword co-ocurrance network, authors' collaboration, source network and the subject category co-ocurrance network. The linkage of the keywords in the keyword co-ocurrance network indicates that both appeared in one paper, and the same for the authors, which means they cooperated in one paper, at least. Similar to the linkage of sources or to other subject categories.

Despite the noticeable increase in the last decades, KIBS research is still an emerging theoretical field. The division of KIBS into four clusters brought coherence to its analysis. These clusters reflect the key dimensions that allow a better understanding about the conceptual definition of KIBS, the interaction with other firms and its role in the economy. This study aimed to find the most important keywords, researchers, scientific journals, subject categories and the development process of hot topics in the field of KIBS. After identifying how the topic is defined in the international literature and the progress achieved in the research field, in a first moment, and evaluating/measuring the research productivity, key authors and scientific journals with the highest impact on this research field, and the networks of association between the respective institutions and countries of origin, some characteristics of these networks were analysed. It allowed us to identify topics and dimensions which are related to KIBS in order to support future research.

In the subject category co-occurrence network, the hot categories were plus Business and Economics (according to our redefinition), Strategy, Operation Research and Management Studies, Geography and Environmental studies, Engineering and Information and Library Science. As one can see, KIBS research is applied in many areas, therefore researchers could do more empirical analysis in other industries except for IT services, communication, and computer services. It may be creative to apply KIBS theory to some different areas, for instance, an emerging area in the literature is the tourism sector.

With respect to keywords, we found that the relationship between the studies became more and more close. As the academics in the service innovation field, gradually turned into a research system (Zhu and Guan, 2013). According to the authors, some hot topics were focused on for a long time, such as customer orientation and telecommunication, and others were changeable with years, market or information process over the period 2004–5, globalisation and collaboration over the period 2006–7, then the focus were to innovation process and service innovation model over the period 2008–9, and shifted into internet and network effects over the period 2010–11. This study searched for analyse the research situation, and found the research focus of the field of innovation and knowledge. Few of the papers on the sample used subject category to establish networks and interaction between KIBS and the client. These findings can be useful to give directions to future research.

In our research, we found that, geographically, the highest number of publications on KIBS field, in leading international journals, is found in Europe (especially, England, Italy and Spain), and followed by Canada, USA and Asian countries (with special emphasis on China). For instance, we did not find any publication of African researchers and only one article by Latin American researchers (from Brazil). Co-authoring relationships from different institutions in one country were found but rarely international co-authorships. Only two articles in international co-authorship, we highlight University of Manchester (England), Bocconi University (Italy) and Suffolk University (USA). Manchester University (England) is the institutions with more co-authorship relationships, with other four publications with researchers from University of Leeds (England). It seems to be possible to conclude that internationalisation is a still weak feature in KIBS research. In addition, as the collaboration between KIBS and other firms brings recognised benefits to the latter (Wong and He, 2005) as well as for the whole economy (Shi *et al.*, 2014), it would also be beneficial to take this collaborative research to an international level. Furthermore, internationalisation is a topic that seems to gain prominence in the literature on KIBS (Doloreux and Laperriere, 2014). In the light of these results, internationalisation will be a dimension to be explored in future investigations.

This study uses only the ISI Web of Science database (so we did not consider other important databases) and involves articles published in journals exclusively allocated to the categories of business and economics. Despite its limitations, this study is one of the first attempts to systematically map the research on KIBS using bibliometric tools. Several different bibliometric methods can be used to analyse the same sample and compare the results of different means, as well as studying literature in different periods or using different databases to find different research focuses.

The analysis of 140 scientific articles contributes to the literature on KIBS, and the structure form on the analysis provides a solid basis for how to conceptualise KIBS in future research.

Notes

1 "VOSviewer can (for example) be used to construct maps of authors or journals based on cocitation data or to construct maps of keywords based on co-occurrence data. The program offers a viewer that allows bibliometric maps to be examined in full detail. VOSviewer can display a map in various different ways, each emphasizing a different aspect of the map. It has functionality for zooming, scrolling, and searching, which facilitates the detailed examination of a map. The viewing capabilities of VOSviewer are especially useful for maps containing at least a moderately large number of items (e.g.,

at least 100 items). Most computer programs that are used for bibliometric mapping do not display such maps in a satisfactory way" (van Eck and Waltman, 2010, p. 524).

2 CitNetExplorer is a software tool for visualizing and analyzing citation networks of scientific publications. The tool allows citation networks to be imported directly from the Web of Science database. Citation networks can be explored interactively, for instance by drilling down into a network and by identifying clusters of closely related publications.

3 The academic community usually recognises ISI journals as "certified journals", and the ones bearing a prominent role in scientific knowledge diffusion.

4 Last updated on May 6, 2015.

5 Impact factor is a quantitative measure citation-based of the importance and significance of a scientific journal Garfield, E. (1979) Is citation analysis a legitimate evaluation tool? *Scientometrics*, 1, 359–75. Considering impact factor as a gross approximation of the reputation and overall scientific standing of academic journals in which articles have been published, we included 2013 impact factor of journals referred.

6 Although there is a potential danger for mistakes arising from changes in the authors' names.

7 It refers to the first author of the paper.

References

Aarikka-Stenroos, L. and Jaakkola, A. E. (2012) Value co-creation in knowledge intensive business services: a dyadic perspective on the joint problem solving process. *Industrial Marketing Management*, 41, 15–26.

Abecassis-Moedas, C., Ben Mahmoud-Jouini, S., Dell'era, C., Maceau, D. and Verganti, R. (2012) Key resources and internationalization modes of creative knowledge-intensive business services: the case of design consultancies. *Creativity and Innovation Management*, 21, 315–31.

Abreu, M., Grinevich, V., Kitson, M. and Savona, M. (2010) Policies to enhance the 'hidden innovation' in services: evidence and lessons from the UK. *Service Industries Journal*, 30, 99–118.

Alvarez-Gonzalez, J. A. and Gonzalez-Morales, M. O. (2014) The role of knowledge-intensive business services in Spanish local tourist production systems. *Tourism Economics*, 20, 355–71.

Amara, N., Landry, R. and Doloreux, D. (2009) Patterns of innovation in knowledge-intensive business services. *Service Industries Journal*, 29, 407–30.

Amara, N., Landry, R. and Traore, N. (2008) Managing the protection of innovations in knowledge-intensive business services. *Research Policy*, 37, 1530–47.

Andersson, M. and Hellerstedt, K. (2009) Location attributes and start-ups in knowledge-intensive business services. *Industry and Innovation*, 16, 103–21.

Antonelli, C. (1998) Localized technological change, new information technology and the knowledge-based economy: the European evidence. *Journal of Evolutionary Economics*, 8, 177–98.

Aslesen, H. W. and Isaksen, A. (2007) Knowledge intensive business services and urban industrial development. *Service Industries Journal*, 27, 321–38.

Ávilla, L., Barros, I., Madruga, L. and Júnior, V. (2014) Características das publicações sobre Empreendedorismo (Social) no Web of Science no período 2002–11. *Administração Pública e Gestão Social*, 62, 88–100.

Bader, M. A. (2008) Managing intellectual property in the financial services industry sector: learning from Swiss Re. *Technovation*, 28, 196–207.

Bengtsson, L. and Dabhilkar, M. (2009) Manufacturing outsourcing and its effect on plant performance-lessons for KIBS outsourcing. *Journal of Evolutionary Economics*, 19, 231–57.

Bessant, J. and Rush, H. (1995) Building bridges for innovation: the role of consultants in technology transfer. *Research Policy*, 24, 97–114.

Bettencourt, L. A., Ostrom, A. L., Brown, S. W. and Roundtree, R. I. (2002) Client co-production in knowledge-intensive business services. *California Management Review*, 44, 100-+.

Bettiol, M., Di Maria, E. and Grandinetti, R. (2012) Codification and creativity: knowledge management strategies in KIBS. *Journal of Knowledge Management*, 16, 550–62.

Borodako, K., Berbeka, J. and Rudnicki, M. (2014) The potential of local KIBS companies as a determinant of tourism development in Krakow. *Tourism Economics*, 20, 1337–48.

Carmona-Lavado, A., Cuevas-Rodriguez, G. and Cabello-Medina, C. (2013) Service innovativeness and innovation success in technology-based knowledge-intensive business services: an intellectual capital approach. *Industry and Innovation*, 20, 133–56.

Chae, B. (2012) A framework for new solution development: an adaptive search perspective. *Service Industries Journal*, 32, 127–49.

Consoli, D. and Elche-Hortelano, D. (2010) Variety in the knowledge base of knowledge intensive business services. *Research Policy*, 39, 1303–10.

Corrocher, N. and Cusmano, L. (2014) The 'KIBS engine' of regional innovation systems: Empirical evidence from European regions. *Regional Studies*, 48, 1212–26.

Corrocher, N., Cusmano, L. and Morrison, A. (2009) Modes of innovation in knowledge-intensive business services evidence from Lombardy. *Journal of Evolutionary Economics*, 19, 173–96.

De Marchi, V. (2012) Environmental innovation and R and D cooperation: empirical evidence from Spanish manufacturing firms. *Research Policy*, 41, 614–23.

Di Maria, E., Bettiol, M., De Marchi, V. and Grandinetti, R. (2012) Developing and managing distant markets: the case of KIBS. *Economia Politica*, 29, 361–79.

Doloreux, D. and LaPerriere, A. (2014) Internationalisation and innovation in the knowledge-intensive business services. *Service Business*, 8, 635–57.

Doloreux, D. and Mattsson, H. (2008) To what extent do sectors 'socialize' innovation differently? Mapping cooperative linkages in knowledge-intensive industries in the Ottawa region. *Industry and Innovation*, 15, 351–70.

Doloreux, D. and Shearmur, R. (2012) Collaboration, information and the geography of innovation in knowledge intensive business services. *Journal of Economic Geography*, 12, 79–105.

Dragos, C., Dinu, V., Pop, C. and Dabija, D. (2014) Scientometric approach of productivity in scholarly economics and business. *Economic Research-Ekonomska Istrazivanja*, 27, 496–507.

Drejer, I. (2004) Identifying innovation in surveys of services: a Schumpeterian perspective. *Research Policy*, 33, 551–62.

Fernandes, C. I. and Ferreira, J. J. M. (2013) Knowledge spillovers: cooperation between universities and KIBS. *R & D Management*, 43, 461–72.

Ferreira, J. M., Ferreira, F. F., Fernandes, C. M. A. S., Jalali, M., Raposo, M. R. and Marques, C. (2015) What do we [not] know about technology entrepreneurship research? *International Entrepreneurship and Management Journal*, 1–21.

Freel, M. (2006) Patterns of technological innovation in knowledge intensive business services. *Industry and Innovation*, 13, 335–58.

Gallouj, F. Z. and Weinstein, O. (1997) Innovation in services. *Research Policy*, 26, 537–56.

Garfield, E. (1979) Is citation analysis a legitimate evaluation tool? *Scientometrics*, 1, 359–75.

Gimzauskiene, E. and Staliuniene, J. D. (2010) Model of core competence ranking in audit business. *Inzinerine Ekonomika-Engineering Economics*, 21, 128–35.

Grimshaw, D. and Miozzo, M. (2006) Institutional effects on the IT outsourcing market: analysing clients, suppliers and staff transfer in Germany and the UK. *Organization Studies*, 27, 1229–59.

Grimshaw, D. and Miozzo, M. (2009) New human resource management practices in knowledge-intensive business services firms: the case of outsourcing with staff transfer. *Human Relations*, 62, 1521–50.

Hakanen, T. (2014) Co-creating integrated solutions within business networks: the KAM team as knowledge integrator. *Industrial Marketing Management*, 43, 1195–203.

Hauknes, J. (1998) Services in innovation – innovation in services. *SI4S Final Report to the European Commission*. Oslo, TSER program, STEP Group.

Hauknes, J. and Knell, M. (2009) Embodied knowledge and sectoral linkages: an input-output approach to the interaction of high- and low-tech industries. *Research Policy*, 38, 459–69.

He, Z. L. and Wong, P. K. (2009) Knowledge interaction with manufacturing clients and innovation of knowledge-intensive business services firms. *Innovation-Management Policy & Practice*, 11, 264–78.

Hertog, P. D. (2000) Knowledge-intensive business services as co-producers of innovation. *International Journal of Innovation Management*, 04, 491–528.

Hipp, C. and Grupp, H. (2005) Innovation in the service sector: the demand for service-specific innovation measurement concepts and typologies. *Research Policy*, 34, 517–35.

Hoyler, M., Freytag, T. and Mager, C. (2008) Connecting Rhine-Main: the production of multi-scalar polycentricities through knowledge-intensive business services. *Regional Studies*, 42, 1095–111.

Hu, T. S., Lin, C. Y. and Chang, S. L. (2013) Knowledge intensive business services and client innovation. *Service Industries Journal*, 33, 1435–55.

Huang, C. Y. and Ji, L. (2013) Knowledge-intensive business services and economic growth with endogenous market structure. *Journal of Macroeconomics*, 38, 95–106.

Huggins, R. and Johnston, A. (2012) Knowledge alliances and innovation performance: an empirical perspective on the role of network resources. *International Journal of Technology Management*, 57, 245–65.

Klerkx, L. and Leeuwis, C. (2008) Balancing multiple interests: embedding innovation intermediation in the agricultural knowledge infrastructure. *Technovation*, 28, 364–78.

Koch, A. and Stahlecker, T. (2006) Regional innovation systems and the foundation of knowledge intensive business services. A comparative study in Bremen, Munich, and Stuttgart, Germany. *European Planning Studies*, 14, 123–45.

Koch, A. and Strotmann, H. (2006) Impact of functional integration and spatial proximity on the post-entry performance of knowledge intensive business service firms. *International Small Business Journal*, 24, 610–34.

Koschatzky, K. and Stahlecker, T. (2010) The emergence of new modes of RD services in Germany. *Service Industries Journal*, 30, 685–700.

Larsen, J. N. (2001) Knowledge, human resources and social practice: the knowledge-intensive business service firm as a distributed knowledge system. *Service Industries Journal*, 21, 81–102.

Manning, S., Ricart, J. E., Rique, M. S. R. and Lewin, A. Y. (2010) From blind spots to hotspots: How knowledge services clusters develop and attract foreign investment. *Journal of International Management*, 16, 369–82.

Mas-Tur, A. and Soriano, D. R. (2014) The level of innovation among young innovative companies: the impacts of knowledge-intensive services use, firm characteristics and the entrepreneur attributes. *Service Business*, 8, 51–63.

Mas-Verdú, F., Wensley, A., Alba, M. and Garcia Alvarez-Coque, J. M. (2011) How much does KIBS contribute to the generation and diffusion of innovation? *Service Business*, 5, 195–212.

Miles, I. (2005) Knowledge intensive business services: prospects and policies. *Foresight*, 7, 39–63.

Miles, I. (2008) Patterns of innovation in service industries. *IBM Systems Journal*, 47, 115–28.

Miles, I., Andersen, B., Boden, M. and Howells, J. (2000) Service production and intellectual property. *International Journal of Technology Management*, 20, 95–115.
Miles, I., Katrinos, N., Flanagan, K., Bilderbeek, R., Den Hertog, P. and Huntink, W. (1995) Knowledge-intensive business services: users, carriers and sources of innovation. *EIMS Publication No 15,* EIMS, Luxembourg.
Miozzo, M. and Grimshaw, D. (2005) Modularity and innovation in knowledge-intensive business services: IT outsourcing in Germany and the UK. *Research Policy*, 34, 1419–39.
Miozzo, M. and Soete, L. (2001) Internationalization of services: a technological perspective *Technological Forecasting and Social Change*, 67, 159–85.
Muller, E. and Zenker, A. (2001) Business services as actors of knowledge transformation: the role of KIBS in regional and national innovation systems. *Research Policy*, 30, 1501–16.
Murray, J. Y., Kotabe, M. and Westjohn, S. A. (2009) Global sourcing strategy and performance of knowledge-intensive business services: a two-stage strategic fit model. *Journal of International Marketing*, 17, 90–105.
Najafi-Tavani, Z., Giroud, A. and Andersson, U. (2014) The interplay of networking activities and internal knowledge actions for subsidiary influence within MNCs. *Journal of World Business*, 49, 122–31.
Nonaka, I. and Takeuchi, H. (eds) (1995) *The knowledge-creating company*, New York: Oxford University Press.
O'Farrell, P. N. and Moffat, L. A. R. (1995) Business services and their impact upon client performance – an exploratory interregional analysis. *Regional Studies*, 29, 111–24.
Palacios-Marques, D., Gil-Pechuan, I. and Lim, S. (2011) Improving human capital through knowledge management practices in knowledge-intensive business services. *Service Business*, 5, 99–112.
Pavitt, K. (1984) Sectoral patterns of technical change: towards a taxonomy and a theory. *Research Policy*, 13, 343–73.
Peiker, W., Pflanz, K., Kujath, H. J. and Kulke, E. (2012) The heterogeneity of internationalisation in knowledge intensive business services. *Zeitschrift Fur Wirtschaftsgeographie*, 56, 209–25.
Philips, W., Lee, H., Ghobadian, A., O'Regan, N. and James, P. (2015) Social innovation and social entrepreneurship: a systematic review. *Group and Organization Management*, 1–34.
Plaza, B., Galvez-Galvez, C. and Gonzalez-Flores, A. (2011) Orchestrating innovation networks in e-tourism: a case study. *African Journal of Business Management*, 5, 464–80.
Ramos-Rodríguez, A. R. and Ruíz-Navarro, J. (2004) Changes in the intellectual structure of strategic management research: a bibliometric study of the Strategic Management Journal, 1980–2000. *Strategic Management Journal*, 25, 981–1004.
Ramsey, E., Ibbotson, P., Bell, J. and McCole, P. (2005) Internet-based business among knowledge intensive business services: some Irish regional evidence. *Service Industries Journal*, 25, 525–45.
Santos-Vijande, M. L., Diaz-Martin, A. M., Suarez-Alvarez, L. and Del Rio-Lanza, A. B. (2013a) An integrated service recovery system (ISRS): influence on knowledge-intensive business services performance. *European Journal of Marketing*, 47, 934–63.
Santos-Vijande, M. L., Gonzalez-Mieres, C. and Lopez-Sanchez, J. A. (2013b) An assessment of innovativeness in KIBS: implications on KIBS' co-creation culture, innovation capability, and performance. *Journal of Business & Industrial Marketing*, 28, 86–101.
Shearmur, R. and Doloreux, D. (2009) Place, space and distance: towards a geography of Knowledge-intensive business services innovation. *Industry and Innovation*, 16, 79–102.

Shi, X., WU, Y. R. and Zhao, D. T. (2014) Knowledge intensive business services and their impact on innovation in China. *Service Business*, 8, 479–98.

Simmie, J. and Strambach, S. (2006) The contribution of KIBS to innovation in cities: an evolutionary and institutional perspective. *Journal of Knowledge Management*, 10, 26–40.

Skjolsvik, T., Lowendahl, B. R., Kvalshaugen, R. and Fosstenlokken, S. M. (2007) Choosing to learn and learning to choose: strategies for client co-production and knowledge development. *California Management Review*, 49, 110+.

Smith, L. (1981) Citation analysis. *Library Trends*, 30, 83–106.

Stahl, M. J., Leap, T. L. and Wei, Z. Z. (1988) Publication in leading management journals as a measure of institutional research productivity. *Academy of Management Journal*, 31, 707–20.

Strambach, S. (1994) Knowledge-intensive business services in the Rhine-Neckar area. *Tijdschrift Voor Economische En Sociale Geografie*, 85, 354–65.

Tether, B. S. and Hipp, C. (2002) Knowledge intensive, technical and other services: patterns of competitiveness and innovation compared. *Technology Analysis & Strategic Management*, 14, 163–82.

Toivanen, H. (2014) The shift from theory to innovation: the evolution of Brazilian research frontiers 2005–2011. *Technology Analysis & Strategic Management*, 26, 105–19.

Toivonen, M. and Tuominen, T. (2009) Emergence of innovations in services. *Service Industries Journal*, 29, 887–902.

Tomlinson, M. (1999) The learning economy and embodied knowledge flows in Great Britain. *Journal of Evolutionary Economics*, 9, 431–51.

Tseng, C. Y., Pai, D. C. and Hung, C. H. (2011) Knowledge absorptive capacity and innovation performance in KIBS. *Journal of Knowledge Management*, 15, 971–83.

van Eck, N. and Waltman, L. (2010) Software survey: VOSviewer, a computer program for bibliometric mapping. *Scientometrics*, 84, 523–38.

Viljamaa, A., Kolehmainen, J. and Kuusisto, J. (2010) For and against? An exploration of inadvertent influences of policies on KIBS industries in the Finnish policy setting. *Service Industries Journal*, 30, 71–84.

Wallin, M. W. (2012) The bibliometric structure of spin-off literature. *Innovation*, 14, 162–77.

Windrun, P. and Tomlinson, M. (1999) Knowledge-intensive services and internal competitiveness: a four country comparison. *Technology Analysis and Strategic Management*, 11, 391–408.

Wong, P. K. and He, Z. L. (2005) A comparative study of innovation behaviour in Singapore's KIBS and manufacturing firms. *Service Industries Journal*, 25, 23–42.

Wood, P. (2005) A service-informed approach to regional innovation – or adaptation? *Service Industries Journal*, 25, 429–45.

Yam, R. C. M., Lo, W., Tang, E. P. Y. and Lau, A. K. W. (2011) Analysis of sources of innovation, technological innovation capabilities, and performance: an empirical study of Hong Kong manufacturing industries. *Research Policy*, 40, 391–402.

Zhu, W. and Guan, J. (2013) A bibliometric study of service innovation research: based on complex network analysis. *Scientometrics*, 94, 1195–216.

Zitt, M. and Bassecoulard, E. (1994) Development of a method for detection and trend analysis of research fronts built by lexical or cocitation analysis. *Scientometrics*, 30, 333–51.

2 KIBS and the dynamics of industrial clusters

A Complex Adaptive Systems approach

Benoit Desmarchelier, Faridadh Djellal and Faiz Gallouj

This chapter aims at intertwining – in a theoretical and operational way – three strands of literature: (1) innovation through Knowledge Intensive Business Services (KIBS thereafter), (2) industrial clusters' dynamics, and (3) Complex Adaptive Systems.

KIBS are services which are processing, generating, and diffusing knowledge within the economy, and as such they are largely regarded as important (co-)producers of innovations (Miles *et al.*, 1995; Gadrey and Gallouj, 1998; Den Hertog, 2000; 2002; Gallouj, 2002), as well as a promising engine for economic growth (Desmarchelier *et al.*, 2013a) and a key component of regional and national innovation systems (Muller and Zenker, 2001) and of technological and sectoral systems of innovation alike. Typical KIBS activities are training services, R&D, engineering services and consultancy in its various forms (technical or not). KIBS include both traditional professional services (such as legal services, audit and accountancy, market research, personnel services, management consultancy, etc.) and new technology based services. According to Miles *et al.*, (1995), regarding "*their relation to new technology*, compared to the latter, the former are "*users rather than agents in development and diffusion*" (p. 27) of new technologies. Universities are often not included into the broad category of KIBS (Muller and Zenker, 2001; Miles *et al.*, 1995). They have indeed many functions (e.g. teaching and fundamental/academic research), which are not directly oriented towards businesses' technological (and non-technological) needs. However, some of their functions clearly fit into KIBS purposes, especially but not exclusively new technology based KIBS (ex. technical training, technical consultancy, business funded R-D, establishment of research centers in partnership with businesses) and industrial clusters' studies very often highlight their central role in explaining clusters emergence (Saxenian, 1994; Audretsch and Feldman, 1996a; 1996b). The present study itself also underlines universities' role in favoring the emergence of new technologies within an industrial cluster. In this paper we therefore include universities and research bodies within the KIBS category and the empirical part is mainly focused on such types of KIBS.

More generally, KIBS central role within successful industrial clusters has been emphasised since the birth of this latter concept. Indeed, in Porter's words "*clusters are geographic concentrations of interconnected companies, specialized suppliers, service providers, firms in related industries, and associated institutions (e.g. universities, standard agencies, trade associations) in a particular field that compete but also cooperate*" (Porter, 2000, p. 15). In this definition, KIBS enter mainly into the "*associated institutions*" category, as it includes "*universities, think-tanks, vocational training providers*" (p. 17).

An important and highly debated question in economic geography is how to explain the dynamics of these clusters, i.e. their emergence and evolution through time (Frenken *et al.*, 2015; Boschma and Fornhal, 2011). Two main theories are generally explored, without being confronted[1]: the cluster life cycle theory (Menzel and Fornahl, 2010; Shin and Hassink, 2011; Audretsch and Feldman, 1996) – which adopts mainly an aggregate[2] point of view – and a network-based approach (Saxenian, 1994). Surprisingly, these two theories pay little attention to KIBS as a potential driver of clusters' dynamics.

We show in this paper that taking KIBS into account requires considering an alternative and integrative approach that conciliates these two theories. In particular, we argue that Complex Adaptive Systems – or CAS – (Martin and Sunley, 2011; Holland, 2012) constitute a promising candidate for such a synthesis. Since these systems are mainly encountered into the theoretical literature (Dilaver *et al.*, 2014; Albino *et al.*, 2003; Squazzoni and Boero, 2002; Boero *et al.*, 2004), we choose to justify our theoretical stance by studying KIBS' leading role within an existing industrial cluster, conceived as a CAS: Skywin (aeronautics in Wallonia region, Belgium).

The remaining of the paper is organised in two parts: we begin by discussing about the competing theories of clusters dynamics and we advocate for the CAS approach, then we conduct our empirical analysis in order to illustrate the usefulness of this theoretical stance.

Life cycle theory vs. network-based approach: the need for a synthesis

From product and industry life cycles to cluster life cycle

The life cycle hypothesis in economics and management literature is, in its original form, a descriptive model aiming at synthesizing in a coherent manner a wide variety of stylised facts about the evolution through time of the marketed products.

Pioneers of the product life cycle (PLC) theory, Utterback and Abernathy (1975) portray product evolution in three successive steps: (1) the "uncoordinated process, within which firms undertake mainly product innovations aiming at improving their technical performance, (2) the "segmental process, where firms modify minor characteristics for increasing product variety and earning market shares, then (3) the "systemic process" where innovation efforts focus on reducing production costs. These three phases are also and more often labeled: "fluid, "transition" and "specific" phases (Abernathy and Utterback, 1978). Even though authors claim that "*there is reason to believe that in any cases the progression may stop for long periods, or even reverse*" (Utterback, and Abernathy, p. 645), they insist on the high degree of predictability/determinism in the way products evolve through time. Klepper (1996, 1997) systematises this PLC into a model of industry life cycle within which firms' entries, exit, growth and innovations in "*technologically progressive industries*" (p. 564) are the driving forces behind the PLC.

The general character of this theory has been challenged by the advent of service economies. In particular, Barras (1986) points out the existence of a "reverse product cycle" (RPC) within service sectors. In this view, service firms acquire innovations (mainly information technologies) coming from manufacturing sectors first for improving the efficiency of service operations. Then comes the stage of service quality improvement, and eventually the production of totally new services. The PLC is supposed to be reversed as far as process innovations precede product innovation in the cycle. Likewise, this life cycle theory of the innovation dynamics in services proved to be incomplete. Gallouj (1998) argues that it reflects a "*technological bias*" (p. 128): indeed in Barras model (1986), services cannot innovate by themselves as their innovations mainly come from the use of the so-called "enabling technologies, i.e. information technologies. Non-technological forms of innovation which are important in services and which concern not only the organisation and the process but also the product are not taken into account by the RPC theory. In contrast, Gallouj (1998) finds that many KIBS perform several types of non-technological innovation including "ad hoc" innovations, i.e. custom made innovations adapted to their clients' needs.

Gallouj and Weinstein (1997) proposed a more complete view of firms' innovations (whether they originate from industry or services). Adopting a characteristics-based approach of the product – good or service – conceived as a set of technical and service characteristics[3], they identify six different modes of innovation: radical, improvement, incremental, ad hoc, recombinative and formalisation (see Gallouj and Weinstein, 1997 for details). These six modes are not exclusive to each other nor a priori ordered in a pre-determined sequence. Another important point as regards the life cycle theory is that, according to the authors, the PLC encompasses only "one point of entry" for innovations: the technical characteristics of the product. It follows that the life cycle conception offers a limited and deterministic view of innovation dynamics. Nevertheless, it is a popular metaphor for reporting industrial clusters' evolutions.

A first exploration of the life cycle theory applied to industrial clusters is undertaken by Audretsch and Feldman (1996a; 1996b). According to them, the main driver of firms' agglomeration is the low transferability of tacit knowledge through long distances. Following Klepper's industry life cycle, they postulate that tacit/ localised knowledge is important in early developments of a given industry, fostering a certain degree of firms' agglomeration. However, as the industry becomes mature, a dominant design emerges and the product becomes standardised. Firms thus mainly rely on codified knowledge and information, which are easy to share in long distances. The initial clustering is thus replaced by a movement of firms' dispersion when the industry reaches maturity.

Even though KIBS are not explicitly mentioned by Audretsch and Feldman (1996a; 1996b), universities are seen by these authors as an important source of tacit knowledge and are thus a key focal point for early clusters' developments. Moreover, knowledge codification process appears to be the driving force of the life cycle. Arguably, services and KIBS in particular are major actors in knowledge processing (Gallouj, 2002) and transmission (Miles *et al.*, 1995; Lau and Lo, 2015) and should thus be regarded as key actors in explaining clusters' evolution. This service-friendly theory is questioned by Klepper (2010), who rather emphasises the role of a "spinoff process, that is of the emergence of "*firms with one or more founders that previously worked at another firm*" (p. 16).

Menzel and Fornahl (2010) for their part also propose a knowledge-driven clusters' life cycle theory, summarised in Figure 2.1. Two dimensions of the cluster are considered: the number of employees and the heterogeneity of "accessible knowledge." The main driver of the cluster life cycle, addressed in terms of number of employees, is a gradual process of knowledge homogenisation among the members of the cluster. Although similar to Audretsch and Feldman's approach (1996a; 1996b) by the role it attributes to the nature of knowledge in the dynamics of the cluster, this theory has the advantage to avoid too deterministic evolutions from emergence to death, since clusters can always enter into loops of self-sustainment, successive cycles of growth and decline, or even re-orient themselves through a process of "transformation". This adaptability is determined by the degree of knowledge heterogeneity and by the openness to new comers of incumbent firms' networks.

Interestingly, cluster life cycle theory has moved away from the original product and industry life cycles, as it becomes less deterministic and more influenced by local drivers, notably clusters' ability to maintain a healthy degree of knowledge heterogeneity (especially through new intrant firms or new "imported" technologies).

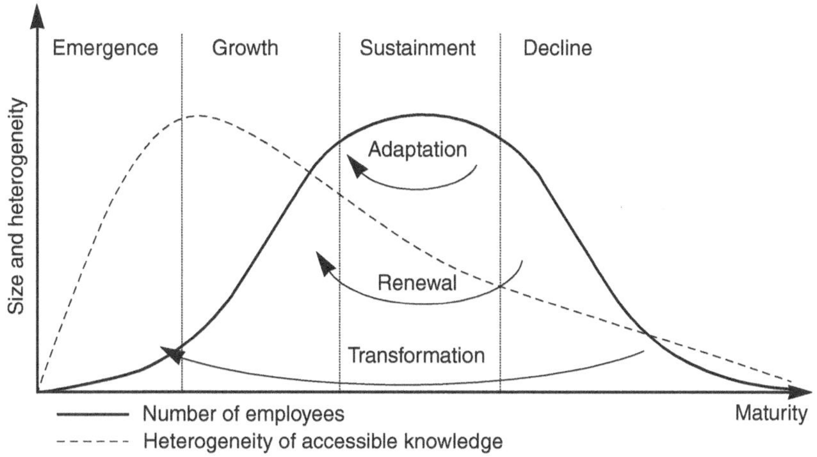

Figure 2.1 Knowledge-based cluster life cycle (from Menzel and Fornahl, 2010 p. 218)

At the opposite of Audretsch and Feldman (1996a; 1996b), for whom clusters' life cycles are "shaped" by the industries that they belong to, Menzel and Fornahl (2010) consider clusters as more independent entities. However, we argue that this approach remains too restrictive and deterministic since, at least for incumbent cluster agents, it considers only one kind of innovation trajectory, i.e. only one kind of knowledge processing mode, namely: formalisation (Gallouj, 2002). Actually, "renewal, "adaptation" and "transformation, i.e. the reverse innovation trajectory or knowledge processing mode (namely differentiation/localisation) (Figure 2.1) are only possible through "external knowledge" (Menzel and Fornahl, 2010, p. 229), thus through exogenous/unexplained factors.

More generally (and beyond Menzel and Fornahl's contribution) another point of criticism towards the life cycle theory is that it gives too few importance to cluster's actors in explaining aggregate dynamics. Indeed, this approach generally considers the cluster – i.e. an aggregation of heterogeneous actors – as a relevant decision maker. Following Martin and Sunley (2011), one can wonder whether "products, technologies, industries and clusters [can] be treated as if they are the economic equivalent of biological organisms" (p. 1301). Besides, even though universities are sometimes cited in early clusters' dynamics (Audretsch and Feldman, 1996a; 1996b), the main actors mentioned are very often the "firms, but we neither know which primary sector of activity they belong to (or whether they all belong to the same sector) nor the nature of the interactions they entertain between each other. Klepper (2010) "spinoff process" is clearer on this point, since spinoffs generally belong to the same sector as the original company – or as the research team in the case of university spin-offs – and are, at first, of smaller size. The exclusive focus on firms is not satisfactory for addressing the clusters dynamics. Indeed, according to Porter (2000), firms are also supported by a number of *"associated institutions"* within clusters, mainly *"universities, think-tanks, vocational training providers*" (p. 17). Regarding the account for the diversity of the actors involved, the network-based approaches are obviously more appropriate.

Networks and clusters dynamics

An alternative explanation of clusters' dynamics is focusing on their internal organisation in the form of networks of interacting entities. According to Newman (2003, p. 2), "a network is a set of items [called] vertices or sometimes nodes, with connections between them, called edges. Systems taking the form of networks abound in the world." Within clusters, nodes are companies and supporting institutions and the edges are all kind of relations between these actors: common investments in R&D, involvements in the same production processes, common patents or shared resources, etc. However, the network is not only a structure, it is also a mode of coordination that fits between market and hierarchy. From an innovation perspective, the network is considered as a coordination mode that is more effective than both market and hierarchy. Indeed resorting to the market assumes the establishment of explicit contracts, while in the field of research and innovation, projects are highly complex and uncertain. This makes it difficult to establish explicit contracts, which furthermore raise the risk that strategic secrets might be divulged (Hakansson, 1989; Callon, 1991; Hakansson and Johansson, 1993). The hierarchy for its part reduces transaction costs but involves the risk of bureaucratisation, which (as already foreseen by Schumpeter) may be prejudicial to innovation. In this network tradition, one can mention here Saxenian's (1994) seminal work, comparing the "*network*" or system-based Silicon Valley and the "*independent firms-based*" Route 128. Saxenian argues that this is the prevalence of horizontal networks between firms and research institutions (e.g. Stanford) in Silicon Valley that allowed this cluster to successfully switch from semiconductors to microcomputers during the 1980s, whereas independent firms in Route 128 failed to adapt to the new technological conditions of that time. An horizontal network is in Saxenian's words a set of actors who "*deepen their own capabilities by specializing*" (p. 4), thus whose links are different from just input-output flows.

An interesting observation here is that, in a network perspective, clusters exist and develop because of a specialisation process, which is the opposite of the knowledge homogenisation generally invoked by the life cycle literature. However, KIBS can be drivers of both dynamics: formalisation of existing tacit knowledge or generation of custom-made (specialised) knowledge (Gallouj, 2002). It follows that if we recognise that KIBS are active members of industrial clusters, we have to acknowledge that both dynamics are possible. This remark about KIBS advocates for an integrative approach recognizing the influence of actors' interactions on the direction taken over time by the cluster as a whole. We argue in the following that Complex Adaptive Systems (CAS) allows integrating, within a single framework, both the aggregate perspective of the life cycle theory and the micro (or multi-agents based) perspective of the network-based approach, without falling into deterministic predictions.

Complex Adaptive Systems: towards and integration of network and life cycle perspectives

Martin and Sunley (2011) recently proposed to consider industrial clusters as a particular type of CAS. According to them, a CAS is a system "*made up of numerous components with functions and inter-relationships that imbue the system as a whole with a particular identity and a degree of connectivity or connectedness*" (p. 1303). Furthermore, a CAS is "characterized by non-linear dynamics because of various feedbacks and self-reinforcing interactions amongst component (…). It is also characterised by emergence and self-organisation." However, reading from these authors, it is not clear what improvements these CAS bring to Saxenian's network-based framework (Saxenian, 1994) nor to the life cycle theory discussed above. Indeed, Martin and Sunley (2011) use a typology of "*meta-models*" covering the various forms of CAS dynamics (inspired by Cumming and Collier, 2005). These meta-models range from deterministic (traditional) life cycles to totally random walks. Life cycle trajectories are envisaged as special case of CAS among others[4]. Among the proposed models, Martin and Sunley argue that clusters dynamics are well depicted by the so-called "*adaptive life-cycle model*"[5] and they try to adapt it to the cluster dynamics. The resulting "*modified cluster adaptive cycle*" that they propose is reproduced in Figure 2.2. Arguably, this "*meta-model*" is very similar to the knowledge-based life cycle proposed by Menzel and Fornahl (2010), as we can easily draw a parallel between their respective alternative trajectories: "*constant cluster mutation*" in Martin and Sunley (2011) stands for "*adaptation*" in Menzel and Fornahl (2010), similarly "*cluster stabilization*" stands for "*renewal*", and "*cluster re-orientation*" stands for "*transformation*". However, the two cycles are not equivalent: in Menzel and Fornahl (2010), knowledge heterogeneity between firms and other actors explains the emergence of a cluster, and the process of knowledge homogenisation drives cluster's evolution. In Martin and Sunley (2011), there is no general mechanism of evolution, since there is no general principle explaining why a cluster shifts from one phase to another. Instead, these authors propose a descriptive list of potential drivers. For instance, cluster re-emergence is possible thanks to "sufficient resources, inherited capabilities and competencies" (p. 1313) left after a phase of decline, or a constant mutation comes from "high rates of spin-offs" (p. 1313). Apart from a chance factor, there is no explanation of why the rate of spin-offs is high or why the remaining capabilities are enough and up-to-date. Another weakness of their model is that, despite their definition of a CAS, they do not precisely ground clusters' dynamics in a network-based view of the actors, and the actors are not considered as heterogeneous entities.

Although we point out limitations of Martin and Sunley's (2011) adaptive cycle, we find very relevant their proposition to rely on CAS for conceptualizing clusters' functioning and dynamics. Rather than trying to classify such systems, we consider that a general definition and a list of properties can justify this point of view.

According to Holland (2012) a CAS "*consists of a multitude of interacting components called agents [...]. The agents are diverse rather than standardized,*

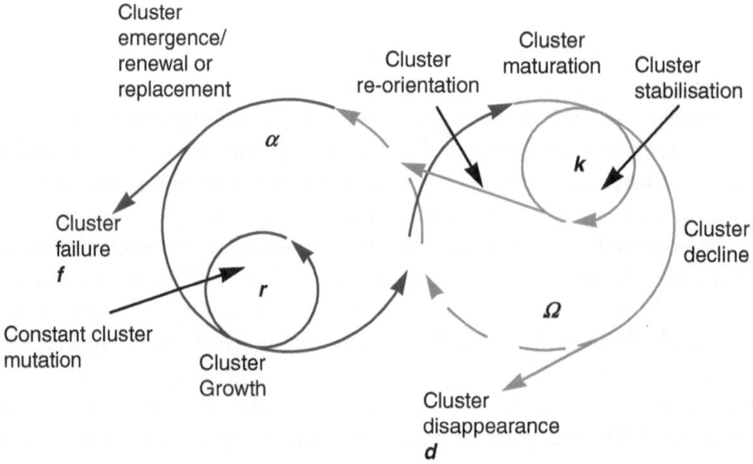

Figure 2.2 Martin and Sunley (2011) "modified cluster adaptive cycle" (p. 1312)

and both their behavior and their structure change as they interact" (p. 57). Furthermore CAS display the following three main features:

1 "*There is no universal competitor or global optimum in a CAS*" (p. 58).
2 "*Innovation is a regular feature of CAS*" (p. 58).
3 "*In a CAS, anticipations change the course of the system*" (p. 60).

Clusters' network structure has already been documented by many authors, including Porter (1998; 2000) and Saxenian (1994). All of them focus on the diversities of the "agents" involved: firms, universities, think-tanks, etc. It might thus be argued that, as structures, clusters are examples of CAS. But do they share CAS properties?

Applied to clusters, the first characteristic mentioned above implies (1) that networks of agents can be found in many technological or market niches, and (2) that cooperation between specialised agents can always allow for improvements. The remark about niches is particularly relevant for clusters, since the clustering phenomenon reflects a tendency towards regional specialisations in very distinctive activities including vine production, sportswear, semiconductors, etc.

Regarding the second characteristic, evidence shows that, within clusters, agents are specialised and that they cooperate in order to be more innovative (Porter, 1998; 2000; Saxenian, 1994). Innovations can take various forms, without following any pre-determined sequence (Gallouj and Weinstein, 1997).

Finally with regards to the third characteristics, it can be underlined that within a cluster, every agent can anticipate/forecast new technological or market opportunities, although their anticipation is imperfect because of bounded rationality (Frenken, 2006; Desmarchelier *et al.*, 2013b). This characteristic is important, because it contradicts the very conception of a from birth to death pre-determined cluster cycle. In addition, unlike Menzel and Fornahl (2010), who introduce exogenous factors as the main drivers likely to change the course of a system, in a CAS approach, clusters adapt because of their agents' individual anticipations.

In conclusion, and as we will try to confirm it in the empirical part of this work, these characteristics seem to fit well with what is known about clusters functioning. Then, how does the conception of clusters as CAS change the way we understand their dynamics?

Our literature review identified the very reason of clusters' existence: the knowledge-seeking behaviour of the firms. They seek knowledge from other firms or from other types of agents – notably KIBS, including universities. But we also identified an important difference between life cycle theory and network-based theory, regarding the way they address the knowledge dynamics within clusters: the life cycle theory postulates a knowledge homogenisation process, whereas the network-based approach postulates a specialisation process. The first CAS property (i.e. no global optimum) fits well with the idea of specialised agents, but the "anticipation" and "innovation" properties are not imposing any type of pre-determined process. Agents are heterogeneous, and KIBS may allow for both homogenisation and specialisation trajectories, admitting that a general/cluster-level trajectory can be found. The two other CAS properties (i.e. innovation and anticipation) indicate that the alternative routes in Menzel and Fornahl (2011) (i.e. adaptation, renewal and transformation) are the rule rather than the exception in CAS dynamics. It follows that a proper deterministic life cycle is likely to be the reflect of a degenerative cluster (ex. Route 128 in Saxenian, 1994).

Clusters have already been modelled as CAS thanks to agent-based modelling[6] (Dilaver *et al.*, 2014; Boero *et al.*, 2004; Squazzoni and Boero, 2002; Albino et al., 2003). Important question to tackle within this perspective are how specialised agents tie to each other and then how these ties evolve. In this respect, Boero *et al.*, (2004) propose various matching strategies. As an example – for a specific agent – the strategy could be: "*look at the first agent with different technology/ techno-organizational asset you meet*" (p. 12). These theoretical efforts are welcome, but the building of relevant models has to rely on a set of well-established stylised facts (Borrill and Tesfatsion, 2011). Discussing about clusters' dynamics and their main drivers supposes to decide in a first step which aggregate indicators/variables (number of actors, quantity produced, number of patents, R&D expenditures, etc.) and which underlying networks to consider. Unfortunately, to our knowledge, there is still no such empirical study within a CAS framework. Indeed, although theoretically appealing, the CAS approach remains hard to put into practice. We thus propose, in the remaining of the paper, a strategy for operationalizing the CAS framework.

Clusters as CAS: empirical example and insights about clusters dynamics

In order to conduct an empirical investigation, we have to make several choices: (a) Which industry to study? (b) Which cluster to focus on within this industry? (c) What kind of networks are we looking at? After making our choices, we explore the dynamics of the selected empirical network of agents and we draw general conclusions about clusters dynamics.

Which industry to study?

The "*non-universal competitor*" principle (Holland, 2012) advocates for the study of sectors with complex technology landscapes[7], because they are the most likely to offer many niches and thus many opportunities for clusters to emerge. According to Arthur (2009), all products display a tree-like recursiveness: "*the technology is the trunk, the main assemblies the main branches, their subassemblies the sub-branches, and so on, with the elemental parts the furthest twigs ... The depth of this hierarchy is the number of branches from trunk to some representative twigs*" (p. 38). The more complex a product is, the more it relies on a complex technology – i.e. a technology with a high depth. Arthur (2009), among others (Frenken, 2006; Niosi and Zhegu, 2005), argues that the aircraft industry relies on very complex technologies, composed of many subparts.

The resulting hierarchy between producers of various airplanes' subparts is represented in Figure 2.3, from Niosi and Zhegu (2005). Following the CAS approach,

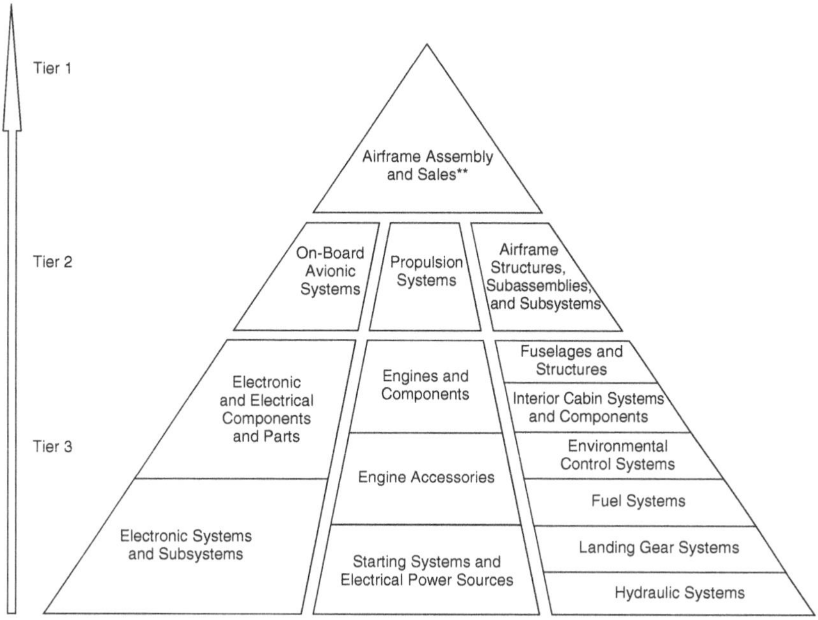

Figure 2.3 Aircraft producers' pyramid (from Niosi and Zhegu, 2005 p.8)

clusters can be found in any layer of this hierarchy. The biggest clusters include agents involved in top layers, and more particularly the "prime contractors" or "airframe assemblers": Bombardier in Montréal, Airbus in Toulouse, Boeing in Seattle (Niosi and Zhegu, 2005) or Lockheed Martin in Los Angeles (Scott, 1990).

Which cluster to study?

For every sector, it is common to find contributions focusing on successful/first class clusters: Route 128, Silicon Valley, Detroit, Los Angeles (Saxenian, 1994; Klepper, 2010; Scott, 1990). The most important ones for the aircraft industry have already been mentioned (Montréal, Toulouse, Seattle, Los Angeles) and won't be considered for the present study. Indeed, we rather choose to focus on a niche cluster, for highlighting the difference between the CAS and the more traditional/aggregate (or life cycle) approach of the clustering phenomenon.

In the traditional approach, authors generally study the geographical concentration of employment or companies by industry (Shin and Hassink, 2011; Niosi and Zhegu, 2005; Scott, 1990), or the geographical concentration of production (Shin and Hassink, 2011) or of the innovation activity (Audretsch and Feldman, 1996a; 1996b) for a given industry. Interestingly, there are very few accounts of the actual linkages between these actors, although Saxenian (1994) reported – with the example of the Route 128 – that spatial proximity does not necessarily impliy strong cooperation. One can also highlight a very limited account for actors' diversity within this aggregate approach, since only "firms" are generally mentioned.

Figure 2.4a displays the most recent account for the number of employees in aerospace industry within all European regions. We observe that aerospace employment is widespread, although quite concentrated in Western Europe and Russia. This dispersion is higher than what is observed in the United States and is generally explained by political reasons, notably the need to ensure countries independence (Niosi and Zhegu, 2005). In an aggregate view, we would consider studying the Southwest of France, Northern Germany or South England. However as far as we favor the CAS approach, we rather choose a small cluster. Figure 2.4b displays a map of Belgium and the regional concentration of employment in aerospace. We distinguish two relatively major poles within the southern regions – surrounding the cities of Mons and Liège – and two smaller poles near Brussels and Leuven. In 2006, all the actors from these 4 poles joined to create an official association called Skywin.[8] We propose to study this cluster.

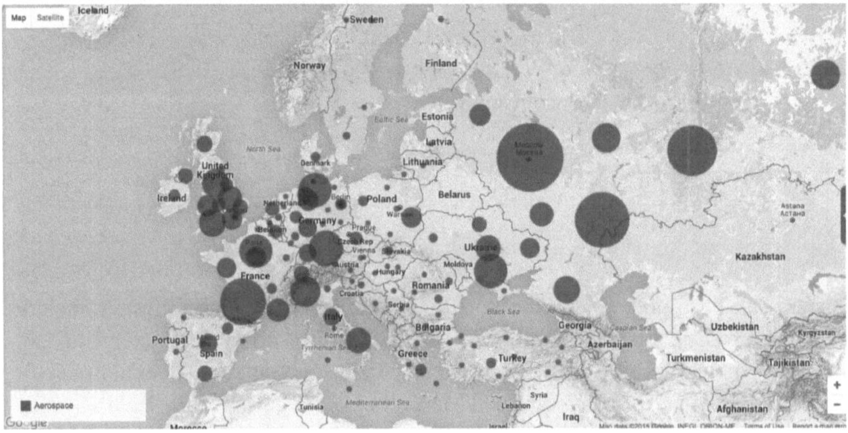

Figure 2.4a Number of employees in the aerospace industry in Europe in 2011[9]

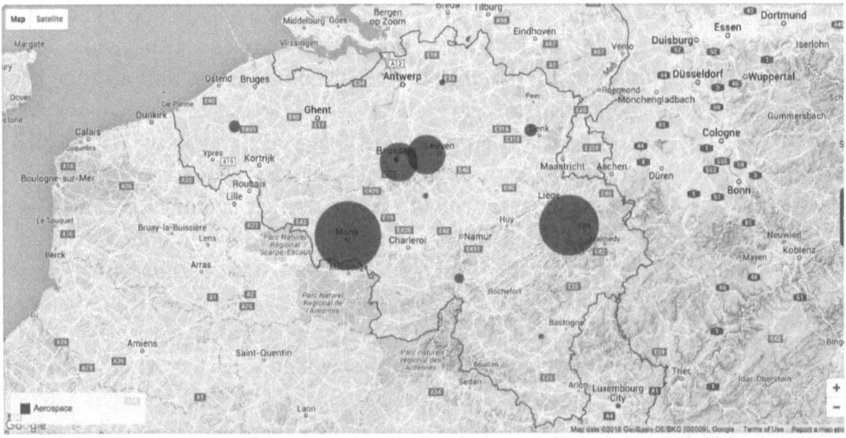

Figure 2.4b Number of employees in the aerospace industry in Belgium in 2011

What kind of networks are we looking at?

As we have already pointed out, a CAS is primarily a network of heterogeneous and specialised agents, which can be described in the following terms: (1) it emerges within (technological or market) niches, (2) it is strongly oriented towards innovation and (3) it is quick to adapt to changes in its environment thanks to its agents' anticipations and innovations (Holland, 2012). Skywin is presented as "*a group of companies, training centers and research units engaged in public and private partnership and building synergies around common and innovative projects*".[10] These projects fit into six main axes that the agents anticipate to be of strategic importance for their future development[11]:

1 Composite materials and processes
2 Metallic materials and processes
3 Embedded systems
4 Airport services
5 Space applications and systems
6 Modelling and simulation

These six themes reflect the niche position of Skywin within the aircraft producers' pyramid (Figure 2.3), as they mainly fit into some of the third tier activities: fuselage and structure for the first two axes and the electronic systems for the third axe. Interestingly, the fourth one – airport services – is not part of the aircraft production process and it responds to a potential market in developing countries.[12] This exemplifies clusters ability to re-orient their activity through time. The sixth axe – "modeling and simulation" – arguably applies in every parts of the pyramid (Figure 2.3) since simulation is generally involved in the conception phase of any airplane components. Finally, the fifth axe on "space applications and systems" reveals a specialisation relevant for the space industry (not considered in the present work).

Skywin is thus a group of heterogeneous agents, and the main interactions we should look at are those taking place within these "common and innovative projects". An exhaustive list is provided by the cluster website,[13] which covers 46 common projects undertaken collectively between 2006 and 2014. The following informations are provided for every project: the agents involved – classified into two categories ("industries" and "research bodies" [14]) – as well as the total budget (in millions of Euros) and the duration of the project (start and end years).

We use these informations for creating a bi-partite relational database in which two sets of nodes are linked (cf. the definition of a network): the agents and the projects in which they are involved. The structure of this database is presented in Figure 2.5a. Rows list the actors (industries or research bodies) and columns stand for the projects. Each time an actor is involved into a project, we report the value of this project into the corresponding cell. Another strategy could have been to report binary numbers, but using monetary values represent the advantage of allowing to assess for the relative importance of the various projects. These values represent links, which last for only few years. This information appears at the top of Figure 2.5a, in the row labelled "T". For instance, this row indicates that the project SW_3WSA (Wallonia World Wide Space Applications) lasts from 2006 to 2011 included – that is from year 1 to year 6 of the lifetime of the cluster.

A way to represent this database is to draw a network (Figure 2.5b)[15]. In this figure, red ellipses represent the projects and the blue ones represent the agents, whether they are "industries" or "research bodies". Links represent somehow the involvement within a project or more exactly the level of expected involvement, as it may be expressed by the budget allocated to the project[16]. The thicker and darker they are, the higher the project's financial value (and the expected involvement) is. This Figure does not distinguish between time periods: it summarises all the interactions that took part within Skywin from 2006 to 2014.

Studying clusters dynamics through innovation networks

In Figure 2.5b, the agents are not directly linked, that's why we modify the network on the basis of the hypothesis that agents who are participating to the same project are in fact directly linked. We also consider the time dimension and obtain as a result the 9 configurations of the network for each year from 2006 to 2014 displayed in Appendix (Figures 2.A1 to 2.A9). Table 2.1 summarises some descriptive statistics of these configurations.

In a traditional – aggregate – perspective, authors focus on the number of actors within a given geographical area. In Table 2.1, we rather report the actors who actively cooperate on a set of common projects. Looking at first at the number of these actors – and abstracting from the European economic crisis during the period – we may consider that Skywin enters into a phase of decline starting from 2010. This general movement goes hand in hand with a decrease in the average degree and networks' densities, which means that the agents who take part to innovation projects are more and more loosely connected. In other words, the cluster loses its attractiveness for potential newcomers: knowledge-seeking agents cannot find enough useful knowledge for exploring their technological or market landscapes, or simply the niches within which Skywin is evolving are not promising enough. Overall, these evolutions are consistent with the life cycle theory.

However, several elements contradict this pessimistic and deterministic conclusion. (1) The configurations of the network displayed in Appendix (Figures 2.A1 to 2.A9) are particularly dynamic throughout the considered time span. From year to year actors are leaving and others are entering into the network, following projects life cycles. Active actors are thus changing: looking at the network's configurations, we count in total 92 different active agents from 2006 to 2014, although no more than 67 were operating at the same time. This suggests a positive rate of turn-over among these agents. (2) There remain plenty of opportunities for partnerships, as innovation networks' densities always evolve within a range comprised between 17 per cent and 31 per cent.[20] In this respect, Skywin accounts for a total of 117 members.[21] Considering the 92 active agents, this highlights both a high rate of cooperation between Skywin members and the existence

	Units=Millions of Euros															
	T	1-6	1-2	2-5	2-7	1-3	2-5	1-3	2-5	3-5	2	3-7	3-5	3-8	4-7	3-9
	Projects	SW_3WSA	SW_AIRCAP	SW_APC	SW_HM+	SW_RADAC	SW_TELECOM	SW_TLS	SW_CM Experts	SW_PLM	SW_VIRCAP	SW_E-COM	SW_EASIPM	SW_ICS	SW_MULTIPHY	SW_PAU
Actors	Spacebel	3														
	Vitrociset	3														
	Creaction	3														
	Ulyces	3														
	Amos	3													7,1	
	Walphot	3														
	IonicSoftwar	3														
	WSLux	3														
	APC	3														
	Aquapole	3														
	CSL	3														
	Cetic	3			7,9		6,3									
	Advanced Coating		1													
	Sonaca			19,3	7,9				1,2	1,3						
	SABCA			19,3												
	Techspace Aero			19,3	7,9							9,2		6,5		
	e-Xstream			19,3												
	Optim			19,3								9,2				
	Samtech			19,3	7,9										7,1	
	SD&A			19,3												
	Sobelcomp			19,3												
	Cenaero			19,3	7,9								0,4	6,5	7,1	
	CRIF			19,3												
	UCL			19,3	7,9		6,3					9,2			7,1	3,2
	ULB			19,3	7,9		6,3					9,2		6,5		
	ULg			19,3	7,9		6,3					9,2		6,5	7,1	3,2
	UMH			19,3	7,9		6,3						0,4		7,1	3,2
	ERM			19,3										6,5		
	Etca				7,9											
	ThalesAS				7,9		6,3									
	GDTECH				7,9							9,2			7,1	
	Open Engineering				7,9										7,1	
	Cissoid				7,9		6,3									

Figure 2.5a Relational database built from Skywin's innovation projects

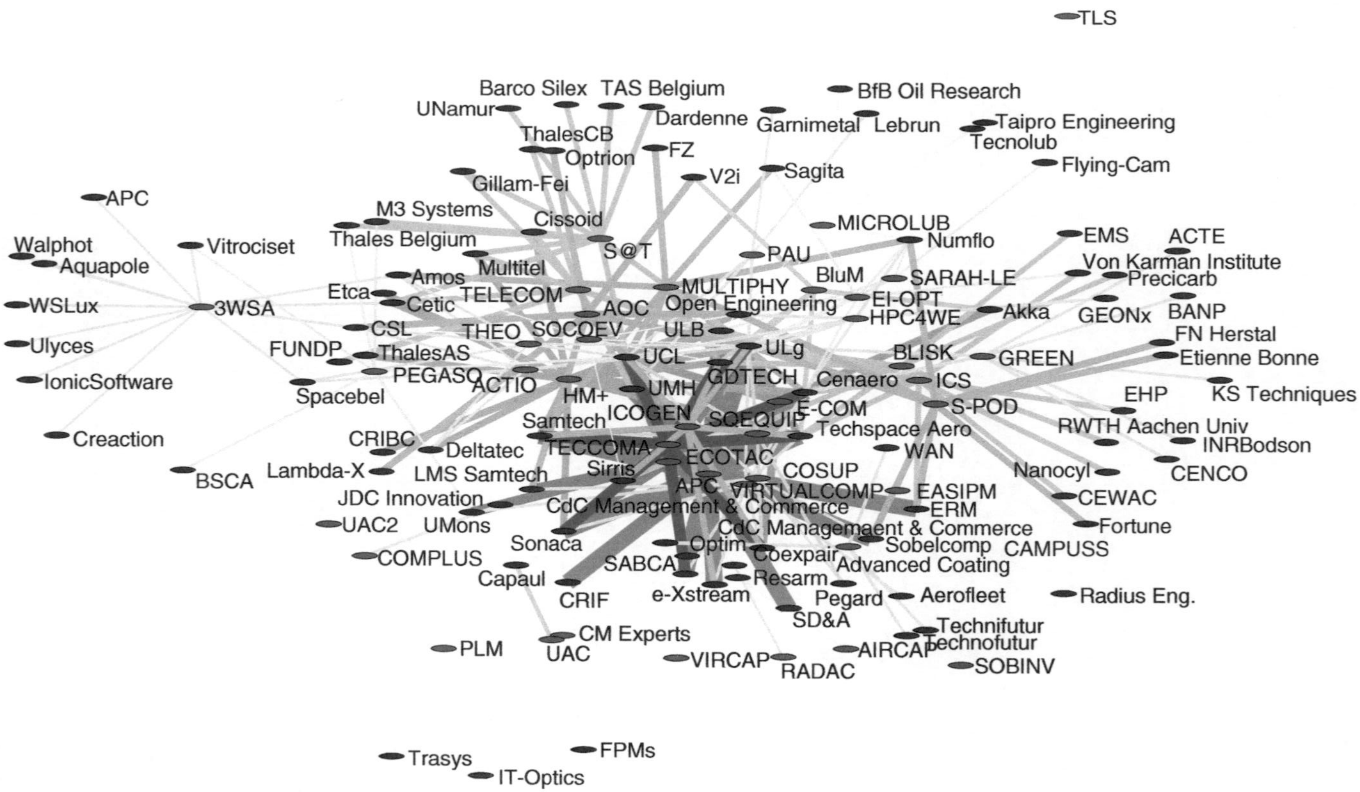

Figure 2.5b Skywin bi-partite innovation network (Light gray vertices stand for projects, and dark gray vertices stand for the agents)

Table 2.1 Properties of Skywin temporal networks

	Number of actors	*Av. degree*[17]	*Density (no loops allowed)*[18]	*Av. path length*[19] *among reachable pairs*
2006	15	7.33	0.52	1.00000
2007	44	13.54	0.31	1.72740
2008	52	13.57	0.26	1.88690
2009	61	14.46	0.24	1.90169
2010	67	14.68	0.22	1.90629
2011	61	13.97	0.23	1.84570
2012	62	13.45	0.22	1.76497
2013	60	10.53	0.18	1.82214
2014	56	9.53	0.17	1.79285

of unused opportunities. (3) The path length values in Table 2.1 are always much smaller than the number of agents within the network. This reveals the presence of a small-world effect (Newman, 2003): because of a high rate of overlap between the members of different projects, it becomes easy for any agent – even for a newcomer – to obtain information or knowledge coming from other projects. These short path lengths can thus be considered as a good proxy for evaluating the advantages of taking part in this cluster. Considering the relative stability of this measure in Table 1, we cannot sustain the declining hypothesis that results from the sole observation of the number of active agents.

We argue that, in a knowledge-seeking perspective, agents benefit from the existence of a small-world effect, which itself emerges from a certain degree of overlap between the various projects. It follows that the core agents of a given cluster are those who allow for overlaps to occur. A good way to identify them is to compute agents' degree centralities, i.e. for a given agent i, to count the number of links he/she has with the other agents in the network. We can also take into account the fact that different links are not equivalent, in the sense that – at least for our current networks – the financial values of the various projects are different. In order to take these links' values into account, we compute the weighted degree centrality measure proposed by Opsahl *et al.*, (2010).

Be $\alpha \in [0;1]$ a tuning parameter, k_i the number of links connected to the agent i and s_i the average weight (or value) of these links, then $C_D^{W\alpha}(i)$, the weighted degree centrality of the agent i is given by the following equation:

$$C_D^{W\alpha}(i) = k_i^{1-\alpha} \times s_i^{\alpha}$$

The more α is important, the more we attribute importance to links' values in computing agents' centralities. Top 10 weighted degree centralities for every year and for various α values are reported in Appendix (Tables 2A.1, 2A.2, and 2A.3). We observe that, in virtually all cases, the two most central actors belong to the "research bodies" category. The University of Liège and the Catholic University of Leuven are particularly central. Industrial firms are also well represented in these rankings and, even though we observe the recurrence of national leaders in the aerospace industry (e.g.. Sonaca), they are much more "volatile" than universities in the sense that their relative positions are less stable and that there is a much higher rate of turnover among firms within the top 10.

These results give interesting insights about the drivers of clusters' attractiveness – measured in our case by a small-world effect within clusters' innovation networks. We show that this attractiveness relies on the presence of a stable core of highly connected Knowledge Intensive Business Services (universities or research bodies in general). Arguably, what determines if a cluster is declining is not the age of the cluster as a whole, nor the number of the (active) agents it includes, but the quality and connectivity of the Knowledge Intensive Business Services in its core part. When looking at the number of innovation-active agents, we could say that Skywin is entering into a phase of decline, but a closer look to its innovation networks' properties reveals an attractive cluster and thus show potential for new phases of growth.

Conclusion

Clusters' dynamics are generally understood as the evolution of an aggregate indicator, like the number of firms operating in a given geographical area. Two competing theories aim at explaining the dynamics of this indicator: the life cycle theory and the network-based approach. Both consider knowledge processing as the main driver but in an opposite way. For the tenants of the life cycle, clusters evolve through a process of knowledge homogenisation among their members, whereas the network-based approach considers that knowledge becomes more and more specialised. We argue that KIBS play a major role in both of these directions, and we thus advocate for an alternative/ synthesizing approach.

Such synthesis should combine the aggregate point of view of the life cycle theory with the actor-centered network-based approach, while avoiding their deterministic predictions. Complex Adaptive Systems are a promising candidate for such a purpose. In order to consider their implications for clusters' dynamics, we conducted within this framework an empirical analysis on a given industrial cluster (the aeronautics cluster in Belgium).

We discovered that this cluster's innovation networks exhibit a small-world effect. This implies that any agent who takes part into an innovation project of this cluster can easily benefit from knowledge and information generated in another ongoing project. We argue that this effect is an interesting proxy of a cluster's attractiveness and an appropriate aggregate variable for studying clusters' dynamics as it shows cluster's potential for further growth. We also demonstrate that KIBS are the main responsible for the emergence of this small-world effect in innovation networks.

Appendix 2A

Table 2A.1 Weighted degree centralities ($\alpha = 0.2$)

2006			*2007*			*2008*		
Agents	*Type*	*Cwd(i)*	*Agents*	*Type*	*Cwd(i)*	*Agents*	*Type*	*Cwd(i)*
Aquapole	Research	6.80948	CetIndustryc	Research	15.4418	ULg	Research	18.7524
CSL	Research	6.80948	UCL	Research	14.7705	ULB	Research	17.8305
Spacebel	Industry	6.30957	ULB	Research	14.7705	UCL	Research	17.2989
CetIndustryc	Research	6.30957	ULg	Research	14.7705	UMH	Research	17.0241
CreactIndustryon	Industry	6.30957	UMH	Research	14.7705	Techspace Aero	Industry	16.1466
Ulyces	Industry	6.30957	Sonaca	Industry	14.213	Cenaero	Research	15.4421
Amos	Industry	6.30957	Techspace Aero	Industry	12.9557	CetIndustryc	Research	15.4418
Walphot	Industry	6.30957	Samtech	Industry	12.9557	Sonaca	Industry	14.6256
IndustryonIndustrycSoftware	Industry	6.30957	Cenaero	Research	12.9557	Samtech	Industry	12.9557
WSLux	Industry	6.30957	ThalesAS	Industry	11.2266	GDTECH	Industry	11.3936
APC	Research	6.30957	CIndustryssoIndustryd	Industry	11.2266	ThalesAS	Industry	11.2266

(continued)

Table 2A.1 Weighted degree centralities ($\alpha = 0.2$) (*continued*)

2009			*2010*			*2011*		
Agents	*Type*	*Cwd(i)*	*Agents*	*Type*	*Cwd(i)*	*Agents*	*Type*	*Cwd(i)*
ULg	Research	25.3146	ULg	Research	27.7763	ULg	Research	25.8309
Cenaero	Research	20.4387	UCL	Research	22.4516	UCL	Research	20.8232
UCL	Research	19.8468	Cenaero	Research	20.4387	Cenaero	Research	20.3615
ULB	Research	19.4721	ULB	Research	19.4721	Techspace Aero	Industry	19.1457
Techspace Aero	Industry	19.2091	Techspace Aero	Industry	19.2091	GDTECH	Industry	17.7651
UMH	Research	18.2283	CetIndustryc	Research	18.2331	CetIndustryc	Research	16.7783
Sonaca	Industry	15.9253	UMH	Research	18.2283	ULB	Research	14.2981
CetIndustryc	Research	15.4418	Sonaca	Industry	15.9253	Sonaca	Industry	13.124
GDTECH	Industry	15.1836	GDTECH	Industry	15.1836	CIndustryssoIndustryd	Industry	12.6856
Samtech	Industry	14.6959	Samtech	Industry	14.6959	MultIndustrytel	Research	12.6856
Amos	Industry	11.3936	CIndustryssoIndustryd	Industry	14.2038	UMH	Research	12.1846

2012			*2013*			*2014*		
Agents	*Type*	*Cwd(i)*	*Agents*	*Type*	*Cwd(i)*	*Agents*	*Type*	*Cwd(i)*
ULg	Research	28.9931	ULg	Research	26.9847	ULg	Research	25.3146
Cenaero	Research	21.9441	UCL	Research	18.8493	UCL	Research	20.5866
UCL	Research	20.8232	Cenaero	Research	16.0135	Cenaero	Research	13.4368
Techspace Aero	Industry	19.1457	Techspace Aero	Industry	14.6372	GDTECH	Industry	13.2441
GDTECH	Industry	17.7651	GDTECH	Industry	11.3942	Open EngIndustryneer-Industryng	Industry	11.1082
ULB	Research	14.2981	LMS Samtech	Industry	10.2474	Techspace Aero	Industry	9.8875
Open EngIndustryneerIndustryng	Industry	13.4635	Sobelcomp	Industry	9.96496	Sobelcomp	Industry	9.51092
Sonaca	Industry	13.124	CoexpaIndustryr	Industry	9.24402	CoexpaIndustryr	Industry	8.79849
CetIndustryc	Research	12.6856	UMons	Research	9.05126	UMons	Research	8.58552
CIndustryssoIndustryd	Industry	12.6856	M3 Systems	Industry	8.48205	Numflo	Industry	8.48344
MultIndustrytel	Research	12.6856	Thales BelgIndustryum	Industry	8.00911	CSL	Research	8.24125

Table 2A.2 Weighted degree centralities ($\alpha = 0.5$)

2006			*2007*			*2008*		
Agents	*Type*	*Cwd(i)*	*Agents*	*Type*	*Cwd(i)*	*Agents*	*Type*	*Cwd(i)*
Aquapole	Research	3.31662	UCL	Research	6.32456	ULg	Research	7.68115
CSL	Research	3.31662	ULB	Research	6.32456	ULB	Research	7.4162
Spacebel	Industry	3.16228	ULg	Research	6.32456	UCL	Research	7.2111
CetIndustryc	Research	3.16228	UMH	Research	6.32456	UMH	Research	6.9282
CreactIndustryon	Industry	3.16228	CetIndustryc	Research	6	Techspace Aero	Industry	6.7082
Ulyces	Industry	3.16228	Sonaca	Industry	5.74456	Cenaero	Research	6.32456
Amos	Industry	3.16228	Techspace Aero	Industry	5.47723	CetIndustryc	Research	6
Walphot	Industry	3.16228	Samtech	Industry	5.47723	Sonaca	Industry	5.83095
IndustryonIndustrycSoftware	Industry	3.16228	Cenaero	Research	5.47723	Samtech	Industry	5.47723
WSLux	Industry	3.16228	ThalesAS	Industry	5.09902	ThalesAS	Industry	5.09902
APC	Research	3.16228	CIndustryssoIndustryd	Industry	5.09902	CIndustryssoIndustryd	Industry	5.09902

2009			2010			2011		
Agents	*Type*	*Cwd(i)*	*Agents*	*Type*	*Cwd(i)*	*Agents*	*Type*	*Cwd(i)*
ULg	Research	9.69536	ULg	Research	10.247	ULg	Research	9.59166
UCL	Research	8.12404	UCL	Research	8.77496	UCL	Research	8.12404
Cenaero	Research	8.06226	Cenaero	Research	8.06226	Techspace Aero	Industry	7.74597
Techspace Aero	Industry	7.81025	Techspace Aero	Industry	7.81025	Cenaero	Research	7.68115
ULB	Research	7.74597	ULB	Research	7.74597	GDTECH	Industry	7.34847
UMH	Research	7.48331	UMH	Research	7.48331	CetIndustryc	Research	6.08276
Sonaca	Industry	6.48074	CetIndustryc	Research	6.85565	ULB	Research	5.83095
GDTECH	Industry	6.40312	Sonaca	Industry	6.48074	Sonaca	Industry	5.65685
Samtech	Industry	6.245	GDTECH	Industry	6.40312	UMH	Research	5.38516
CetIndustryc	Research	6	Samtech	Industry	6.245	CIndustryssoIndustryd	Industry	5.19615
ThalesAS	Industry	5.09902	CIndustryssoIndustryd	Industry	6.08276	MultIndustrytel	Research	5.19615

(*continued*)

Table 2A.2 Weighted degree centralities ($\alpha = 0.5$) (*continued*)

2012			*2013*			*2014*		
Agents	*Type*	*Cwd(i)*	*Agents*	*Type*	*Cwd(i)*	*Agents*	*Type*	*Cwd(i)*
ULg	Research	10.247	ULg	Research	9.27362	ULg	Research	9.69536
UCL	Research	8.12404	UCL	Research	7.14143	UCL	Research	8.544
Cenaero	Research	8	Techspace Aero	Industry	6.55744	GDTECH	Industry	6.63325
Techspace Aero	Industry	7.74597	Cenaero	Research	6.245	Cenaero	Research	6
GDTECH	Industry	7.34847	GDTECH	Industry	5.2915	Open EngIndustryneer-Industryng	Industry	5.38516
ULB	Research	5.83095	LMS Samtech	Industry	4.79583	Techspace Aero	Industry	5.2915
Sonaca	Industry	5.65685	CoexpaIndustryr	Industry	4.47214	CoexpaIndustryr	Industry	4.89898
Open EngIndustryneerIndustryng	Industry	5.65685	Sobelcomp	Industry	4.47214	CSL	Research	4.69042
UMH	Research	5.38516	UMons	Research	4.24264	Numflo	Industry	4.47214
CetIndustryc	Research	5.19615	M3 Systems	Industry	4	Samtech	Industry	4.47214
CIndustryssoIndustryd	Industry	5.19615	Sonaca	Industry	4	Sobelcomp	Industry	4.3589

Table 2A.3 Weighted degree centralities (α = 0.8)

2006			2007			2008		
Agents	*Type*	*Cwd(i)*	*Agents*	*Type*	*Cwd(i)*	*Agents*	*Type*	*Cwd(i)*
Aquapole	Research	1.61539	UCL	Research	2.7081	ULg	Research	3.14626
CSL	Research	1.61539	ULB	Research	2.7081	ULB	Research	3.08461
Spacebel	Industry	1.58489	ULg	Research	2.7081	UCL	Research	3.00598
CetIndustryc	Research	1.58489	UMH	Research	2.7081	UMH	Research	2.81953
CreactIndustryon	Industry	1.58489	CetIndustryc	Research	2.33133	Techspace Aero	Industry	2.78696
Ulyces	Industry	1.58489	Sonaca	Industry	2.32181	Cenaero	Research	2.59032
Amos	Industry	1.58489	ThalesAS	Industry	2.31594	CetIndustryc	Research	2.33133
Walphot	Industry	1.58489	CIndustryssoIndustryd	Industry	2.31594	Sonaca	Industry	2.32469
IndustryonIndustrycSoftware	Industry	1.58489	MultIndustrytel	Research	2.31594	ThalesAS	Industry	2.31594
WSLux	Industry	1.58489	Techspace Aero	Industry	2.31558	CIndustryssoIndustryd	Industry	2.31594
APC	Research	1.58489	Samtech	Industry	2.31558	MultIndustrytel	Research	2.31594

(*continued*)

Table 2A.3 Weighted degree centralities (α = 0.8) (*continued*)

2009			*2010*			*2011*		
Agents	*Type*	*Cwd(i)*	*Agents*	*Type*	*Cwd(i)*	*Agents*	*Type*	*Cwd(i)*
ULg	Research	3.71328	ULg	Research	3.7802	ULg	Research	3.56163
UCL	Research	3.32547	UCL	Research	3.4296	UCL	Research	3.16954
Cenaero	Research	3.18024	Cenaero	Research	3.18024	Techspace Aero	Industry	3.13386
Techspace Aero	Industry	3.17558	Techspace Aero	Industry	3.17558	GDTECH	Industry	3.03966
ULB	Research	3.08134	ULB	Research	3.08134	Cenaero	Research	2.89763
UMH	Research	3.07215	UMH	Research	3.07215	Sonaca	Industry	2.43828
GDTECH	Industry	2.70028	GDTECH	Industry	2.70028	UMH	Research	2.38005
Samtech	Industry	2.6538	Samtech	Industry	2.6538	ULB	Research	2.37793
Sonaca	Industry	2.63731	Sonaca	Industry	2.63731	Samtech	Industry	2.2444
CetIndustryc	Research	2.33133	CIndustryssoIndustryd	Industry	2.60494	Open EngIndustryneer-Industryng	Industry	2.2444
ThalesAS	Industry	2.31594	MultIndustrytel	Research	2.60494	CetIndustryc	Research	2.20523

2012			*2013*			*2014*		
Agents	*Type*	*Cwd(i)*	*Agents*	*Type*	*Cwd(i)*	*Agents*	*Type*	*Cwd(i)*
ULg	Research	3.62155	ULg	Research	3.18699	ULg	Research	3.71328
UCL	Research	3.16954	Techspace Aero	Industry	2.93772	UCL	Research	3.54599
Techspace Aero	Industry	3.13386	UCL	Research	2.70567	GDTECH	Industry	3.32224
GDTECH	Industry	3.03966	GDTECH	Industry	2.45739	Techspace Aero	Industry	2.83186
Cenaero	Research	2.91651	Cenaero	Research	2.43545	CoexpaIndustryr	Industry	2.72774
Sonaca	Industry	2.43828	LMS Samtech	Industry	2.24447	Cenaero	Research	2.6792
UMH	Research	2.38005	CoexpaIndustryr	Industry	2.16356	CSL	Research	2.6695
ULB	Research	2.37793	Sonaca	Industry	2.06913	Open EngIndustryneer-Industryng	Industry	2.61069
Open EngIndustryneerIndustryng	Industry	2.3768	SABCA	Industry	2.06913	Amos	Industry	2.50622
Samtech	Industry	2.2444	Sobelcomp	Industry	2.00703	Samtech	Industry	2.47352
CetIndustryc	Research	2.12841	UMons	Research	1.98867	Spacebel	Industry	2.40225

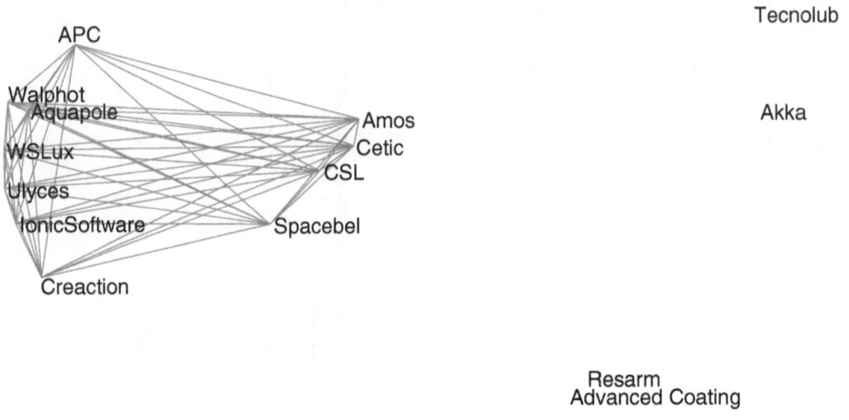

Figure 2.A1 Skywin's innovation network in 2006

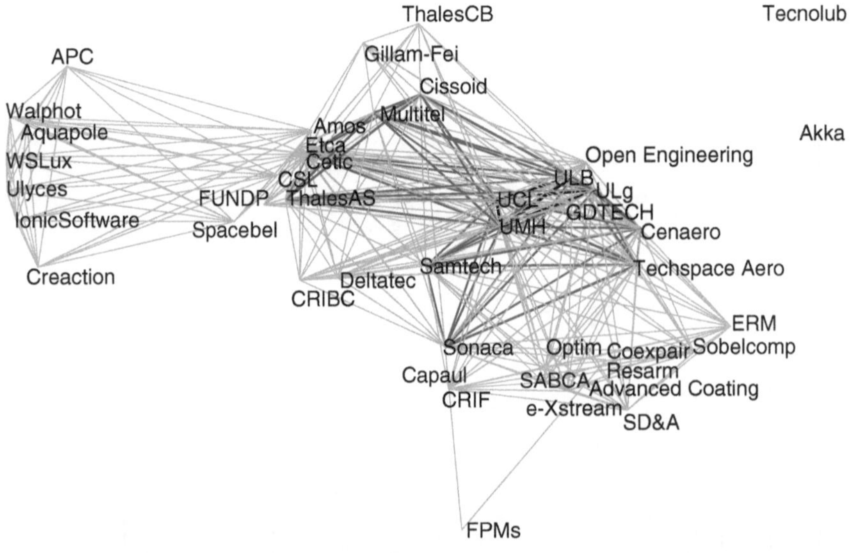

Figure 2.A2 Skywin's innovation network in 2007

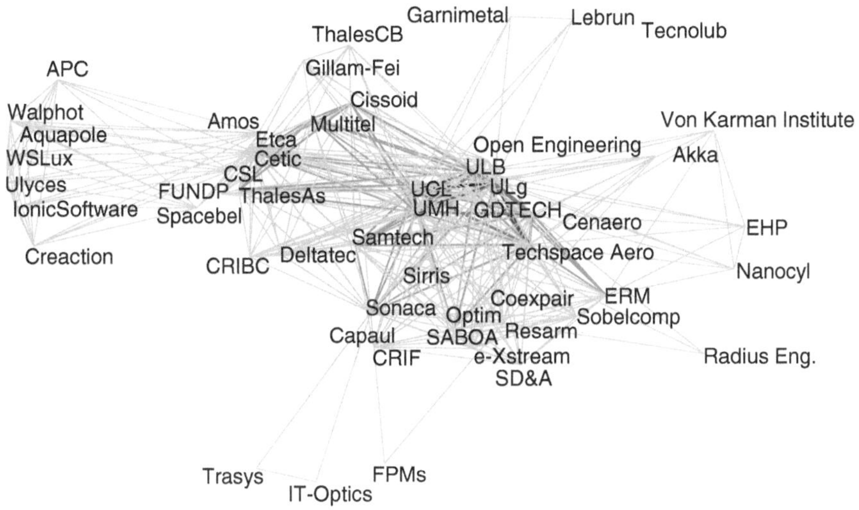

Figure 2.A3 Skywin's innovation network in 2008

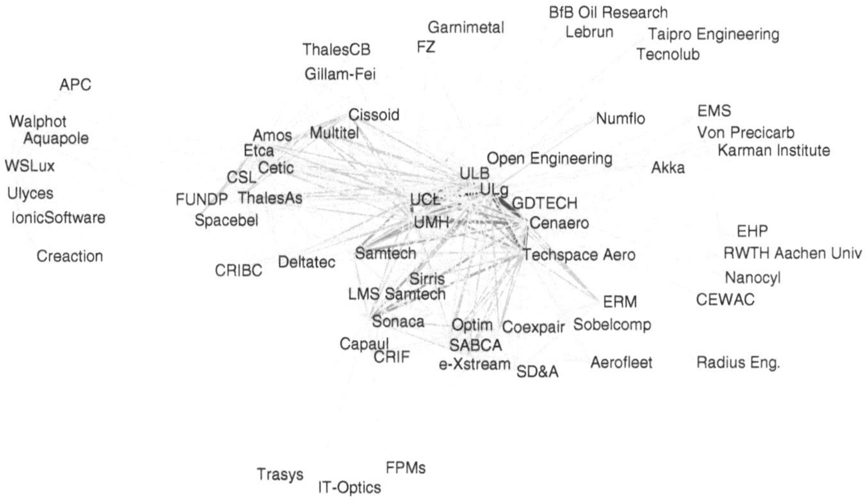

Figure 2.A4 Skywin's innovation network in 2009

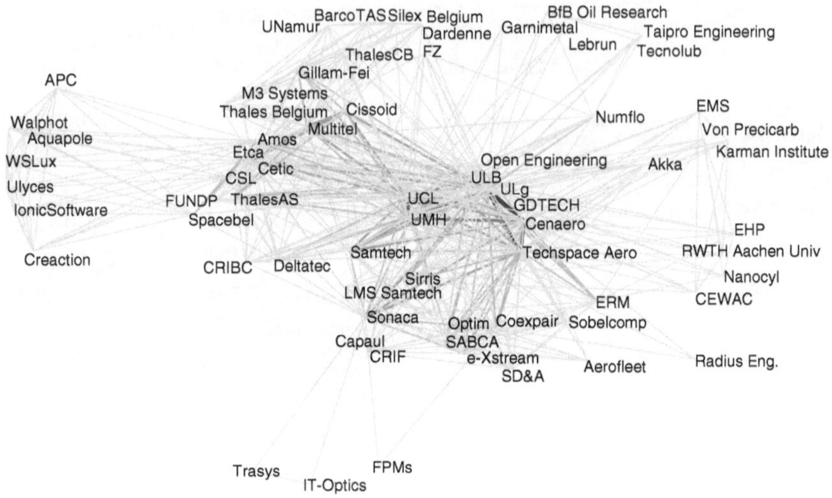

Figure 2.A5 Skywin's innovation network in 2010

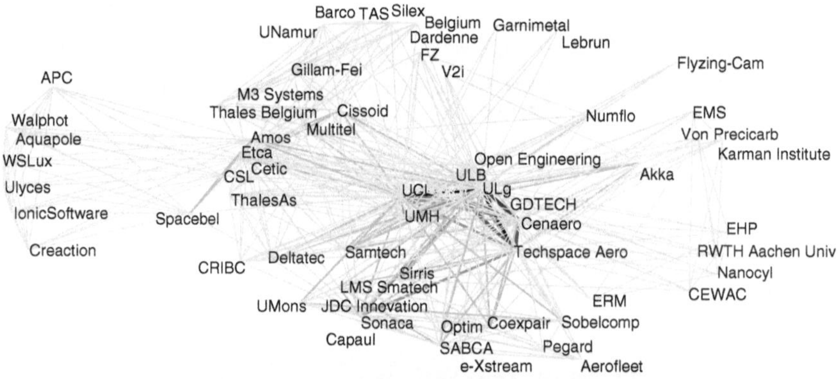

Figure 2.A6 Skywin's innovation network in 2011

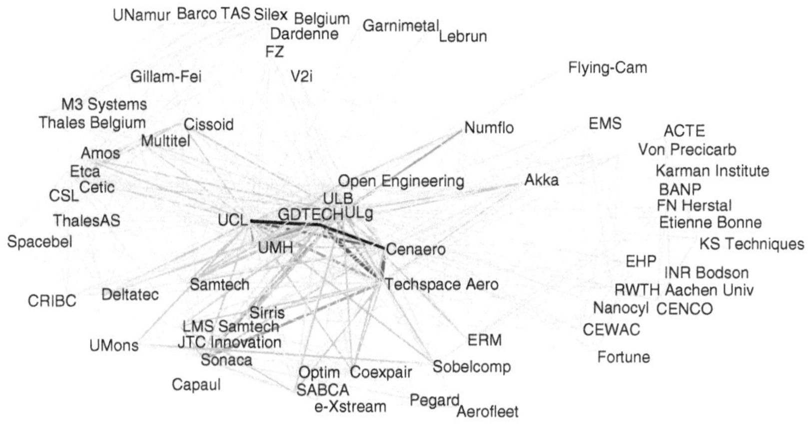

Figure 2.A7 Skywin's innovation network in 2012

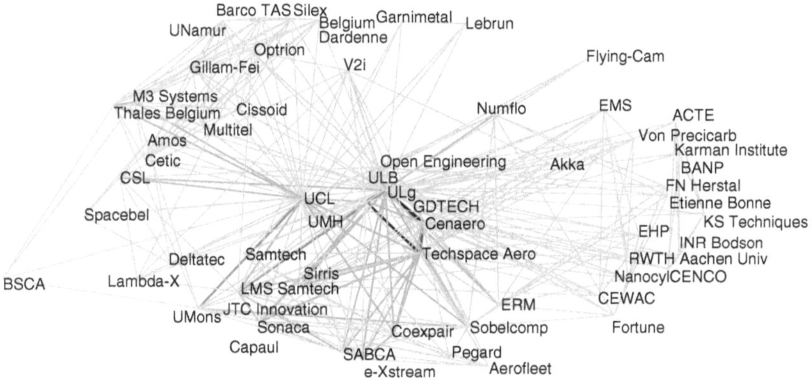

Figure 2.A8 Skywin's innovation network in 2013

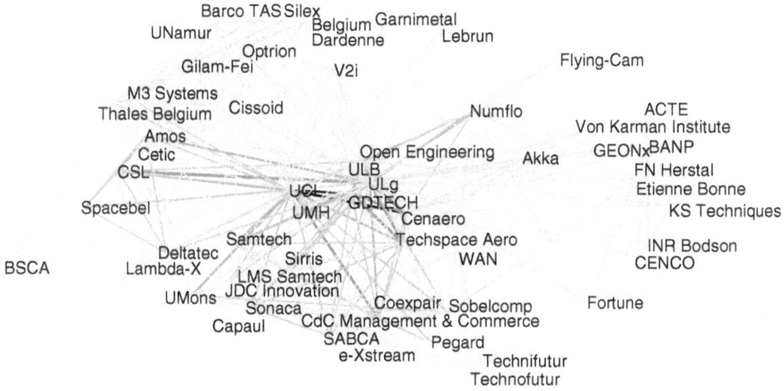

Figure 2.A9 Skywin's innovation network in 2014

Notes

1 A notable exception is to be found in Martin and Sunley (2011), whose contribution will be discussed in this chapter.
2 This means that the age of the cluster is proxied by just one (or a limited number of) variable(s), which can be for example the number of employees or the number of firms, etc.
3 In the Lancasterian tradition (Lancaster, 1966, Saviotti and Metcalfe, 1984).
4 The typology of meta-models of CAS includes the following meta-models (types of complex systems): life cycle, random walk, replacement, limitation, succession, adaptive cycle, evolutionary.
5 The "adaptive cycle model of the evolution of a complex system" (Martin and Sunley, 2011 p.1307) is similar to the "modified adaptive cycle" represented in Figure 2.2, minus the alternative trajectories of "failure", "constant cluster mutation", "cluster disappearance", "cluster stabilization" and "cluster re-orientation".
6 Applied to economics, agent-based modelling is "a computational approach that aims to explain economic systems by modeling them as societies of intelligent software agents. The individual agents make autonomous decisions, but their actual behaviors are constrained by available resources, other individuals' behaviors, and institutions" (Osinga *et al.*, 2011).
7 Kauffman et al (2000) define a technology landscape as a set of values attributed to all the various possible "production recipes" (p. 8), which are represented as vertices of a "directed graph". A production recipe "encompasses all the deliberate organizational and technical practices which, when performed together, result in the production of a specific good" (p. 4). Technology landscape is a metaphor originated from biology (Kauffman, 1993) for representing the choice of economic agents when they have to decide what to produce and how to produce it. It states that agents' initial choice has long term incidence on their adaptability since it constraints their innovation capabilities, this is the reason why production recipes are embedded into a directed graph: it is not possible to switch easily from a recipe to another.
8 www.skywin.be/?q=en (last access: 10 Feb. 2015)
9 www.clusterobservatory.eu (last access: 10 Feb. 2015, the numbers are for 2011)
10 www.skywin.be/?q=en/mission_and_strategy (last access: 10 Feb. 2015)
11 www.skywin.be/?q=en/mission_and_strategy (last access: 10 Feb. 2015)
12 As an example, it has been reported that 25.2% of flights in Mainland China have been delayed in 2012, 14.9% of them for an "unidentified reason". Beijing News, 23 May 2013: www.bjd.com.cn/10beijingnews/201305/23/t20130523_3774403.html
13 www.skywin.be/sites/default/files/kcfinder/1–vincent.marchal%40skywin.be/files/Fiches%20projets%20Skywin%20Calls%201–9.pdf (last access: 10 Feb. 2015)
14 These "research bodies" include universities, training centers and private research institutions. We thus assimilate them to KIBS.
15 Network graphs are obtained by using Pajek, a software for network analysis freely available at the following web-page: http://mrvar.fdv.uni-lj.si/pajek/ (last access: 13 September 2015).
16 It should be noted that as far as there is no information on how the budget is allocated between the different partners of a given project, the whole budget is associated with each of them.
17 The average degree of a network gives the average number of links per agent within the network. For instance, the agents taking part to innovation projects in 2008 were, on average, linked to 13.57 agents. More formally, let N be the number of agents, and k_{ij} stands for the link between agents i and j, then the average degree is given by:

$$\frac{1}{N}\sum_{i=1}^{N}\sum_{\substack{j=1\\ j\neq i}}^{N} k_{ij}$$

18 Network density is the ratio between the actual number of links and the maximum possible in a hypothetical situation where each agent is connected to all the others. For instance, a density equal to 0.52 means that 52 per cent of the connection possibilities are exploited by the agents. In an undirected graph, the maximum number of connections is equal to $\frac{N(N-1)}{2}$, with N the total number of actors. If K stands for the number of observed links, then the degree of the network is given by: $K \,/\, \frac{N(N-1)}{2}$

19 The path length between two agents i and j is the shortest distance between them (i.e. the shortest sequence of vertices). The average path length of a network is obtained by averaging all the path length between all the reachable agents of this network. Isolated agents are thus not taken into account.

20 We omit the density of the network configuration in 2006 (date of the creation of the cluster) because its relatively high value is explained by the fact that the cluster included only one project with several actors.

21 www.skywin.be/?q=en/members

References

Abernathy, W. J. and Utterback, J. M. (1978), Patterns of industrial innovation, *Technology Review*, 80 (7), pp. 40–47.

Albino V., Carbonara N., Giannoccaro I. (2003), Coordination mechanisms based on cooperation and competition within industrial districts: an agent-based computational approach, *The Journal of Artificial Societies and Social Simulation*, 6. http://jasss.soc.surrey.ac.uk/6/4/3.html.

Arthur W.B. (2009), *The nature of technology. What is it and how it evolves.* Free Press.

Audretsch D.B., Feldman M.P. (1996a), Innovative clusters and the industry life cycle, *Review of Industrial Organization*, 11, pp. 253–73.

Audretsch D.B., Feldman M.P. (1996b), R&D spillovers and the geography of innovation and production, *The American Economic Review*, 86, pp. 630–40.

Barras R. (1986), Towards a theory of innovation in services, *Research Policy*, 15, pp. 161–73.

Boero R., Castellani M., Squazzoni F. (2004), Micro behavioral attitudes and macro technological adaptation in industrial districts: an agent-based prototype, *The Journal of Artificial Societies and Social Simulation*, 7. http://jasss.soc.surrey.ac.uk/7/2/1.html.

Borrill P., Tesfatsion L. (2011), Agent-based modeling: the right mathematics for the social sciences?, in Davis J. B. and Hands D. W. (eds) *The Elgar companion to recent economic methodology*, pp. 223–55. Edward Elgar.

Boschma R., Fornahl D. (2011), Cluster evolution and a roadmap for future research, *Regional Studies*, 45, pp. 1295–98.

Callon M. (1991), Réseaux technico-économiques et irréversibilité, in Boyer, R. Chavance, and Godard, O. (eds) *Figures de l'irréversibilité en économie*, Paris, Edition de l'EHESS, pp.195–230.

Cumming G.S., Collier J. (2005), Change and identity in complex systems, *Ecology and Society*, 10.

Den Hertog P. (2000), Knowledge-intensive business services as co-producers of innovation, *International Journal of Innovation Management*, 4, pp. 491–528.

Den Hertog P. (2002), Co-producers of innovation: on the role of knowledge-intensive business services in innovation, in Gadrey J. and Gallouj F. (eds) *Productivity, innovation and knowledge in services*, pp. 223–55. Edward Elgar.

Desmarchelier B., Djellal F., Gallouj F. (2013a), Knowledge-intensive business services and long term growth, *Structural Change and Economic Dynamics*, 25, pp. 188–205.

Desmarchelier B., Djellal F., Gallouj F. (2013b), Environmental policies and eco-innovation by service firms: an agent-based model, *Technological Forecasting and Social Change*, 80, pp. 1395–1408.

Dilaver O., Bleda M., Uyarra E. (2014), Entrepreneurship and the emergence of industrial clusters, *Complexity*, 19, pp. 14–29.

Frenken K. (2006), *Innovation, evolution and complexity theory*. Edward Elgar.

Frenken K., Cefis E., Stam E. (2015), Industrial dynamics and clusters: a survey, *Regional Studies*, 49, pp. 10–27.

Gadrey J. Gallouj F. (1998), The provider-customer interface in business and professional services, *The Service Industries Journal*, 18 (2), pp. 1–15.

Gallouj F. (1998), Innovating in reverse: services and the reverse product cycle, *European Journal of Innovation Management*, 1, pp. 123–38.

Gallouj F. (2002), Knowledge-intensive business services: processing knowledge and producing innovation, in Gadrey J. and Gallouj F. (eds) *Productivity, innovation and knowledge in services*, pp. 256–84. Edward Elgar.

Gallouj F., Weinstein O. (1997), Innovation in services, *Research Policy*, 26, pp. 537–56.

Hakansson H. (1989), *Corporate technological behavior, cooperation and networks*, Routledge.

Hakansson H., Johanson J. (1993), The network as a governance structure: interfirm cooperation beyond market and hierarchie's, in Grabher G. (ed.), *The embedded firm: on the socio-economics of industrial networks*, Routledge, pp. 35–51.

Holland J.H. (2012), *Signals and boundaries. Building blocks of complex adaptive systems*, The MIT Press.

Kauffman S. (1993), *The origins of order. Self-organization and selection in evolution*, Oxford University Press.

Kauffman S., Lobo J., Macready W.G. (2000), Optimal search on a technology landscape, *Journal of Economic Behavior and Organization*, 43, pp. 141–66.

Klepper S. (1996), Entry, exit, and innovation over the product life cycle, *The American Economic Review*, 86, pp. 562–83.

Klepper S. (1997), Industry life cycles, *Industrial and Corporate Change*, 6(1), pp. 145–81.

Klepper S. (2010), The origin and growth of industry clusters: the making of Silicon Valley and Detroit, *Journal of Urban Economics*, 67, pp. 15–32.

Lancaster K.J. (1966), A new approach to consumer theory, *Journal of Political Economy*, 74, pp. 132–57.

Lau A.K.W., Lo W. (2015), Regional innovation system, absorptive capacity and innovation performance: an empirical study, *Technological Forecasting and Social Change*, 92, pp. 99–114.

Martin R., Sunley P. (2011), Conceptualizing cluster evolution: beyond the life cycle model?, *Regional Studies*, 45, pp. 1299–1318.

Menzel M., Fornahl D. (2010), Cluster life cycles – dimensions and rationales of cluster evolution, *Industrial and Corporate Change*, 19, pp. 205–38.

Miles I., Kastrinos N., Flanagan K., Bilderbeek R., den Hertog P., Huntink W., Bouman M. (1995), Knowledge-intensive business services. Users, carriers and sources of innovation, *Report to DG13 SPRINT–EIMS*

Muller E., Zenker A. (2001), Business services as actors of knowledge transformation: the role of KIBS in regional and national innovation systems, *Research Policy*, 30, pp. 1501–16.

Newman M.E.J. (2003), The structure and function of complex networks, *SIAM Review*, 45, pp. 167–256.

Niosi J., Zhegu M. (2005), Aerospace clusters: local or global knowledge spillovers?, *Industry and Innovation*, 12, pp. 1–25.

Opsahl T., Agneessens F., Skvoretz J. (2010), Node centrality in weighted networks: generalizing degree and shortest paths, *Social Networks*, 32, pp. 245–51.

Osinga S., Hofstede G.J., Verwaart T. (2011), *Emergent results of artificial economics*, Springer.

Porter M.E. (1998), Clusters and the new economics of competition, *Harvard Business Review*, November–December, pp. 77–90.

Porter M.E. (2000), Location, competition, and economic development: local clusters in a global economy, *Economic Development Quarterly*, 14, pp. 15–34.

Saviotti P.P., Metcalfe J.S. (1984), A theoretical approach to the construction of technological output indicators, *Research Policy*, 13, pp. 141–51.

Saxenian A. (1994), *Regional advantage. Culture and competition in Silicon Valley and Route 128*, Harvard University Press.

Scott A.J. (1990), The technopoles of Southern California, *Environment and Planning A*, 22, pp. 1575–1605.

Shin D.H., Hassink R. (2011), Cluster life cycles: the case of the shipbuilding industry cluster in South Korea, *Regional Studies*, 45, pp. 1387–1402.

Squazzoni F., Boero R. (2002), Economic performance, inter-firms relations and local institutional engineering in a computational prototype of industrial district, *The Journal of Artificial Societies and Social Simulation*, 5. http://jasss.soc.surrey.ac.uk/5/1/1.html.

Utterback J.M., Abernathy W.J. (1975), A dynamic model of process and product innovation, *The International Journal of Management Science*, 3, pp. 639–56.

3 Bad news travels fast

The role of informal networks for SME-KIBS cooperation

Daniel Feser and Till Proeger

Introduction

Two dominant developments have been identified as substantial determinants of regional competitiveness in recent decades. First, learning and knowledge creation have gained significant relevance for firms' individual performance, as well as overall economic growth. Therefore, the capabilities of a region's firms in terms of successfully establishing and developing knowledge intensive innovative processes increasingly determine regional levels of competitiveness (Cooke and Leydesdorff, 2006; Lundvall, 1992). In this context, cooperation within networks that foster the dissemination of knowledge has been discussed as a positive influence on regional innovation systems (RIS) (Cantner *et al.*, 2010). Second, the qualitative and quantitative relevance of the service industries – and particularly that of Knowledge Intensive Business Services (KIBS) – has increased across sectors due to higher competitive pressure in the globalised economy. Accordingly, the innovativeness of the service sector and the degree of its interaction with firms have been identified as key determinants of sectoral and regional competitiveness (Evangelista *et al.*, 2013; Gallouj and Savona, 2009).

In this chapter, we present a barrier precluding the broad implementation of KIBS that is particularly problematic in peripheral RIS. We investigate the role of informal networks for SMEs' decisions to cooperate with KIBS and focus on the dissemination of negative attitudes towards external partners within informal networks precluding innovative cooperation. Analyzing a sample of interviews with CEOs, we find that informal networks are the one central determinant of the choice whether and with whom to cooperate. Due to the lack of alternative sources of information, networks of informal contacts largely shape the degree of external cooperation within the region investigated. Despite being rational from a transaction cost perspective, the dominance of informal networks has the counter-productive side effect of fueling skeptical attitudes among CEOs by disseminating primarily negative experiences and attitudes towards KIBS, which can be explained by drawing upon behavioural research. To counter this effect, measures for regional policy-makers determined to increase innovative cooperation among SMEs and KIBS are presented.

The remainder of this chapter is structured as follows. Section 2 gives a short literature review. Section 3 presents the methodology, before section 4 provides our results, case studies and behavioural interpretation. Finally, the results are linked to the implications for regional policy in peripheral RIS in the concluding section 5.

Literature review

Cooperation with KIBS[1] has a substantial positive impact on client industries (Camacho and Rodriguez, 2007; Kox and Rubalcaba, 2007) Furthermore, following the seminal contribution by Cooke (1992), KIBS are considered to positively influence RIS (Muller and Zenker, 2001; Probert *et al.*, 2013; Strambach, 2002), whereby the literature is mostly focused on KIBS within metropolitan regions (e.g. Aslesen and Isaksen, 2007; Wood, 2002).

Particularly for small and medium-sized enterprises (SMEs)[2] lacking internal resources, intermediate organisations providing external knowledge are an essential source to preserve and increase innovative capacities (Howells, 2006). Consequently, for regional economic structures primarily comprising SMEs, a well-established network of institutions providing and disseminating research-based knowledge holds pivotal relevance for the respective region's competitiveness (Pinto *et al.*, 2012; Tether and Tajar, 2008), since KIBS provide a means of efficiently transferring innovative knowledge that is otherwise unavailable to SMEs (Muller and Doloreux, 2009). Thus, it can be suggested that the establishment and operation of knowledge-sharing networks of SMEs and KIBS are the core requirement for regional policy aimed at increasing regional competitive advantages, particularly in peripheral regions (Muller and Zenker, 2001). However, a number of determinants for the successful use of KIBS' are considered problematic for peripheral regions (Karlsen *et al.*, 2011), particularly the lack of supporting infrastructure and access to human and social capital required for successful KIBS-SMEs cooperation (Shearmur and Doloreux, 2009; Tödtling and Trippl, 2005). These factors necessarily reduce the level of innovative cooperation achieved in peripheral regions. Extending these studies, we focus on another barrier to cooperation by investigating the role of informal networks and their potential negative side effects in peripheral regions.

Data and methodology

Following Eisenhardt (1989) and Eisenhardt and Graebner (2007) in their approach of designing case studies, we analyse the role of SMEs' informal networks regarding the demand for KIBS and the consequences for RIS. A qualitative approach is appropriate for our research goals since the response rate in surveys is fairly low, particularly for SMEs (Newby *et al.*, 2003), while SME managers favor direct human interaction in research projects with complex content (Bartholomew and Smith, 2006). Based upon the theoretical sampling, the firms were selected to achieve theoretical saturation (Glaser, 1965; Glaser and Strauss, 2008). Using semi-structured questionnaires, we attempt to gain insights for the cooperation between SMEs and KIBS since empirical and theoretical results on KIBS-SME cooperation to date remain limited. Therefore, we use an exploratory research design without examining specific hypotheses. The prerequisite for the selection of interview participants was previous experience in cooperating with KIBS and innovation integration as a substantial part of their business model.

The interviews were conducted between May and September 2014, with a sample size of 19 interviews and duration of 45–90 minutes for each interview. All interviewees except for one – who was the chairman of a board – were CEOs of their respective companies. The interviews were recorded, whereby each participant could request to stop the recording at any time. Only one participant refused to be recorded; nonetheless, his remarks are used for our analysis. Since the details of cooperation with KIBS contain sensitive information, anonymity was guaranteed to all interviewees.

The interviews were transcribed, analysed and compared with additional publically available documents (e.g. published official firm records, newspaper and results from online research) to control for the reliability of the interview results. The questionnaire was structured in three parts: first, open questions about the general experience and attitude towards using KIBS were asked; second, the interviewees were asked to describe the pre-cooperation phase, including the decision process to cooperate with KIBS and their final selection; and third, the outcome of cooperation and the impact on the innovative capacity of the respective SME were discussed, which constituted the significant part of the interviews. Following the interview, the participants received feedback and had the opportunity to criticise and validate the results.[3]

The interview data was analysed using the content analysis put forth by Mayring (2000). To apply this method, the content was reduced to its core parts using a mixed approach with inductive codes for newly mentioned aspects of the interviewees and elements that have been discussed in the respective literature (for an elaboration of our approach, see Fereday and Muir-Cochrane (2006). Based upon the results, a cross case analysis with the sample of interviews was performed, identifying and comparing SMEs' different reactions using KIBS caused by both informal networks and the consequences in terms of firms' capability to innovate.

The sample comprises 19 firms, all located in the southern part of the German federal state of Lower Saxony, a peripheral region constituted primarily of SMEs and only few large international corporations. A number of renowned universities as well as basic and applied research institutions[4] exist, although the link between these institutions and local SMEs appears to be rather weak as graduates and academic staff tend to leave the region due to the lack of job and career opportunities (Süssberger, 2011). Furthermore, small companies in Lower Saxony have been found to be least capable of innovating when compared to the rest of Germany (Berthold *et al.*, 2009). Despite the membership of Goettingen in the metropolitan area comprising the industrial centers of

Hanover, Brunswick and Wolfsburg, the low level of network integration adds to a low level of innovativeness.

We initially gained access to local SMEs using gatekeepers from the local chamber of crafts (LCC) and regional economic support (RES) which often were involved in publicly funded innovation projects with a regional focus. Subsequently, the sample was extended by following interviewees' recommendations for other CEOs (Probert *et al.*, 2013). The sample is characterised by its heterogeneity, covering the whole range from self-employed persons up to a firm with 150 employees. Furthermore, the qualification level of the companies considerably differed; for instance, Firm D was founded as a research spin-off and almost exclusively employs academic staff, in contrast to Firm B, which employs no staff with an academic background whatsoever. For the majority of the firms, the national market is most important, while the firms' focus varies regarding regional, national and international markets. The companies are in most cases family owned, with the exception of those belonging to local shareholders. Table 3.1 provides an overview of our sample and the firms' specifics. Note that we used the same set of interview responses in Feser and Proeger (2015), discussing other aspects of KIBS.

In the following section, we present our results regarding the role of informal networks for SMEs' demand for KIBS and subsequently focus on the impact of negative experiences with KIBS and its dissemination within informal networks.

The role of informal networks for SME-KIBS cooperation

SMEs aiming to establish cooperation with external partners face substantial uncertainty concerning the expected gains of a cooperation and information asymmetries regarding the specialised capabilities of KIBS. Consequently, the knowledge intensity of cooperating with KIBS makes it difficult to evaluate the quality of the service ex ante for potential clients. To reduce search costs, the SMEs in our sample primarily relied on their CEOs' informal networks, i.e. friends, business contacts and family members who had prior experience in using KIBS. The information retrieved through these sources determined the decision whether to use KIBS in the first place and largely determined which KIBS was ultimately chosen.

Despite being rational from a transaction cost view in a situation with little reliable information and high search costs, retrieving information using an individual informal network has potential drawbacks. SMEs in a peripheral region with a low density of firms necessarily have rather few contacts, which limits the information and experiences with KIBS that can be accessed through the network. This might lead to an excessive interpretation of individual experiences in using KIBS or limit cooperation to the few external contacts already established among the members of the informal network.

In the following, we describe the results of our interviews with CEOs and discuss the role of informal networks in shaping the decision whether to cooperate

Table 3.1 Overview of the sample of SMEs

Firm Coding	*Classification*	*Size*	*Sector*	*Market*	*Gate-keeper*	*Firm location*	*Ownership structure*
A	Service	Small	Crafts	Regional	LCC	Urban	Family
B	Industry	Micro	Manufacturing	Regional	-	Urban	Family
C	Service	Micro	Trade	Germany	-	Urban	Family
D	Service Industry	Small	Biotech	Worldwide	RES	Urban	Family
E	Service	Micro	Health	Regional	LCC	Urban	Family
F	Industry	Medium	Engineering	Regional	LCC	Rural	Family
G	Industry	Medium	Casting	Germany	-	Rural	Local Shareholders
H	Service	Small	Construction	Regional	LCC	Urban	Family
I	Service	Small	IT	Worldwide	RES	Urban	Local Shareholders
J	Service	Small	Biotech	Worldwide	RES	Urban	Local Shareholders
K	Service	Small	Medicine	Worldwide	RES	Urban	Local Shareholders
L	Service	Micro	Publishing	Worldwide	-	Urban	Family
M	Service Industry	Medium	Steel	Worldwide	LCC	Urban	Family
N	Industry	Medium	Engineering	Worldwide	LCC	Rural	Family
O	Service	Micro	Consulting	Germany	RES	Urban	Family
P	Industry	Medium	Car Industry	Worldwide	-	Rural	Local Shareholders
Q	Industry	Medium	Engineering	Worldwide	-	Rural	Family
R	Industry	Medium	Crafts	Worldwide	-	Rural	Local Shareholders
S	Service	Medium	Biotech	Worldwide	RES	Urban	Family

with KIBS, as well as which external partner is chosen. We thus emphasise the central role of informal networks in shaping CEOs' decision processes in peripheral regions. Subsequently, by presenting three case studies representative of our sample, we highlight the dismal effects of relying on informal networks for innovative cooperation, resulting from the excessive dissemination of negative experiences. We explain this effect using behavioural evidence on the "negativity bias" in human interaction and subsequently conclude by presenting policy implications for the design of RIS in peripheral regions.

Pre-cooperation decision

Two factors are regularly discussed in the literature on innovative cooperation as impeding SME-KIBS cooperation: the risk aversion by SME managers and financial constraints (Beck and Demirguc-Kunt, 2006; Cohen and Levinthal, 1990). However, both factors do not play a major role for the SMEs interviewed. While all interviewees considered risk as a relevant factor, it was also acknowledged as part of the regular business environment and a natural part of looking for external partners. Regarding the costs, firms recognise the costs of using KIBS and consider this a significant aspect of the decision to cooperate with KIBS. Nevertheless, the financing of the projects is not seen as problematic since they had a positive expected outcome and financial assets were available. Consequently, neither risk nor cost constraints play a significant role for the CEOs in our sample.

Instead, we find that the decision whether to cooperate with KIBS in the first place is strongly influenced by the personal attitudes of the SME manager regarding KIBS in general. The shift from using internal knowledge to external sources for creating innovation involves substantial uncertainty about the cooperating partners' behaviour and a high degree of knowledge intensity. Thus, SMEs perceive difficulties in evaluating the prospective outcome of an external cooperation, which reinforces the need to gain information through informal networks of friends and professional acquaintances in other regional firms. Within these networks of CEOs, a rather skeptical perspective on KIBS prevails. We find that this attitude is connected with stories and often stereotypical anecdotes that CEOs' have heard from and about other cooperating firms. Combined with news from the local and national media, this influences and even drives the attitudes of CEOs regarding KIBS. This effect is stronger for the field of p-KIBS, where our respondents often use negative experiences discussed within informal networks as their central argument not to cooperate with KIBS. One prominent story and sentiment hindering cooperation for SMEs in our sample – regardless of prior experience with KIBS – is the widespread feeling of being an "easy target" for KIBS exploiting informational asymmetries during the cooperation. This sentiment is based upon individual experiences by friends, acquaintances and business partners and it strongly hampers the willingness to cooperate with KIBS prior to any cooperation. Another widely disseminated story describes an SME that suffered substantial financial harm due to a cooperation with a KIBS and consequently vowed to quit external cooperation with the specific

KIBS sub-sectors in the future. While obviously not being representative, this story is repeatedly discussed and disseminated within the informal networks of SMEs in the region investigated.

Overall, uncertainties in the pre-cooperation phase substantially influence the demand for KIBS and lead to the crucial role of communication about previous experiences with KIBS within informal networks. Both affirmative and dismissive stories about KIBS strongly influence SMEs choices, while financial limitations and general risk aversion only play a minor role for the firms' behaviour prior to external cooperation.

Selection process

The selection process cannot be strictly separated from the decision process. Consequently, financial and risk assessment issues also influence the selection process to a certain degree. However, similar to the decision process, the information gained in informal networks dominates SMEs' choices, since the acquisition of reliable information about the quality of potential partners is the major difficulty. Using informal networks, CEOs state that they succeed in gaining an effective measure for the selection of KIBS.

The process of choosing external partners is challenging for SMEs since the matching of individual demands and the service supplied is complex in terms of information availability. In standard services, prices serve as a signal for quality, which makes it fairly easy for SMEs to choose the preferred service. By contrast, for knowledge intensive services in a business-to-business market, the services used are experience or even credence goods, whereby the quality of the services can only be confirmed when investing substantial search costs (see Feser and Proeger (2015) for a comprehensive discussion). The CEOs regularly stated that the actual quality and specifics of the cooperation with a KIBS were hardly understood before the start of the cooperation and indeed only partly after the cooperation. Consequently, the dominant strategy when considering a novel cooperation was to choose an external firm with which the CEOs already had positive experiences. If none was available, the difficulties in finding a suitable cooperation partner prompted CEOs to use their informal networks as the key instrument for selecting a KIBS.

The choice of a KIBS was primarily based upon the recommendations of family members, friends and business partners. Typically, the selection process started with finding someone in the firms' informal network who had experience with a specific class of KIBS. If this proved impossible, members of the informal network were asked to give recommendations concerning whom to contact. On each occasion, this information was double-checked with other sources to acquire second opinions. When a firm was found that appeared trustworthy, the SMEs assessed its reliability by checking with other companies currently cooperating with the specific KIBS. The final decision was often based upon the competence and perceived trustworthiness of the KIBS' CEO. While the specific duration and success of the selection process was highly dependent on the individual CEO's network, in all cases the informal network was accepted as lowering the informational barriers, which gave CEOs a higher level of trust in the cooperating KIBS. In some cases, even when CEOs were skeptical about external cooperation ex ante, a credible recommendation could turn their uneasiness into a cooperation that led to the successful integration of external knowledge.

Case studies on the role of informal and formal networks in SME-KIBS cooperation

While the predominant role of informal networks for the cooperation with KIBS was pointed out by all respondents in our sample, the effect on the innovative outcome of the knowledge intensive service substantially varied. We selected three case studies to exemplify the influence of informal network on the SMEs' use of KIBS. In the first case, despite initial skepticism, informal recommendations lead to cooperation, which resulted in negative experiences with KIBS. This subsequently led to the complete avoidance of the specific KIBS sub-sector and the further dissemination of the frustration with KIBS by the respective CEO (Firm I). Our second example describes the refusal to use KIBS merely based upon hearsay by colleagues' and business partners' negative experiences leading to the total exclusion of KIBS use (Firm F). By contrast, the third case study illustrates the use of an informal network following the membership in a formal network, which fostered communication with KIBS in the pre-cooperation phase and resulted in satisfactory outcomes for the respective SME (Firm H).

Negative bias confirmed

Firm I was founded in 1983 as an IT specialist, which is active in the business-to-business sector. The small-sized firm mainly focuses its products and services on international markets. Operating in niche markets, Firm I offers full and flexible support for printing machines. Based upon the firm's reputation, a large share of its customers are family businesses with a long-term cooperation history. Rough competition influences the firm's strategy and requires constant adaption to the dynamic environment, because their competitors are larger corporations. Innovation is thus seen as crucial part of its business culture; for instance, Firm I has participated in several innovation competitions.

The interviewee stated that external knowledge is used regularly and more specifically explained that cooperation in the field of t- and p-KIBS is a daily routine for the firm. Struggling with the competitive environment, the need for support from external consulting regarding strategic questions was thus identified at an early stage. However, based upon stories from business partners and acquaintances who spread negative experiences, the CEO had a skeptical attitude towards consulting companies. Despite being convinced that strategic consulting was required, the CEO thus feared that a prospective cooperation could have a strong impact on the firm's future and structure, with potentially negative consequences for the employees. Underpinning the relevance for the firm, mainly the CEO – involving only few employees – conducted the selection process of KIBS and was active in the cooperation. This procedure took place twice with different consulting firms, on each occasion without satisfying results for Firm I. Both cooperation efforts were resumed without the expected outcome and interrupted before a new strategy was implemented, because the CEO lost the confidence in the positive impact for his firm following the cooperation. Quitting the cooperation led to the firm's decision to avoid cooperation with this specific KIBS sector altogether in the future. In both cases, the selection of the KIBS had been made following the recommendations of the firm's informal network and the CEO based the selection on established business contacts, which were double-checked with additional information. The final decision about cooperating was mainly based upon the CEO's trust in the cooperation partner's competence, which faded during the cooperation with both consulting firms. The actual experiences thus led to the confirmation of the previous negative bias against consulting firms that it had spread within the informal network. Subsequently, the CEO's disappointment about his failed attempts at external cooperation was shared within his informal network. His strong feelings are thus likely to preclude cooperation by firms within his network that are unclear about the potential merits of an innovative cooperation. Accordingly, the negative experience of a single firm may reinforce the skepticism of a larger number of other firms regarding a specific KIBS sector. Indeed, our second example serves as an example of this issue.

Negative bias precludes cooperation

Founded in 1930, Firm F, which is situated in a rural environment, operates in the industrial engineering sector in the third generation. The family business mainly serves as a supplier to machine builders and constantly enhances its innovative capacity to comply with the demanding requirements of its customer firms. To extend its knowledge basis, the SME uses external sources based upon informal contacts. Nevertheless, firm F only cooperates with KIBS to a very limited extent since their benefits are evaluated critically.

Negative attitudes in the informal network of Firm F have substantially influenced its demand for KIBS, and particularly for P-KIBS. A nearby firm had started a cooperation with an accounting KIBS to restructure accounting processes, which ended negatively with substantial financial losses. The CEO of Firm F thus drew the consequences for the firm not to cooperate with KIBS operating in accounting at all. Although the exact circumstances were not transferrable and it may appear unreasonable to avoid an entire class of KIBS, the respective CEO overemphasised the negative experiences by a member of his informal network and suspended all future cooperation with KIBS.

The interviewee explicitly confirmed that his contacts to business partners and friends led to an atmosphere of skepticism, which precludes the decision to cooperate with KIBS. The central argument is the potentially problematic interaction between internal and external knowledge sources and the failure of communication between academic-based KIBS and SMEs. These factors lead to the skeptical attitude of Firm F, and particularly for business sectors where the firm's knowledge basis is low. Hence, the fear of potential dismal outcomes of cooperation combined with negative experiences spread in informal networks determines the CEO's decisions. The widespread perception of SMEs being an "easy victim" has an informal network function as catalysts of negative information about SME-KIBS cooperation. The lack of personal experience further supports the belief in negative stories about KIBS and contributes to premature rejections of innovative cooperation.

Cooperation fostered by a formal network

The third example is a family business that has operated since 1919 with a focus on regional markets in the fourth generation. The small business began as standard painter firm and has been searching for new business opportunities and customers since 2000. Market liberalisation and new foreign competitors forced Firm H to search for new strategic perspectives, although its regular demand for KIBS was limited to outsourcing activities in the legal and accounting sector.

The first attempt to cooperate with an external knowledge intensive institution was in the sector of energy consulting, to enable the firm to offer additional services for house owners. Firm H cooperated with a public institution and received staff training concerning the consulting of customers in energy-related issues. Nonetheless, given that the change in the firm's strategy failed to increase the number of customers, the additional service was terminated and the cooperation with the institution in the field of energy advising was ended.

The next attempt at innovative cooperation started with a business contact from a formal network. The SME's initial motive for joining the horizontal network was to improve the firm's professional training. Being familiar with a t-KIBS from the network, the barriers to contact the firm on an informal level and the risk connected to cooperation were considered low. Although the CEO had already left the respective network, the reason for contacting the KIBS was the meeting of both CEOs at a network event. The result of the cooperation was a new strategic concept for the firm and a publication used as a marketing instrument to attract new customers. The established trust in the cooperation partner was essential: since the CEO already personally knew the KIBS's CEO, the risk seemed acceptable. Following the successful innovative cooperation, Firm H plans to continue working with the respective t-KIBS.

The case of Firm H illustrates that individual contacts established in formal networks can substantially simplify the contact to KIBS. Despite negative experiences with previous cooperation, the interviewee did not shun communicating with KIBS connected via a formal network. The network thus worked as an informational platform connecting KIBS and SMEs. These conditions generated a situation in which trust could be established between SMEs and KIBS through the meeting of the respective CEOs. Thus, unlike the two prior examples, Firm H did not rely on biased information drawn from informal networks; instead, personal contact to a KIBS could be established via a regional formal network that fostered innovative cooperation.

Interpretation: the "negativity bias" in informal networks

Our core finding regarding the effect of informal networks – as exemplified in the first two case studies – is that negative experiences are transmitted quickly within informal networks and often preclude future innovative cooperation by increasing existing doubts and skepticism regarding KIBS. While this result is based upon a limited number of SMEs interviewed, it is supported by a broad strand of literature from behavioural research in Psychology and Experimental Economics. We would argue that informal networks are particularly prone to the "negativity bias" in human interaction and – more specifically – communication.

This bias describes the human tendency to overrate negative events and news, whereby "*bad impressions and bad stereotypes are quicker to form and more resistant to disconfirmation than good ones*" (Baumeister *et al.*, 2001, p. 323). Behavioral research has discussed different aspects and explanations for this phenomenon.[5] While no unanimous psychological explanation has been presented to date, it has been established that negative events receive more attention in individual reasoning and feeling, which translates into a substantial negativity bias in all forms of human communication.

Rozin and Royzman (2001) suggest that a central driver of the negativity bias in human interaction lies in its contagiousness. Thus, negative or worrisome information dominates more positive issues in individual interaction and is involuntarily transmitted with a higher priority than other pieces of information. Evolutionary approaches can offer good explanations for the utility and superiority of this mechanism; however, its dominance can also – as in our case – lead to a structural failure to use new ideas and innovative concepts once an informal network has incorporated and disseminated a number of negative experiences. Since these are weighed more heavily than positive examples of cooperation in informal networks without a systematic presentation of best-case examples, skepticism is likely to increase over time.

We would thus argue that since the dominance of single negative experiences with KIBS in informal networks dominate positive examples, regional policy-makers need to counter this effect with a systematic use of positive examples disseminated within formal networks. Accordingly, a balance between negative and positive news regarding KIBS can be established helping to compensate the negativity effect.

Conclusion: consequences for regional policy

The essential role of KIBS in RIS is regularly highlighted and discussed in the literature regarding the competiveness of entire regions. While the research emphasises the positive impacts of KIBS on sectors, firms and regions, the client firm's internal perspective largely remains unclear at present. We run interviews with CEOs of SMEs to analyse the role of informal networks in the demand for innovative cooperation with KIBS. All interviewees stressed the relevance of informal networks for selecting KIBS to reduce the information asymmetries between SMEs and KIBS. We thus find that informal networks play a predominant role in the pre-cooperation and selection process, and particularly for peripheral RIS, where formal network structures are less well established. Since there are fewer firms compared to metropolitan areas, the informal networks structures are similarly weakly developed, which limits SMEs' potential sources for recommendations when looking for KIBS. This aggravates the problem whereby a few bad experiences in combination with negative media coverage can swiftly be disseminated and lead to an overall refusal to cooperate with KIBS among the local business community. This is accentuated by the negativity bias, which fosters the spreading of worst-case examples within informal networks.

This effect is likely to contribute to a weak demand for KIBS, which disturbs the circulation of knowledge and can impede the innovation processes and thus regional competitiveness in peripheral RIS.

We would argue that formal network structures can offer a way to reduce the uncertainties for SMEs otherwise relying on the information from informal networks. Therefore, based upon our investigation, we would consider formal networks as the key element for increasing the innovative capabilities of a peripheral region. We would offer three different suggestions for policy-makers to foster the innovative capability of peripheral regions. First, in peripheral regions without any formal networking efforts, the initial implementation of formal networks should be pursued. Building these structures should be conducted in close cooperation with the targeted firms to fulfill their specific requirements for public support. Networks should regularly be reshaped regarding potentially changing demands among the participating firms.

Furthermore, strengthening existing formal networks by making efforts to include additional firms and organise additional activities in peripheral regions should be used to build mutual trust between SMEs and KIBS. A local chamber of crafts/commerce – such as those existing in many countries such as Spain and Germany – should organise formal networks and activate previously disinterested firms. This can improve the communication between firms and KIBS as information ceases to be exclusively related through informal networks.

Thirdly, the promotion of best practice examples by formal networks structures can help to reduce the widespread skepticism about using KIBS. However, this goal hinges on the network's credibility among participating firms, whereby it is vital that the organisation running the network is perceived as impartial by firms to prevent the impression that the networks are used as a mere promotional tool by KIBS trying to gain additional customers. This impression would reduce the credibility of the information provided through the network and thus jeopardise the networking goals; rather, the networks need to be perceived as impartial institutions that are financially independent from KIBS' business interests. This impartiality can be achieved provided that networking institutions are publicly funded and required to work in favor of SME interests, as in chambers of crafts/commerce.

Let us finally comment on some limitations of our study. Given our explorative approach, further research needs to add and specify in greater detail the quantitative and qualitative effect to which the communication in informal networks influences SMEs' decisions to cooperate with KIBS. Bearing in mind that our results are based upon a German sample, other examples from peripheral regions are required. However, the region of the South of Lower Saxony exemplifies a region of weak network structures between academic research institutions and SMEs in a peripheral regional innovation system with few large corporations driving innovative efforts. Given the importance of formal network structures, a further development of regional policy instruments is desirable to learn how to best build regional institutions to create and implement effective network structures. In particular, organisational structures of networks – whether private, public or public-private – providing a neutral SME-KIBS communication platform should be addressed in further research.

Acknowledgement

Financial support for conducting the interviews from the PraxisResearcher program funded by the University of Goettingen and the iENG project (grant number 03EK3517A), funded by the Federal Ministry of Education and Research, is gratefully acknowledged.

Notes

1 We use the seminal definition of KIBS provided by Miles *et al.* (1995). KIBS are the part of the service sector with an above average share of academic staff (defined as a higher than 11 per cent of employees with academic degree or above 4.5 per cent of employees with academic degree in Engineering or Science) and supplying professional business-to-business services. Strambach (2008) divided the KIBS sector in two sub-groups: professional (p-), which includes e.g. marketing, legal and accounting, and technological (t-) KIBS with information and technology (ICT) services, engineering and technological consulting services. Despite the heterogeneity of the KIBS sector, the involvement in learning processes of customer firms by supporting the transfer and implementation of knowledge from the KIBS (Strambach (2008). Hence, being a major contributor in the innovation process, KIBS role was largely discussed in the literature, e.g. by Castaldi *et al.* (2013); Tödtling *et al.* (2006).

2 We apply the definition of Eurostat: SMEs include enterprises with fewer than 250 employees, medium–sized between 50 and 250, small 10 to 49, and micro enterprises fewer than 10 employees.

3 Additional information on the content and analysis of the interviews is available from the authors upon request.

4 For example, the University of Goettingen, the Clausthal University of Technology and the University of Applied Sciences and Arts Holzminden are situated in the south of Lower Saxony. Moreover, the research institutes of the German Primate Center (DPZ), Max Planck Institute for Biophysical Chemistry (MPI) and the German German Aerospace Center (DLR) are located in the same region.

5 For an introductory reading to the broad literature regarding the negativity bias, its causes and implications, we refer to Rozin and Royzmann (2001), Vaish et al. (2008) and – more recently – Hsee et al. (2014).

References

Aslesen, H.W., Isaksen, A., 2007. Knowledge-intensive business services and urban industrial development. *The Service Industries Journal* 27 (3), 321–38.

Bartholomew, S., Smith, A.D., 2006. Improving survey response rates from chief executive officers in small firms: the importance of social networks. *Entrepreneurship Theory and Practice* 30 (1), 83–96.

Baumeister, R.F., Bratslavsky, E., Finkenauer, C., Vohs, K.D., 2001. Bad is stronger than good. *Review of General Psychology* 5 (4), 323–70.

Beck, T., Demirguc-Kunt, A., 2006. Small and medium-size enterprises: access to finance as a growth constraint. *Journal of Banking and Finance* 30 (11), 2931–43.

Berthold, N., Kögel, D., Kullas, M., 2009. *Innovation in den Bundesländern. Ergebnisse einer repräsentativen Unternehmensumfrage*, Würzburg.

Camacho, J.A., Rodriguez, M., 2007. Integration and diffusion of KIS for industry performance, in: Rubalcaba, L., Kox, H.L. (Eds), *Business services in European economic growth*. Palgrave Macmillan, Basingstoke, New York, pp. 128–43.

Cantner, U., Meder, A., ter Wal, A.L.J., 2010. Innovator networks and regional knowledge base. *Technovation* 30 (9–10), 496–507.

Castaldi, C., Faber, J., Kishna, M.J., 2013. Co-innovation by KIBS in environmental services: a knowledge based perspective. *International Journal of Innovation Management* 17 (05), 1–17.

Cohen, W.M., Levinthal, D.A., 1990. Absorptive capacity: a new perspective on learning and innovation. *Administrative Science Quarterly* 35 (1), 128–52.

Cooke, P., 1992. Regional innovation systems: competitive regulation in the new Europe. *Geoforum* 23 (3), 365–82.

Cooke, P., Leydesdorff, L., 2006. Regional development in the knowledge–based economy: the construction of advantage. *Journal of Technology Transfer* 31, 5–15.

Eisenhardt, K., 1989. Theories from case study research. *The Academy of Management Review* 14 (4), 532–50.

Eisenhardt, K., Graebner, M.E., 2007. Theory building from cases: opportunities and challenges. *Academy of Management Journal* 50 (1), 25–32.

Evangelista, R., Lucchese, M., Meliciani, V., 2013. Business services, innovation and sectoral growth. *Structural Change and Economic Dynamics* 25, 119–32.

Fereday, J., Muir-Cochrane, E., 2006. Demonstrating rigor using thematic analysis: a hybrid approach of inductive and deductive coding and theme development. *International Journal of Qualitative Methods* 5 (1), 80–92.

Feser, D., Proeger, T., 2015. Knowledge-intensive business services as credence goods. A demand-side approach. *cege Discussion Papers* (232).

Gallouj, F., Savona, M., 2009. Innovation in services: a review of the debate and a research agenda. *Journal of Evolutionary Economics* 19 (2), 149–72.

Glaser, B.G., 1965. The constant comparative method of qualitative analysis. *Social Problems* 12 (4), 436–45.

Glaser, B.G., Strauss, A.L., 2008. *The discovery of grounded theory: strategies for qualitative research*. Recording for the Blind and Dyslexic, Princeton, NJ.

Howells, J., 2006. Intermediation and the role of intermediaries in innovation. *Research Policy* 35 (5), 715–28.

Hsee, C.K., Rottenstreich, Y., Tang, J., 2014. Asymmetries between positives and negatives. *Social and Personality Psychology Compass* 8 (12), 699–707.

Karlsen, J., Isaksen, A., Spilling, O.R., 2011. The challenge of constructing regional advantages in peripheral areas: the case of marine biotechnology in Tromsø, Norway. *Entrepreneurship and Regional Development* 23 (3–4), 235–57.

Kox, H.L., Rubalcaba, L., 2007. The contribution of business services to economic growth, in: Rubalcaba, L., Kox, H.L. (eds), *Business services in European economic growth*. Palgrave Macmillan, Basingstoke, New York, pp. 97–115.

Lundvall, B. Å., 1992. *National innovation systems: towards a theory of innovation and interactive learning*. Pinter Publishers, London.
Mayring, P., 2000. Qualitative content analysis. *Forum: Qualitative Social Research* 1 (2), Art. 20.
Muller, E., Doloreux, D., 2009. What we should know about knowledge-intensive business services. *Technology in Society* 31 (1), 64–72.
Muller, E., Zenker, A., 2001. Business services as actors of knowledge transformation: the role of KIBS in regional and national innovation systems. *Research Policy* 30 (9), 1501–16.
Newby, R., Watson, J., Woodliff, D., 2003. SME survey methodology: response rates, data quality, and cost effectiveness. *Entrepreneurship Theory and Practice* 28 (2), 163–72.
Pinto, H., Fernandez-Esquinas, M., Uyarra, E., 2012. Universities and knowledge-intensive business services (KIBS) as sources of knowledge for innovative firms in peripheral regions. *Regional Studies*, 1–19.
Probert, J., Connell, D., Mina, A., 2013. R&D service firms: the hidden engine of the high-tech economy? *Research Policy* 42 (6–7), 1274–85.
Rozin, P., Royzmann, E.B., 2001. Negativity bias, negative dominance, and contagion. *Personality and Social Psychology Review* 5 (4), 296–320.
Shearmur, R., Doloreux, D., 2009. Place, space and distance: towards a geography of knowledge-intensive business services innovation. *Industry and Innovation* 16 (1), 79–102.
Strambach, S., 2002. Change in the innovation process: new knowledge production and competitive cities. The case of Stuttgart. *European Planning Studies* 10 (2), 215–31.
Strambach, S., 2008. Knowledge-intensive business services (KIBS) as drivers of multilevel knowledge dynamics. *International Journal of Service Technology and Management* 10 (2–4), 152–74.
Süssberger, H., 2011. Unterstützung der regionalen Wirtschaft durch universitäre Ausbildung und Forschung am Beispiel der Georg-August-Universität, in: Udmurt State University (Ed.), *International cooperation: integration of education areas*. Proceedings of the 2nd International Conference November 17–19, Izhevsk, pp. 24–28.
Tether, B.S., Tajar, A., 2008. Beyond industry-university links: sourcing knowledge for innovation from consultants, private research organisations and the public science-base. *Research Policy* 37 (6–7), 1079–95.
Tödtling, F., Lehner, P., Trippl, M., 2006. Innovation in knowledge intensive industries: the nature and geography of knowledge links. *European Planning Studies,* 14 (8), 1035–58.
Tödtling, F., Trippl, M., 2005. One size fits all? *Research Policy* 34 (8), 1203–19.
Vaish, A., Grossmann, T., Woodward, A., 2008. Not all emotions are created equal: the negativity bias in social-emotional development. *Psychological Bulletin*, 134 (3), 383–403.
Wood, P. (Ed), 2002. *Consultancy and innovation: the business service revolution in Europe*. Routledge, London, New York.

Part II

KIBS and their context

4 Does the geographic distribution of Knowledge Intensive Business Services affect the use of services for innovation?

Empirical evidence from Quebec KIBS manufacturers

David Doloreux and Richard Shearmur

Introduction

Innovation is an open process, with collaboration and outsourcing increasingly recognised as key innovative behaviours (Huizingh, 2011). This has led to scrutiny of the role and contribution of external partners to innovation. Amongst these external actors, KIBS (Knowledge Intensive Business Services) have been highlighted as important vectors of knowledge creation and transmission, and as facilitators and carriers of innovation (Sundbo and Toivonen, 2012; Martinez-Fernandez, 2010; Doloreux and Shearmur, 2012; Miles, 2008; Den Hertog, 2000).

A related dimension of innovative behaviour is the geographical concentration of economic activities, which is believed to enhance the probability of collaboration, and to facilitate the transmission of tacit information and know-how between agents (Asheim *et al.*, 2011; Asheim and Gertler, 2005; Cooke *et al.*, 2004). The geographic agglomeration and clustering of economic agents is thought to facilitate more intensive interactions between proximate groups of interconnected firms and associated institutions: it thus enhances processes that are fundamental to open innovation (Asheim *et al.*, 2011; Isaksen, 2009; Wolfe, 2009).

This chapter investigates whether manufacturing establishments located in KIBS-rich environments are more likely to use KIBS than those located away from KIBS clusters. Based upon a sample of 804 manufacturing establishments, this question is approached in three stages. The first stage enquires whether manufacturers are more likely to use services when located in KIBS-rich environments. Given the association between KIBS-use and innovation, we than ask the same question of *innovative* manufacturers. Finally, we examine three different types of innovator (product, process, organisational) to see whether the connection between KIBS-use and geographic proximity differs between innovators.

This chapter attempts to overcome a number of limitations in the literature on innovation, services and regional clusters. First, despite KIBS being recognised as contributors to innovation in client firms (Miles, 2012), most empirical research has focused on KIBS as innovators. KIBS as contributors to innovation have been studied qualitatively (e.g. Howells *et al.*, 2003) but there are, as yet, few quantitative studies (Yam *et al.*, 2011; Doloreux and Shearmur, 2012). Second, and related to this, studies focussed on clusters and agglomeration have explored the spatial co-location of KIBS providers and KIBS users, without directly measuring whether services within the agglomeration are used by local users (Rodriguez *et al.*, 2012). Indeed, it is often assumed (as opposed to demonstrated) that innovators turn to KIBS providers if they are available (Yam *et al.*, 2011) – which is not in keeping with studies of KIBS' market extent (that do not, however, focus on innovators, Tether *et al.*, 2012). Finally, work on open innovation itself – for instance that on communities of practice (Amin and Roberts, 2008) – suggests that geographic agglomeration is not necessary for interactions.

The structure of the chapter is of as follows. In the second section, we review several contributions on innovation and clusters. In the third section, we describe the methodology and variables. In the fourth section, we present the result and in the last section, we summarise our main conclusions and discuss policy implications.

Literature review

Innovation in heterogeneous geographic cluster contexts

Innovation is a broad term with multiple meanings characterised by different dimensions and conceptualised in different ways. In particular, innovation can be understood both as a process and as an outcome. Innovation as process includes all activities from idea generation to implementation, commercialisation and diffusion (Rogers, 2003), as well as the feedback, information-gathering and collaboration that accompanies the process. Innovation as outcome details the type of innovation (e.g. product, process, managerial, marketing), the environment relative to which the outcome is innovative (the firm, the local market, the world), and certain other characteristics of the innovation such as the degree of novelty (radical or incremental). Accordingly, innovation as an outcome is defined as 'the development and implementation of a new or significantly improved product (good or service), or process, a new marketing method, or a new organisational method business practice, workplace organisation or external relations' (OECD, 2007: 46).

Different views of innovation capture different activities and ways of integrating knowledge to the process. Early models viewed innovation as a linear sequence of distinct stages or functional activities (design, production, commercialisation, marketing) (Kline and Rosenberg, 1985). There are two basic variations of this linear model. There is the technology-driven model (*technology-push)*, where innovation is primarily linked to the generation and exploitation of technological knowledge developed in R&D. There is the customer model (*demand-pull*), where innovation derives from the exploitation of market knowledge acquired through customers' relations and needs. Both of these models follow a sequential process, either emphasising R&D activities, the market being a receptacle for the results of R&D; or emphasising the market, the market being the source of ideas driving R&D.

Current conceptualisations emphasise the social construction of innovation. This reflects recognition that innovation occurs through processes of interaction between different economic players, including private firms, colleges and HEI, research centres, technology transfer organisations, industrial association, unions and other institutional forms (Chesborough, 2003). From this perspective innovation does not depend exclusively on technological and market capabilities but rather on knowledge integration efforts to mobilise and combine competences, resources and knowledge from other players (Huizingh, 2011). This results in an open model where innovation requires multiple functions – not only R&D – and interactions with heterogeneous economic players and resources within and across firms' boundaries (Sammarra and Biggiero, 2008). Although open innovation strategies at the firm level have been extensively explored over the last decade (Dahlander and Gann, 2010), "the context dependency of open innovation, ... is one of [its] least understood topics" (Huizingh, 2011, p. 2). It is to the geographic context of open innovation that we now turn.

The relationship between innovation and geography has attracted much interest (Shearmur, 2012; Asheim, 2012; Cooke *et al.*, 2011): notwithstanding debates in the field, it is generally accepted that innovation, depending as it does on external interactions, is facilitated by the spatial clustering of agents. Geographic proximity between innovation actors fosters these interactions, and hence facilitates innovation.

Indeed, over the last fifteen years, in parallel with the articulation of the open innovation paradigm, many studies have investigated the association between the geographic concentration of economic activities and innovation (Wolfe, 2009; Currid, 2007; Cooke *et al.*, 2011). There are three main processes at work. First, spatial agglomeration provides benefits and externalities that enhance firms' abilities to absorb and apply knowledge. Following Marshall's (1890) seminal work, the sources of these benefits and externalities include specialised labour, the support and provision that firms in the same industry draw from specialised suppliers and service providers, and finally, the positive externalities (in particular knowledge spillovers) that flow among geographically proximate firms. Innovation is stimulated because of increased circulation of knowledge and employees within and between firms, facilitated by ongoing relationships and frequent face-to-face contacts. When agglomerated, firms perceive more clearly and rapidly new technological possibilities and market needs, and learn more about evolving technologies and machinery availability. Second, agglomeration and clustering enable competitive advantage to be built (Porter, 2003). This includes better and cheaper access to related and supporting industries (such as specialised KIBS), to production factors, and to institutions as well as governments: the policies, institutions and support industries within the cluster adapt to, and enhance, the competitiveness of the principal export industries. Finally, there is a positive effect of concentration on new firm formation: clusters appear to provide an attractive environment for entrepreneurs and new subsidiaries (Wennberg and Linqvist, 2010).

One question behind work on innovation and geography concerns why innovation varies across different geographical clusters (Shearmur, 2012). The general assumption is that location matters because the concentration of information and knowledge resources differs from one cluster to another. Highly innovative clusters are thought to be those which concentrate dynamic industrial activities, universities, basic and applied research centres, assisted by a concentration of business services, and possessing a pool of technical knowledge and specialisation (Teirlinck and Spithoven, 2008).

Another factor that varies between clusters is the institutional features which affect the generation, assimilation and transfer of knowledge (Asheim and Gertler, 2005; Cooke *et al.*, 2004). Here, too, innovation is conceptualised as a process grounded in proximity relations. It is argued that geographic proximity and spatial concentration stimulate interactive learning by facilitating the relations between innovating businesses and the external factors needed for the innovation process.

These various approaches to the connection between geography and innovation all point to the fact that certain clusters are better suited – by their institutions, KIBS availability, labour mobility, R&D intensity – to innovation than others. Larger clusters within metropolitan and urban regions, as opposed to clusters in peripheral and rural areas, are often pointed to as gathering the prerequisites for regional innovativeness. They gather a multiplicity of actors and a dense infrastructure of supportive institutions and organisations, both embedded within a local culture and regulatory framework (Tödtling and Trippl, 2005; Cooke *et al.*, 2004; Doloreux, 2002). Indeed, it is often assumed that metropolitan areas – and maybe nearby cities that benefit from easy access to metropolitan areas – are key loci of innovation (Glaeser, 2011; Crevoisier and Camagni, 2001).

Notwithstanding this apparent consensus, the idea that there is a tight connection between clusters, large cities, and innovation is not unanimously held (Shearmur, 2012, 2015). A key question relates to the idea that proximity between actors facilitates interaction: whilst few researchers question this, it is the nature of proximity that has come under scrutiny, since proximity need not necessarily be geographic (Torre, 2008; Bathelt, 2011). As the communities of practice literature makes clear (Amin and Roberts, 2008), proximity that leads to sharing and to innovation can be a multiple sorts. Boschma (2005), for instance, outlines five types of proximity each of which can enhance knowledge exchange: (1) cognitive proximity (the way actors perceive, interpret and evaluate the world; (2) organisational proximity (the nature of relations between the actor; (3) social proximity (individuals' level of relationships, including trust based on friendship; (4) institutional proximity (social and cultural norms that regulate knowledge exchange); and, (5) geographic proximity – physical distance between actors. All of these proximities are interrelated and no single type of proximity is a necessary condition for knowledge exchange and innovation.

In particular, there is no reason to believe that geographic clustering is a necessary, nor a sufficient, stimulus to innovation (Huber, 2012). Neither, conversely, does innovation necessarily rely upon local interactions (Ben Letaifa and Rabeau, 2013; Gordon and McCann, 2005; Doloreux, 2004; Shearmur, 2015). These researchers emphasise the importance of distant knowledge linkages to the innovativeness of clusters: linkages with distant actors are indispensable in the development of firms and clusters not only as an outlet for production but also as a conduit for advanced knowledge. From this perspective, a firm's immediate environment does not determine its level of external interaction.

It is the association between proximity to KIBS and KIBS-use that we investigate in this chapter. The geographic approach to innovation suggests that service utilisation is more frequent in geographical clusters of KIBS because of the physical proximity between service users and providers. However, the proximity approach (Torre, 2008, Boschma, 2005) and the empirical results of Shearmur (2012), Doloreux and Shearmur (2012) and MacPherson (2008) call for caution. Indeed, the proximity approach – which allows for interactions to occur between geographically distant interlocutors provided that they have occasional face-to-face contacts – suggests that the collaboration and interaction needed for innovation does not necessarily have to be local or within a cluster. From this perspective there is no reason to believe that businesses located in KIBS-rich environments will have a higher propensity to use KIBS than businesses in KIBS-poor places.

Research issues and expectations

This chapter examines whether recourse to external KIBS by manufacturing establishments is associated with the presence of KIBS in their immediate environment. Thus the first part of the analysis investigates whether manufacturing firms are more likely to use KIBS if they are located in KIBS clusters. We expect that manufacturing establishments located in metropolitan areas, whether innovative or not, will use a wider variety of KIBS than those located outside of metro areas, given the large number and wide variety of KIBS providers there.

The second part examines if *innovative* manufacturing firms use more services when they locate in an environment with high concentrations of KIBS, particularly those more directly associated with the innovation process. We expect that the metropolitan effect will be more marked for innovative manufacturing establishments, particularly those which introduce technological innovations: they will use a wider variety of KIBS in locations where there is a high presence of technical KIBS, the type of KIBS which can provide direct input to the technological innovation process.

The third part explores whether KIBS use by different types of innovator (radical, product, process and organisational innovators) is increased by being located in proximity to KIBS providers. There are two basic ways of determining the KIBS intensity of an establishment's local environment. The most straightforward is to classify the establishment's location according to its position in the urban hierarchy. All empirical analyses have shown – and Quebec is no exception (see Table 4.1) – that KIBS are over-represented both relatively and absolutely in metropolitan areas and large cities. The second way is to measure the actual presence of KIBS in each locality, and classify localities according to the profile of KIBS that can be found locally. The working hypothesis here is that establishments in location with few KIBS will use fewer of them. Given that a variety of KIBS profiles are identified, it is possible that certain local KIBS profiles are more strongly associated with KIBS use amongst innovators than others.

It is important to compare the results of these two approaches since, even though the second approach appears to be more robust, positive results may merely be replicating those that would have been found using the urban hierarchy. It is only if there is a stronger association between KIBS use and KIBS clusters than there is between KIBS use and the urban hierarchy that our results will corroborate the idea that the local KIBS context has an influence on the

Table 4.1 KIBS and manufacturing employment and location quotients in Québec

	KIBS	*LQ*	*Manuf*	*LQ*
Montréal	341,000	1.28	245,000	0.90
Québec	61,500	1.11	34,000	0.59
Central cities, over 50K	33,000	0.74	60,000	1.32
Central cities, under 50K	14,000	0.60	35,000	1.42
Rural central	37,000	0.54	100,000	1.44
Peripheral cities, over 50K	8,750	0.65	14,000	1.02
Peripheral cities, under 50K	15,250	0.63	21,000	0.84
Rural peripheral	10,500	0.42	25,500	0.98

Note: a central city (or rural area) is one located within about 100km -120km of Montréal, Québec or Ottawa.

propensity to use KIBS. We expect, that in all cases, classification of localities by KIBS profile will have better explanatory power than classification according to the urban hierarchy.

Data and methodology

Research data

The data are derived from two sources. The first source is the Census data of Statistics Canada of 2006 to provide contextual information on each region's industrial structure and characteristics. These employment data, which comprise 24 distinct services sectors at the three- and four-digit NAICS level (see Appendix 4A), cover 104 regions in Quebec, which approximate to labour market areas. These regions are of two types. First, there are 43 urban agglomerations – municipalities or groups of municipalities of over 10,000 people that function as integrated labour markets. Second, there are MRC (Municipalité Régional de Compté – like NUTS 3 in Europe). Urban agglomerations are extracted from these MRCs (which are first aggregated if an agglomeration crosses boundaries) and the resulting 61 regions thus constitute rural areas. These areas do not comprise any urban area of over 10,000 people, and are not necessarily independent labour markets: however, except in the immediate vicinity of large metropolitan areas labour markets do not overlap MRC boundaries (Ribichesi and Shearmur, 2008).

The second source is an original survey of manufacturing establishments that was developed to examine the use of different services in innovation development of manufacturing establishments with fewer than 250 employees. The questionnaire was administered by C.A.T.I (Computer Aided Telephone Interviews) during January and February 2011 and yielded 804 responses. The potential universe of manufacturing establishments had been identified from the business directory of the *Centre de Recherche Industrielle du Québec* (CRIQ). From the list, a random sample of 2,000 establishments was selected and 804 manufacturing participated in the survey. The sample mirrors the current structure of the manufacturing sector in the province of Quebec and its distribution across sectors and regions (more information upon request).

The questionnaire comprises three sections. The first section deals with the use of services and includes questions on the frequency of contact, the nature of service delivery and the location of the main service providers. A list of 15 different service functions is listed in the questionnaire (Table 4.2): note that these service functions are not necessarily associated with a particular KIBS sector. Thus, for instance, advice on accessing capital or finance may have been obtained from a bank, a management consultant, or any other type of service provider that gives this sort of advice. The second section comprises questions about different innovation strategies and activities, including sources of information used and types of innovation developed and their novelty. The

Table 4.2 Service types according to their role in the value chain

Category	*Service type*
Identification of high-value knowledge	Identification of technological and equipment requirements Identification of R&D needs Consulting services for access to technology, patents, etc.
Knowledge validation	Consulting services for business plan preparation Assistance with prototype design or technological feasibility tests Consulting services for patent preparation Certification of product or process safety
Implementation	Consulting services for improving management processes Consulting services for implementing a process or bringing a product on line Consulting services for accessing capital or financing Fiscal services
Commercialisation	Consulting services for commercialisation or marketing
Support services	Human resource management services Services offered by lawyer or notary Accounting services

third section comprises general questions on the establishment's demography such as ownership, number of employees, age, and geographical distribution of sales.

Variables and analysis

The dependent variable in the following analyses is the variety, or total number, of services used. This variable can vary from zero to 15, but in order to have sufficient observations in each category it has been collapsed into 7 categories, the highest frequency users being grouped together in the category '6 or more types of service'.

The independent variable of interest to this study is the classification of localities according to the profile of local KIBS providers. Profiles are identified by first performing a principal component analysis of location quotients of individual sectors (24 high-order service sectors, Appendix 4A). Each component identifies service sectors that tend to co-locate in the 104 localities studied. Cluster analysis is then performed on the localities' component scores in order to identify 'industrial profiles', i.e. we identify nine groups of localities that have similar mixes of high-order services (Table 4.3). This classification of functional regions by KIBS

Table 4.3 Characterisation of local KIBS environment, 2006

Cluster analysis: KIBS profile, mean KIBS component scores (appendix A)

CLUSTER	n regions	n obs.	F1	F2	F3	F4	F5	F6	F7	F8
CL1: no KIBS cluster, low to average KIBS provision	44	240	-0.41	-0.36	0.26	-0.18	0.08	-0.20	-0.20	-0.19
CL2: no KIBS cluster, low KIBS provision	8	24	0.38	0.26	**-1.34**	-0.44	-0.48	0.04	-0.09	**-1.30**
CL3: insurance broking and management, marketing	12	73	**0.62**	-0.30	0.19	-0.41	**-0.72**	0.47	0.00	**1.29**
CL4: technical KIBS and multimedia	8	67	-0.58	**2.11**	0.03	-0.37	-0.01	0.50	-0.46	0.60
CL5: data processing	6	42	-0.03	0.25	0.20	0.24	-0.29	-0.19	**2.38**	0.06
CL6: management, marketing and FIRE	4	23	**1.81**	-0.13	-0.22	-0.22	**2.12**	-0.09	-0.57	0.65
CL7: financial authorities	4	14	0.14	-0.06	0.08	**3.08**	0.09	0.19	-0.19	-0.18
Quebec	1	89	**1.80**	**1.29**	-0.36	**1.14**	**2.74**	-0.05	0.00	**0.82**
Montreal	1	222	**2.41**	**1.76**	**1.52**	**3.92**	**0.62**	**-0.78**	0.09	0.02

(continued)

Table 4.3 Characterisation of local KIBS environment, 2006 (continued)

Cluster analysis: descriptive statistics of localities in each cluster

	Median KIBS jobs	Mean KIBS jobs	Median pop.	Mean pop.	% of localities urban	central
CL1: no KIBS clustering, low to average KIBS provision	540	898	18,505	24,886	35.0%	64.0%
CL2: no KIBS cluster, low KIBS provision	338	428	15,978	15,220	25.0%	50.0%
CL3: insurance broking and management, marketing	810	1,318	22,640	31,567	57.5%	50.0%
CL4: technical KIBS and multimedia	1753	3,498	42,655	65,024	87.5%	62.5%
CL5: data processing	748	2,558	30,405	54,788	33.3%	50.0%
CL6: management, marketing and FIRE	2008	1,649	35,523	33,565	0.0%	100.0%
CL7: financial authorities	955	885	18,230	22,599	75.0%	75.0%
Quebec	61 500		704 180			
Montreal	340 575		3 588 520			

profile is then used as an independent explanatory effect in ordered multinomial logistic regressions where the dependent variable is the variety of services used. Given the exceptional nature of metropolitan areas, which dominate in terms of their absolute number of KIBS, these are treated separately. Of the 102 remaining regions, 86 have at least one observation in them. Our analysis therefore captures 88 different local environments within which establishments are located, and these are grouped into nine different KIBS profiles.

Such a classification allows us to see whether there is a connection between clusters (i.e. local KIBS context) and the use of services. However, it is well established that KIBS presence varies in a systematic manner across the urban hierarchy (Table 4.1): there are marked differences between urban and rural localities, between localities in proximity to metropolitan areas and localities further away, and between metropolitan areas and all others (Desmet and Fafchamps, 2005; Polèse and Shearmur, 2006). Thus an alternative hypothesis – that KIBS use varies according to a classification of regions by their size, proximity to metropolitan areas and urban/rural status (see Table 4.1) – is also tested.

The ordered multinomial logistic regressions are structured as follows:

$$U_i = Cluster_i + Controls_i$$

where U_i is the variety of services used by establishment i

$Cluster_i$ is a series of dichotomous variables indicating the type of cluster in which establishment i is located. An alternative specification has $Cluster_i$ indicating the position of establishment i's location in the urban hierarchy. The *KIBS cluster* and *Urban hierarchy* classifications are not entered concurrently because they are too highly correlated.

$Controls_i$ is a series of establishment level variables that are connected with innovation performance, i.e. manufacturing sectors, establishment size (four size classes); age; R-D intensity, and export intensity.

The analysis proceeds as follows. We first describe the different local KIBS profiles, and, in the light of this classification derive explicit hypotheses based on the general ones enunciated above. We then present the results of regression analyses that explore the connection between KIBS use and local KIBS profile, comparing non-innovators with innovators. Similar results are presented that compare different types of innovator. To conclude the empirical part we show that, in all but one case, local KIBS profiles have an equivalent (i.e. absent) or greater explanatory power than the urban hierarchy when it comes to explaining the propensity to use KIBS.

Results

Classifying the local KIBS environment

Each of the 88 territories (which approximate functional regions) in Quebec have a particular supply of KIBS. In some territories, all KIBS are systematically underrepresented, whereas in others, some combinations of KIBS are over-represented and others are absent. The principal component analysis in Appendix 4A identifies those KIBS which tend to co-locate within the 88 localities, i.e. those which tend to be over or under-represented in the same localities. We identify eight components (there are eight Eigenvalues greater than one), i.e. eight groups of KIBS which tend to co-locate.

Each of these eight components has a factor score associated with each of the 88 localities. This factor score is a standardised variable: if it is high and positive it means that the KIBS loaded onto the component tend to be strongly present in the locality. If it is close to zero, this indicates an average presence, and a negative number indicates under-representation.

Each locality has a different KIBS profile, i.e. different over or under representations of the service sectors that load onto each of the eight components. A cluster analysis is performed in order to identify groups of localities that have approximately the same KIBS profile. The Ward hierarchical clustering approach is used: this generates groups of localities in such a way that within-cluster variance is minimised and between cluster variance is maximised. It is hierarchical, which means that any number of clusters between 1 and 88 can be found: our seven cluster solution reflects our target number of about 5–8 clusters (a tractable number), and the fact that moving from 7 to 8 clusters does not greatly increase the semi-partial R2. We isolate Quebec and Montreal from these clusters, giving a total of seven non-metropolitan clusters, plus the two metro areas.

The cluster profiles, together with some information on the types of locality found in each cluster, are presented in table 4.3. The two metropolitan areas, as well as having by far the greatest absolute and relative concentrations of KIBS (Table 4.1), have in common the strong presence of management and marketing (component 1), scientific and technical (component 2), financial authorities (component 4) and, especially Quebec, finance and insurance (component 5) services. Except for the relative absence of leasing (component 6), data processing (component 7) and, for Montreal, insurance broking (component 8), both metropolitan areas have strong concentrations in most types of business service, especially technical and management-related ones.

Outside of metropolitan areas, the remaining 86 localities have varied KIBS profiles. Fifty-two of these localities (clusters 1 and 2) have virtually no KIBS presence, and a further 12 (cluster 3) have few KIBS except for some insurance and a slight presence of management and marketing. Thus, 64 localities have low or slight KIBS presence. Twenty-two have a more marked presence of KIBS in at least some sectors. Four profiles emerge: a group of 8 localities with a strong local presence of technical KIBS and multimedia (Cluster 4), 6 with a strong presence of data processing (cluster 5), 4 with over-representation of management, marketing, financial, insurance and real estate services (cluster 6), and 4 with on overrepresentation of jobs in financial authorities.

Do establishments use more KIBS in KIBS-rich environments?

Table 4.4 presents a series of four regressions (dependent variable = variety of services used) for different sub-samples of the 804 observations. Each of the four regressions is presented in three stages: the geographic effect, the controls, and the complete regression, for which we indicate the chi2 (and its significance level) associated with adding the geographic effect to the controls.

The first regression analyses the entire sample of 804 establishments. There is no evidence, either before or after controls, that manufacturing establishments located in metropolitan areas have recourse to a wider variety of external services than those located elsewhere in Quebec.

However, even though the classification of localities according to their KIBS profiles does not add significantly to the model's explanatory power, there are two non-metropolitan KIBS clusters in which establishments use a wider variety of KIBS: clusters 3 and 6. Although cluster 3 only has a slight overconcentration of KIBS, both of these clusters have in common their specialisation in management and marketing KIBS, and underspecialisation (though marginally so) in technical KIBS.

These KIBS-use patterns are exacerbated if only innovators (i.e. establishments that have introduced at least one internal innovation) are considered. Clusters 3 and 6 stand out as having KIBS variety significantly higher than in Montreal. Although of lower statistical significance, variety is also greater than Montreal's in clusters 1 and 2, both characterised by low KIBS presence, and in Quebec City – a KIBS-rich environment, but far less so than Montreal. These results raise some key questions about whether or not the local environment influences KIBS use by virtue of local KIBS presence. First, the higher variety of KIBS used by establishments in clusters 1 and 2 suggest, paradoxically, that those establishments immediately surrounded by the lowest concentrations of KIBS use a wider variety of KIBS than those establishments surrounded by the highest concentrations (those in Montreal). Furthermore, establishments in Quebec City, Quebec's smaller metropolitan area, use a wider variety of KIBS than establishments in Montreal. Taken together these results are not strongly suggestive of any direct connection between establishments using a wide variety of KIBS and the KIBS environment within which they are located.

The results are clearer if a more restrictive definition of innovator is used: indeed, if an innovator is an establishment that has introduced a first-to-market innovation, then the discrepancies that emerge for all innovators disappear. It appears that innovative establishments use a wider variety of KIBS in localities which have a local presence of management and marketing consultants: whilst these advisors may not be directly associated with technological innovation, they are important in accompanying the establishment throughout the innovation process, and may advise their clients to seek out other specialist advice.

Finally, whatever the definition of innovation, the KIBS clusters classification enters the model significantly: thus, information about the environment in which innovative establishments are located adds significantly (over and above the controls) to our understanding of the variety of external services that they use.

Table 4.4 Local KIBS environment and variety of service use: innovators and non-innovators

	All observations			*Non-innovators*			*All innovators*			*First to market innovators*		
pseudo r2	0.007	0.216	0.229	0.028	0.198	0.230	0.021	0.146	0.178	0.026	0.150	0.190
nul -2LL	3024	3024	3024	853	853	853	2033	2033	2033	1322	1322	1322
model -2LL	3019	2836***	2823***	847	802***	793***	2021	1948***	1927***	1312	1263***	1246***
chi2 associated with addition of KIBS clusters (DF = 7)			13.30			9.16			20.34***			17.36**
n	804	804	804	239	239	239	555	555	555	374	374	374
Local KIBS environment												
CL1: no KIBS clustering, low to average KIBS provision	0.01		0.07	-0.67**		-0.90***	0.26		0.36*	0.25		0.32
CL2: no KIBS cluster, low KIBS provision	0.10		0.43	-0.25		0.06	1.09**		1.05*	1.07		1.05
CL3: some insurance broking and management, marketing	0.14		0.51**	-0.20		-0.23	0.22		0.63**	0.42		0.84**
CL4: technical KIBS and multimedia	-0.22		-0.13	-0.54		-0.63	0.15		0.23	0.08		0.45
CL5: data processing	-0.10		-0.10	0.01		-0.12	0.06		-0.02	-0.33		-0.60

	All observations			Non-innovators			All innovators			First to market innovators		
CL6: management, marketing and FIRE	0.42		0.93**	0.05		-0.20	1.18**		1.85***	0.90		1.63**
CL7: financial authorities	-0.31		-0.06	-0.90		-0.35	-0.09		-0.08	0.75		0.09
Québec (ref = Montréal)	0.30		0.34	-0.25		-0.29	0.52		0.57*	0.58*		0.64*
Pavitt sector												
Sector: Labour		-0.25	-0.25		-0.02	0.18		-0.24	-0.28		0.00	0.01
Research		-0.39**	-0.41**		-0.23	0.07		-0.34	-0.44*		-0.24	-0.37
Scale (ref = specialised & science based)	-0.22	-0.18		-0.01	0.17		-0.26	-0.23		-0.35	-0.30	
Establishment size												
Size1, 0–5		-1.53***	-1.56***		-1.43***	-1.54***		-1.31***	-1.35***		-1.72***	-1.96***
Size2, 5–10		-1.19***	-1.24***		-0.97***	-0.98***		-0.99***	-1.09***		-0.99***	-1.15***
Size3, 10–20 (ref = over 20 employees)		-0.83***	-0.86***		-0.79**	-0.82**		-0.81***	-0.85***		-0.88***	-0.97***
Establishment age												
Age1, pre 1980		-0.89***	-0.89***		-0.37	-0.53		-1.04***	-1.02***		-0.94***	-0.91***

(continued)

Table 4.4 Local KIBS environment and variety of service use: innovators and non-innovators (*continued*)

	All observations		*Non-innovators*		*All innovators*		*First to market innovators*	
Age2, 1980–89	-0.79***	-0.81***	-0.72*	-0.83**	-0.76***	-0.78***	-0.60**	-0.61**
Age3, 1990–99 (ref = founded 2000 or later)	-0.58***	-0.58***	-0.37	-0.52	-0.62***	-0.59**	-0.67**	-0.63**
Research and development employees								
RDemp1, none	-1.60***	-1.66***	-1.95***	-2.03***	-1.18***	-1.31***	-1.54***	-1.70***
RDemp2, 1–10%	-0.78**	-0.85***	-1.25*	-1.27*	-0.49	-0.62**	-0.95**	-1.09***
RDemp3, 10–35% ref = over 35% of employees in R&D)	-0.60**	-0.65**	-0.58	-0.59	-0.52*	-0.59*	-1.14***	-1.23***
Exports								
export1, none	0.15	0.11	0.34	0.23	0.32	0.32	0.07	0.10
export2, 1–10%	0.57**	0.60**	0.84	0.77	0.57*	0.64**	0.31	0.46
export3, 10–50% (ref = over 50% of output exported)	0.28	0.30	0.07	-0.10	0.49	0.56*	0.33	0.44

Note: *** = significant at 99% level, **= significant at 95% level, * = significant at 90% level.

Do certain types of innovator use more KIBS in KIBS-rich environments?

So far we have only distinguished between non-innovators, innovators and more radical innovators. In this section we consider three different types of innovation. Only first-to-market innovations are considered, given that the results of the previous section are more clear-cut for this more restrictive definition of innovation.

If different types of innovation are considered, it can be seen that the results outlined above are driven by product innovators, and, to a lesser extent, by process innovators. For product innovation, location in clusters 3 and 6 strongly augment the variety of external services used. For process innovation, clusters 3 and 6 have the highest coefficients (and the coefficient on cluster 3 is weakly significant). However, if organisational innovators (either marketing or management) are considered, there is no evidence that presence in any type of KIBS environment enhances, or reduces, the variety of external KIBS used. Only for product innovators does the KIBS cluster classification add significantly to the model's explanatory power.

For all types of innovator, *there is no evidence that being in a metropolitan area increases the variety of external KIBS services used.* Even for technological (product and process) innovators, location in an environment rich in technical KIBS does not increase the variety of external services used. However, for technological innovators – and most particularly for product innovators – location in a non-metropolitan environment rich in management and marketing related KIBS increases the variety of KIBS that are used. Somewhat surprisingly, given the more direct functional link between organisational innovation and management and marketing advice, the variety of KIBS used in localities which have a local presence of management and marketing consultants is no higher than in localities which lack their presence.

Does the local KIBS environment explain KIBS-use better than the urban hierarchy?

Studies of the geography of KIBS show that KIBS tend to locate in larger cities, and that there exists a possible shadow effect (i.e. that cities close to metropolitan areas have fewer KIBS than those further away, since local establishments can use non-local metropolitan KIBS) (Desmet and Fafchamps, 2006; Polèse and Shearmur, 2006). The shadow effect is not evident in our data (Table 4.1), but the increase in number and concentration of KIBS with urban size is clear.

It is therefore important to verify that our conclusions – particularly the overarching one that, for innovators, and particularly for technological innovators, the local KIBS environment has a significant impact on the variety of KIBS that are used – are not simply restating the idea that firms in smaller or more isolated localities use fewer KIBS than those in larger or more central ones.

Table 4.6 presents the significance of the increase in explanatory power of the control model after adding the cluster classification (used in Tables 4.3 to 4.5) and after adding the urban hierarchy classification (presented in Table 4.1).

Table 4.5 Local KIBS environment and variety of service use: different types of first-to-market innovation

	Product			*Process*			*Organisational*		
pseudo r2	0.043	0.157	0.214	0.025	0.184	0.206	0.033	0.169	0.199
nul -2LL	962	962	962	715	715	715	517	517	517
model -2LL	951	916***	898***	710	673***	668***	512	490***	485***
chi2 associated with addition of KIBS clusters (DF=7)			18.60***			5.32			5.42
n	279	279	279	210	210	210	153	153	153
Local KIBS environment									
CL1: no KIBS clustering, low to average KIBS provision	0.16		0.17	-0.20		0.17	0.09		0.02
CL2: no KIBS cluster, low KIBS provision	1.33*		1.10	0.60		0.53	0.63		0.27
CL3: some insurance broking and management, marketing	0.60		0.96**	0.18		0.82*	0.44		0.51
CL4: technical KIBS and multimedia	0.18		0.64	-0.16		0.29	0.96		0.86
CL5: data processing	-0.71		-1.12**	-0.27		-0.44	-0.58		-0.77
CL6: management, marketing and FIRE	1.26		1.70**	-0.21		0.75	1.09		1.25

	Product			Process			Organisational		
CL7: financial authorities	0.68		0.02	0.93		0.14	0.63		-0.64
Québec (ref= Montréal)	0.58		0.60	0.59		0.55	0.14		-0.16
Pavitt sector									
Sector: Labour		-0.05	-0.01		0.03	-0.01		0.11	0.20
Research		-0.37	-0.46		-0.41	-0.50		0.13	0.15
Scale (ref= specialised and science based)		-0.27	-0.23		-0.19	-0.10		-0.51	-0.39
Establishment size									
Size1, 0–5		-1.80***	-2.04***		-2.49***	-2.70***		-2.00***	-2.12***
Size2, 5–10		-1.08***	-1.23***		-0.77*	-0.88*		-1.31**	-1.49***
Size3, 10–20 (ref= over 20 employees)		-0.91***	-1.02***		-0.90***	-0.95***		-0.63	-0.79*
Establishment age									
Age1, pre 1980		-0.90***	-0.82**		-0.75*	-0.73*		-0.57	-0.65
Age2, 1980–89		-0.68**	-0.64*		-0.67*	-0.64		-0.15	-0.17
Age3, 1990–99 (ref= founded 2000 or later)		-0.73**	-0.70**		-0.29	-0.26		-0.41	-0.42

(continued)

Table 4.5 Local KIBS environment and variety of service use: different types of first-to-market innovation (*continued*)

	Product		*Process*		*Organisational*	
Research and development employees						
RDemp1, none	-1.60***	-1.78***	-1.96***	-2.11***	-2.23***	-2.28***
RDemp2, 1–10%	-0.89**	-1.04**	-1.35**	-1.45**	-1.73**	-1.79**
RDemp3, 10–35% ref= over 35% of employees in R&D)	-1.15***	-1.30***	-1.53***	-1.64***	-1.18*	-1.15
Exports						
export1, none	0.28	0.32	0.04	0.02	-0.49	-0.35
export2, 1–10%	0.50	0.68	-0.23	-0.16	-0.13	0.12
export3, 10–50% (ref= over 50% of output exported)	0.26	0.46	0.20	0.22	-0.33	-0.22

Note: *** = significant at 99% level, **= significant at 95% level, * = significant at 90% level.

Table 4.6 Variety of service providers: Comparing effect of KIBS clusters with that of urban hierarchy

		All	*Non-innovators*	*All innovators*	*Radical innovators*	*Product innovators*	*Process innovators*	*Organis. innovators*
	n	804	239	555	374	279	210	153
Urban hierarchy (DF = 7)	Increase in pseudo R2	0.009	0.031	0.021	0.027	0.027	0.028	**0.084**
	Chi2	9.18	9.07	13.59	11.52	8.71	7.40	**15.59**
	p(chi2)=0	0.240	0.248	0.059	0.116	0.274	0.388	**0.029**
KIBS clusters (DF = 8)	Increase in pseudo R2	0.013	0.031	**0.032**	**0.040**	**0.057**	0.021	0.030
	Chi2	13.30	9.16	**20.34**	**17.36**	**18.60**	5.32	5.42
	p(chi2)=0	0.102	0.327	**0.009**	**0.027**	**0.017**	0.723	0.712

For the sample as a whole, and for non-innovators, neither classification adds significantly to the control model. For innovators, radical innovators and product innovators, the KIBS classification outperforms the urban hierarchy very substantially. In order to understand the propensity to use a wider variety of external KIBS, it is the local KIBS environment rather than location within the urban hierarchy (adjusted for remoteness) that is most powerful. Neither classification significantly enhances our understanding of variety of KIBS used by process innovators.

However, for organisational innovators, it is the urban hierarchy that outperforms the KIBS environment. Amongst organisational innovators, it is those located in large central cities (i.e. cities of over 50,000 people located within about 100km of a metropolitan area) that use a far wider variety of KIBS than those in any other types of location. In these locations organisational innovators may use a wider variety of KIBS not only because of the relatively high local presence of KIBS (Table 4.1), but also because of easy access to metropolitan areas.

Conclusion

It is currently accepted that establishments need to draw on their internal capacities, but also – and maybe increasingly so – on the external environment in order to innovate. Although it is recognised that the external environment is not necessarily local, there is still a strong tendency to believe that the actors, agents and institutions immediately surrounding an establishment have a strong impact on the innovation process. Thus, being located in a cluster or in a metropolitan area enhances the variety of information and knowledge sources, facilitating open innovation processes.

In this chapter we investigate this idea by examining the variety of different KIBS inputs that manufacturing establishments use. First, the findings reveal that manufacturing establishments – whether innovators or not – located in metropolitan areas do not necessarily use a wider variety of KIBS. Thus, the widely accepted idea that metropolitan areas are associated with open innovation – at least in the sense that local innovators interact with a wider variety of external sources and partners – is rejected when KIBS are considered.

The second finding relates to what is happening outside of metropolitan areas. Here it is suggested that establishments located in places with higher local concentrations of KIBS will use a wider variety of KIBS. This hypothesis is partly verified: it holds true for localities within which management and marketing KIBS cluster, and most particularly for product innovators. However, it does not hold true for localities in which there are higher concentrations of technical KIBS. These results therefore suggest that innovators do not turn to local service providers for technical advice on innovation: this strategic advice is probably sourced from specific KIBS providers with the necessary expertise (Macpherson, 2008; Shearmur, 2015). However, the innovation process also requires general advice on managing the innovation process, on marketing the new products, and, maybe, on strategies to identify the requisite scientific and technical know-how: the fact that innovators in environments with a higher presence of management and marketing advisors use a wider variety of KIBS may reflect a) the local availability of these services and b) the type of advice that these services give the establishments. It is therefore not the local presence of technical KIBS that enhances KIBS use by innovators but the local presence of generalist advisors: it is the local presence of the latter that seems to enhance open innovation processes.

The third finding is that it is indeed the local KIBS environment, and not merely location relative to the urban hierarchy (adjusted for remoteness) that best explains local differences in variety of KIBS that are used. This result is confirmed overall (for all innovators and for first-to-market innovators), and for product innovators. However, organisational innovators' propensity to use a wide variety of KIBS is connected with the urban hierarchy: it is organisational innovators in large cities close to metropolitan areas (and hence with access to services there) that use a wide variety of service providers.

The results are not clear-cut. If KIBS variety is considered to be an indicator of open innovation, then our results suggests that open innovation is just as prevalent in metropolitan areas as it is in smaller cities and rural areas: establishments seek out the services they need irrespective of their presence in a metropolitan area or not.

However, in non-metropolitan areas, open innovation is more prevalent amongst manufacturing establishments that are in localities where management and marketing services are located. This does not mean that establishments in these localities are more innovative, though they may be. It means that establishments in these environments seem to systematically adopt a more open strategy, whereas those in other environments do not do so systematically: in other environments larger numbers of innovators rely more on internal capacities and on a smaller variety of external KIBS providers (Shearmur, 2015).

Our results are similar to those of Tierlinck and Spithoven (2008) and Grillitsch *et al.*, (2013), who come to similar conclusions about Belgium and Austria respectively. They also confirm Macpherson's (2008) observations about the way in which establishments in remote New York state have recourse to external services, which they identify at a distance (using internet) and with which they meet from time to time. The results also corroborate the theoretical approach to understanding innovation in remote areas developed by Shearmur (2015). These studies, despite their differences in detail, all point to the fact that it is not location in metropolitan areas that enhances the variety of KIBS, or of information and knowledge sources, that are used by innovators. Our study does reveal that there are some differences between establishments located in different types of non-metropolitan locality – so the local KIBS environment plays some role – but the idea that presence in a metropolitan region enhances the open innovation process by increasing the variety of information and knowledge sources is not supported by our results, nor by the results from other similar studies.

Acknowledgements

The authors acknowledge the financial support from the Social Science and Humanities Research Council in Canada (SSHRC 410-2011-0108). The usual disclaimers apply.

Appendix 4A

Table 4A.1 Principal components analysis of KIBS location quotients across Quebec localities

KIBS sectors	*F1*	*F2*	*F3*	*F4*	*F5*	*F6*	*F7*	*F8*	*Com.*
Management and scientific consultants	**0.73**	0.22	0.18	0.21	0.02	-0.01	-0.11	-0.18	0.70
Sound recording	**0.72**	0.05	0.08	0.09	-0.12	0.01	0.25	0.14	0.63
Advertising and marketing	**0.70**	0.13	0.10	0.02	0.17	0.19	-0.03	0.12	0.60
Other scientific and technical services	**0.52**	0.02	0.38	-0.09	0.35	-0.07	0.29	0.00	0.63
Real estate	**0.51**	0.26	0.24	0.17	0.04	***0.40***	-0.08	0.29	0.66
Funds and other financial instruments	**0.50**	-0.04	-0.13	0.13	***0.54***	0.03	-0.02	0.20	0.61
Telecommunications	0.16	**0.76**	0.10	0.07	0.20	-0.13	0.13	-0.11	0.71
Architecture, engineering and related services	0.14	**0.74**	0.19	-0.05	0.18	0.19	0.23	0.13	0.74
Television	0.12	**0.72**	-0.12	0.04	-0.34	0.05	-0.26	0.21	0.77
R&D services	0.07	**0.50**	0.14	***0.40***	***0.46***	-0.05	-0.07	-0.07	0.65

KIBS sectors	*F1*	*F2*	*F3*	*F4*	*F5*	*F6*	*F7*	*F8*	*Com.*
Publishing and information services	0.07	**0.47**	0.35	***0.39***	0.29	0.23	0.07	0.02	0.64
Accounting	-0.01	0.12	**0.72**	-0.16	0.09	-0.12	0.10	0.25	0.66
Legal services	0.25	0.07	**0.60**	0.28	-0.14	0.07	0.17	0.27	0.63
Specialised design services	0.36	0.01	**0.54**	0.04	0.26	0.34	0.08	-0.21	0.66
Financial and investment broking	0.35	0.17	**0.49**	0.24	0.35	0.10	-0.09	-0.02	0.59
Film and video	***0.44***	0.16	**0.49**	0.10	-0.14	0.02	**-0.48**	0.00	0.71
Monetory authorities	0.07	-0.01	0.00	**0.85**	0.06	0.05	-0.07	-0.04	0.74
Stockmarkets and related markets	0.17	0.12	0.04	**0.61**	0.06	-0.08	0.29	0.02	0.51
Insurance	0.00	0.19	0.12	0.07	**0.78**	-0.07	-0.03	0.16	0.70
Rental and leasing services	0.11	0.06	-0.06	0.31	0.04	**0.75**	0.05	0.23	0.73
Data processing	0.11	0.13	0.16	0.14	-0.08	0.07	**0.76**	0.00	0.67
Insurance brokers and agents	0.12	0.05	0.21	-0.05	0.19	0.02	0.00	**0.79**	0.73
Financial intermediation	-0.05	0.01	-0.06	0.30	0.13	**-0.77**	-0.03	0.19	0.74
Computer systems design	***0.41***	***0.40***	***0.40***	***0.39***	0.28	0.14	0.23	-0.11	0.81
variance	*3.13*	*2.56*	*2.38*	*2.04*	*1.94*	*1.65*	*1.32*	*1.20*	*16.23*
Proportion of total variance explained									*67.6%*

Note: com= communality (i.e. proportion of the variance of each variable captured by the 8 components)
Component names
F1: management, marketing and real estate
F2: telecoms. engineers, R&D
F3: accounting, legal, broking
F4: financial authorities
F5: insurance and finance
F6: rental and leasing / not financial intermediaries
F7: data processing
F8: insurance broking

References

Amin, A. and S. Roberts (eds), 2008, *Community, economic creativity and organization*, Oxford: Oxford University Press.

Asheim, B., 2012, The changing role of learning regions in the globalizing knowledge economy: a theoretical examination, *Regional Studies*, 46 (8): 993–1004.

Asheim B., H. Lawton-Smith and C. Oughton, 2011, Regional innovation systems: theory, empirics and policy, *Regional Studies*, 45 (7): 876–91.

Asheim, B. T. and M. S. Gertler, 2005, Regional innovation systems and the geographical foundations of innovation, in J. Fagerberg, D. Mowery and R. Nelson (eds), *Oxford handbook of innovation*. London: Oxford University Press, 291–317.

Bathelt, H., 2011, Innovation, learning and knowledge creation in co-localised and distant contexts, in Pike, A., A. Rodriguez-Pose, and J. Tomaney (eds), *Handbook of local and regional development*, London: Routledge, 149–61.

Boschma, R., 2005, Proximity and innovation: a critical assessment, *Regional Studies*, 39 (1): 61–74.

Chesborough, H., 2003, *Open innovation: the new imperative for creating and profiting from technology*, Cambridge (Mass): Harvard Business Press.

Cooke, P, B. T. Asheim, R. Boschma, R. Martin, D. Schwartz and F. Tödtling (eds), 2011, *Handbook of regional innovation and growth*, Cheltenham: Edward Elgar.

Cooke, P., M. Heidenreich and H. J. Braczyk (eds), 2004, *Regional systems of innovation: the role of governance in a globalized world*, London: Routledge.

Crevoisier O. and R. Camagni, 2001, *Les milieux urbains: innovation, systèmes de production et ancrage*. Neuchâtel: EDES.

Currid, E., 2007, *The Warhol economy: how fashion and music drive New York City*, Princeton: Princeton University Press

Dahlander, L. and D. Gann, 2010, How open is innovation? *Research Policy*, 39: 699–709.

Den Hertog, P., 2000, Knowledge intensive business services as co-producers of innovation, *International Journal of Innovation Management*, 4(4): 491–528.

Desmet, K. and M. Fafchamps, 2005, Changes in the spatial concentration of employment across US counties: a sectoral analysis 1972–2000, *Journal of Economic Geography*, 5 (3): 261–84.

Doloreux, D. and R. Shearmur, 2012, How much does KIBS Contribute to R&D Activities of manufacturing firms?, *Economia Politica*, 29 (3): 319–42.

Glaeser, E., 2011, *The triumph of the city: how our greatest invention makes us richer, smarter, greener, healthier and happier*, New York: Penguin Books.

Gordon, I. and P. McCann, 2005, Innovation, agglomeration and regional development, *Journal of Economic Geography*, 5: 523–43.

Grillitsch, M., F. Tödtling and C. Höglinger, 2013, Variety in knowledge sourcing, geography and innovation: evidence from the ICT sector in Austria, *Papers in Regional Science*, early view. DOI: 10.1111/pirs.12050.

Howells J., A. James and K. Malik, 2003, The sourcing of technological knowledge: distributed innovation processes and dynamic change, *Research and Development Management*. 33 (4): 295–309.

Huber, F., 2012, Do clusters really matter for innovation practices in information technology? questioning the significance of technological knowledge spillovers, *Journal of Economic Geography*, 12 (1): 107–26.

Huizingh, E., 2011, Open innovation: state of the art and future perspectives, *Technovation*, 31: 2–9.

MacPherson, A., 2008, Producer service linkage and industrial innovation: results of a twelve year tracking study of New York state manufacturers, *Growth and Change*, 39 (1): 1–23.

Martinez-Fernandez, C., 2010, Knowledge-intensive service activities in the success of the Australian mining industry, *The Service Industries Journal*, 30 (1): 55–70.

Miles I., 2008, Patterns of innovation in service industries, IBM Systems Journal. 47: 115–28

Miles, I., 2012. KIBS and knowledge dynamics in client-supplier interactions, in Di Maria, E., R. Grandinetti, and B. Di Bernardo (eds), *Exploring knowledge intensive business services*, London: Palgrave MacMillan, 13–34.

OECD, 2007. *Innovation and knowledge-intensive service activities*. Paris: OECD Publishing.

Polèse, M. and R. Shearmur, 2006, Growth and location of economic activity: the spatial dynamics of industries in Canada 1971–2001, *Growth & Change*, 37 (3): 362–95.

Porter, M., 2003, The economic performance of regions, *Regional Studies*, 37: 549–78.

Ribichesi, C. and R. Shearmur, 2008, *Les communautés mono industrielles au Québec: portrait et analyse de la vulnérabilité économique des communautés au Québec*, Montréal: INRS-Urbanisation Culture et Société and Développement économique Canada. www.ucs.inrs.ca/sites/default/files/centre_ucs/pdf/MonoIndustrielle.pdf.

Rodriguez, M., J. Camacho and J. Chica, 2012, The knowledge-intensive services – regional innovation nexus: a European perspective, *The Service Industries Journal*, 32 (4): 605–18.

Rogers, E.M., 2003, *Diffusion of innovations*, New York: Free Press.

Sammarra, A. and L. Biggiero, 2008, Heterogeneity and specificity of inter-firm knowledge flows in innovation networks. *Journal of Management Studies*, *45*(4): 800–29.

Shearmur R., 2012, Are cities the font of innovation? A critical review of the literature on cities and innovation, *Cities*, 29.S2, S2–S19.

Shearmur, R., 2015, Far from the madding crowd: slow innovators, information value, and the geography of innovation, *Growth and Change*, 46.3, 434–42.

Shearmur, R. and D. Doloreux, 2013, Innovation and KIBS: the contribution of KIBS to innovation in manufacturing establishments, *Economics of Innovation and New Technology*, 22 (8): 751–74.

Sundbo, J.F. and M. Toivonen, 2012, *User-based innovation in services*, Cheltenham: Edward Elger.

Tether, B.S. and C. Hipp. 2002, Knowledge intensive, technical and other services: patterns of competitiveness and innovation compared. *Technology Analysis & Strategic Management*, 14: 163–82.

Tether, B. Q.Li and A.Mina, 2012, Knowledge-bases, places, spatial configurations and the performance of knowledge intensive professional service firms, *Journal of Economic Geography*, 12 (5): 969–1001.

Tierlinck, P., and A.Spithoven, 2008, The spatial organization of innovation: open innovation, external knowledge relations and urban structure, *Regional Studies*, 42 (5): 689–704.

Todtling F. and M.Trippl, 2005, One size fits all? Towards a differentiated regional innovation policy approach, *Research Policy*, 34: 1203–19.

Torre, A., 2008, On the role played by temporary geographical proximity in knowledge transmission, *Regional Studies*, 42 (6): 869–89.

Wolfe, D., 2009, *21st century cities in Canada: the geography of innovation*, Ottawa: Conference Board of Canada.

Yam, R., W. Lo, E. Tang and A. Lau, 2011, Analysis of sources of innovation, technological innovation capabilities, and performance: an empirical study of Hong Kong manufacturing industries, *Research Policy*, 40: 391–402.

5 Institutions and spin-offs

Determining factors for establishment and early market entry success of innovation based spin-offs from KIBS-firms

Kjersti Vikse Meland and Tatiana A. Iakovleva

Introduction

Spin-off is, in general, a highly successful phenomenon in terms of competitiveness, innovation, growth and has a positive effect on the socio-economic environment. This specific form of entrepreneurship is seen by European policy makers as both a driver and a result of the shift to the knowledge-based economy.[1] Corporate spin-off, which has proven to be a successful mode of entrepreneurship in Norway, has been given little attention by politicians and funding agencies (Nås and Sandven, 2003).

Much of the literature on entrepreneurship has examined the attributes of individuals, the networks of affiliations in which those individuals are enmeshed, the resources they assemble and the openings present in the competitive environment (Hwang and Powell, 2003). The focus has been on capabilities of individuals or organisations to recognise entrepreneurial opportunities. Legal and political conditions that support entrepreneurial behaviour and the wider ecosystem that serves as barriers or promoters of entrepreneurial activity, such as institutions, first received increased attention during the past decade (Hwang and Powell, 2003).

Institutional theory (Scott, 1995; Scott, 2000) suggests that institutions and business environment affect firm birth rate, churching and dynamics. Scott (1995:p.33; 2001:p.48) defines institutions as: 'social structures that have attained a high degree of resilience. They are composed of cultural-cognitive, normative, and regulative elements that, together with associated activities and resources, provide stability and meaning to social life. Institutions operate at different levels of jurisdiction, from the world system to localised interpersonal relationships. Institutions by definition are subject to change processes, both incremental and discontinuous'. Institutional studies examining institutional aspects of entrepreneurship have been criticised for a narrow focus on culture (Busenitz, Gomez, and Spencer, 2000). Many of them have linked Hofstede's (1983) cultural dimension, especially individualism, to examine a country's propensity to engage in entrepreneurial activities. Studies on global diffusion of entrepreneurial institutions (Gereffi and Hempel, 1996) provide support for the notion that culture alone is insufficient to describe cross-country differences in entrepreneurship.

Economic, political and legal institutions play an important role in fostering or prohibiting entrepreneurship and should be considered in future studies (Djankov, McLiesh and Ramalho, 2006). In the spin-off literature, the role of contextual conditions has been largely overlooked given the largely person-centric view that still dominates (Gilsing, van Burg and Romme, 2010). To try to address these gaps, this paper will focus on how institutions at different levels of the national innovation system condition the effectiveness of establishment and early market entry of spin-offs.

A spin-off is defined as a firm whose intellectual capital originates from its parent institution, such as a university, research institute or another company (Chesbrough, 2002; Mustar *et al.*, 2006). Using Fryges and Wright's (2014) typology of spin-offs highlighting *context*, i.e. whether they originated from a university or commercial context, and the *mode* of the spin-off, i.e. whether it is based on new or existent activity, this paper will focus on commercial spin-offs based on innovations, either product or service that is new to the market. Such spin-offs might be either corporate (Bruneel *et al.*, 2013; Clarysse *et al.*, 2011) where a parent company contributes to equity, or employee spin-off (Fryges and Wright, 2014). We focus on corporate innovation based spin-offs.

In this paper we study spin-off processes from one specific kind of firm, i.e. Knowledge Intensive Business Service (KIBS) firms. KIBS firms are a particular part of the whole service sector. They are defined as firms that provide knowledge intensive goods and services for other business firms. KIBS are distinguished as T-KIBS (those with high use of scientific and technological knowledge, such as R&D services, engineering services, computer services) and P-KIBS that are more traditional professional services (such as legal, accountancy, management consultancy and marketing services). Based on empirical data from three different sectors in Norway, the Maritime/Marine, Oil and Gas and the ICT sector, this paper will focus on spin-off processes from T-KIBS firms. The ICT sector is a typical KIBS sector with a large amount of knowledge intensive service firms, whereas the Maritime/Marine and Oil and Gas industries are represented by KIBS companies in certain parts of their value chain, i.e. within architectural/design, technical services, engineering activities and related technical consultancy (technical testing and analysis etc.).

Over the last 15–20 years, interest in Knowledge Intensive Business Services has grown significantly in Europe, both in science and policy (Schricke, Zenker and Stahlecker, 2012). The increased focus on innovation KIBS is related to the efforts of western economies and the European Union to become knowledge-based economies. 'KIBS are likely to become one of the main engines for future growth within the European Union' (European Commission, 2007:p.7). The demand for knowledge intensive services seems to increase with the effort of European economies trying to maintain their competitive position through development into knowledge-based economies (Schricke, Zenker and Stahlecker, 2012). Therefore, we need a better understanding of how institutional factors impact spin-offs from KIBS firms.

Empirical findings show that in regions with many KIBS firms, start-up activities of KIBS are more frequent, where 'both the overall knowledge intensity of the regional workforce and the size of regional market have a positive influence on KIBS start-ups' (Anderson and Hellerstedt, 2009:p.118). Another finding is that regional patterns of KIBS are dependent on the type of KIBS activity (Wood, 2005). Marketing, advertising and service companies specialising in financial businesses are concentrated in core city regions, whereas computer services are concentrated in prosperous regions and technical services appear to be tied to demand from manufacturing and other industries.

To better understand the role and relevance of institutional factors for the establishment and early market entry of spin-offs from T-KIBS, we have used van der Steen's (1999) and Bekkers and van der Steen' (2003) conceptual model that differentiates among four institutional layers of the national system of innovation as a basis.

The aim of this paper is to develop a more complete and structural picture of institutional factors determining the effectiveness of spin-off process from KIBS firms. We address the following research question: *Which institutional factors and mechanisms on sectoral, regional and managerial levels create favourable conditions for the establishment and early market entry success of corporate innovation based spin-offs from KIBS?*

Corporate innovation based spin-offs

Companies can capture value from their innovation activities in two basic ways: through incorporating innovation in their current businesses, or through launching new ventures that exploit innovation in new business arenas (Chesbrough and Rosenbloom, 2002). The spin-off concept is criticised for its profusion of overlapping terms. In an attempt to improve the awareness of spin-off research, Fryges and Wright (2014) developed a typology of spin-offs by highlighting context and mode. Context is distinguished between commercial environments of for-profit corporations and the non-commercial environment associated with universities, i.e. corporate and academic spin-offs. As the mode of spin-off venture, Fryges and Wright (2014) differentiated between spin-offs involving a new or existing activity. In this study we focus on spin-offs from for-profit corporations involving new activities, i.e. corporate spin-offs having innovation as a foundation for their establishments.

Furthermore, a spin-off can also be distinguished as entrepreneurial or incumbent initiated spin-offs (Van de Velde *et al.*, 2007). Entrepreneurial spin-offs is defined as employees leaving the company to establish their own enterprises, whereas incumbent initiated spin-offs means that the parent company seeks to develop a new technology in a separate company. The focus on of our paper is on incumbent or parent initiated spin-offs. The parent firm decides to establish a new firm for commercializing the new technology, instead of expending the parent's scope of activities or abandoning the new technology (Van de Velde *et al.*,

2007). The parent company can capitalise on the innovation by keeping a certain extent of ownership of the spin-off company (Goduscheit and Brendstrup, 2012).

In this study we focus on spin-offs created in the commercial environment of for-profit corporations involving new activities initiated by parent firms who wish to realise their business idea in a new company.

Institutions

To better understand the role and relevance of institutional factors in establishing spin-offs from KIBS firms, we have used Bekkers et al's (2006) theoretical framework as a starting point. The model argues that different institutional layers of a national system of innovation form the selection environment for spin-offs (Nelson, 2001). Because national laws and policy, and management of the spin-off company, lie beyond the scope of our study, we have developed a modified model of Bekkers et al (2006) framework that includes three layers of the national system of innovation: sectoral, regional and managerial institutions.

The first layer refers to institutions at a sectoral level, defined as the sectoral technology regime, including technological opportunity conditions, variation in technological approaches and patenting behaviour. The second layer reflects institutions at regional level, conceptualised as regional clusters. The third level concerns institutions at managerial level, defined as parent company policies towards spin-off, including parent company strategy towards spin-off and implementation of this strategy. The modified model is presented below (Figure 5.1).

The framework of Bekkers et al (2006) has provided a coherent framework for understanding the combined role of various antecedent conditions for spin-off creation and success. Bekkers et al (2006) focuses on spin-offs from Public Research Organisations (PROs) and especially on intellectual property (IP) based spin-offs. Their findings indicate that national laws and policy, as well as sectoral characteristics, affect the establishment of IP-based spin-offs, whereas in the presence of a regional cluster, a PRO company policy regarding spin-offs and the management of the spin-off firm itself, affect success chances once established (Bekkers *et al.*,

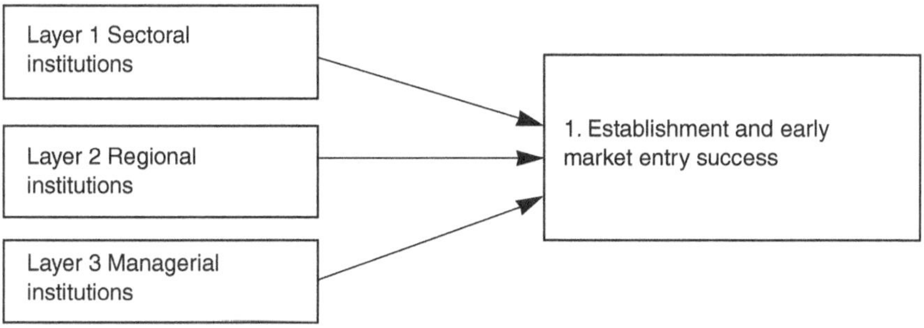

Figure 5.1 Model of institutional layers affecting establishment and early market entry success of corporate spin-offs

2006). Extending this line of research, we suggest that the same factors should be important for corporate spin-offs establishment and early market entry success. In this paper, we focus on how sectoral, regional and managerial institutions affect the establishment and early market entry success of spin-offs from KIBS.

Recent calls in the literature argue for not conflating foundation rates from success rates and to keep the process of spin-off creation analytically separate from its subsequent success or failure (e.g. Djokovic and Souitaris, 2008). We find the concept 'subsequent chances of success' rather unclear, both with regard to the kinds of activities included and to the time period. Therefore, we chose to name the phase 'early market entry success' in an attempt to clarify the concept we are studying. We operationalise the phase 'early market entry success' by activities such as registering of the new enterprise, familiarising potential customers with the product/service idea, involving investors at the early stage, preparing the market by building firms' legitimacy and increasing the visibility of the business, building relationships with potential customers and suppliers and early sales activities (Foss *et al.*, 2011).

Sectoral institutions

Sectors differ in the extent to which they may provide fertile ground for establishing spin-off companies (Gilsing, van Burg, and Romme, 2010). Shane (2001) argues that the characteristics of technological regimes will have systematic effects on whether or not new firms are established to exploit inventions. It is found that sectors with *high technological opportunity conditions* and a *variety of technological approaches* will be more conductive for creation of corporate spin-offs (Shane, 2001).

High technological opportunity conditions exist when technology is rather immature. This is the case for many parts of the industrial value chain in the Norwegian oil and gas and maritime sectors. Within the oil and gas sector, remotely operated vehicle (ROV) technology, used in pipeline inspection, maintenance and repair and maritime operations among others, is an example of a technology having rather high opportunity conditions. New versions of ROVs are being developed continuously focusing, for example, on shorter operation time (faster ROVs) and possibilities for operating in deeper ocean areas.

A variety of technological approaches means that a technological challenge can be solved by various technologies. More environmental friendly propulsion systems can be developed by replacing fuel with natural gas, battery, fuel-cell technology, hydrogen, etc. The reason why a *sectoral technology regime* has a positive effect on spin-off establishment is that incumbents will be able to pursue only a limited number of technological opportunities, given specialised capabilities and scarce resources. Furthermore, it is also related to market entry costs. In the early stages of a new technology, markets are too small to justify investments by large established firms. Instead, independent entrepreneurs with low opportunity costs tend to exploit new markets. These conditions are found typically in sectors with immature technologies, such as software, microelectronics, biotechnology and multimedia (Gilsing, van Burg, and Romme, 2010).

Another sectoral institution assumed to affect the creation of corporate spin-offs is the *sectoral IP regime* (Gilsing, van Burg, and Romme, 2010). The sectoral IP regime refers to the extent to which inventions can be patented and these patents can be effectively defended against infringement. With a patented invention one is assumed to have less aggressive competitors and easier access to investors because the potential of economic rents is larger. Innovative KIBS firms are challenged by the fact that service innovations are hard to protect from imitations (Gallouj and Weinstein, 1997). Intellectual property protection mechanisms in services differ from those in manufacturing (Howells, 2001). Patents are rarely used, while copyright seems to be very relevant for certain KIBS sectors. Miles (2001) argue that KIBS firms have other mechanisms that offer strong protections, such as being a member of a professional association, to document certain quality standards if professional accreditation is necessary. Reputation and secrecy seem to be very important to establish trustful relationships in which knowledge can be transferred and shared (Miles, 2001, p. 97f). Most of the KIBS firms in this study sell services in combination with their own developed and often tailor made equipment. Patents, therefore, are assumed to be of relevance for these firms.

Thus, we expect that sectoral institutions in the form of the technology regime and a strong IP regime would be important for both establishment and early market entry success for innovation based spin-offs from KIBS.

Regional institutions

The *cluster* literature has often pointed to regional factors, such as the availability of state of the art knowledge, experience, capital, talent and housing as factors supporting the development of regional clusters in the form of attracting specialised companies to the cluster (Bekkers *et al.*, 2006). We refer here to clusters as geographic concentrations of interconnected companies, specialised suppliers, service providers, firms in related industries and associated institutions (e.g. universities, standards agencies, trade associations) in a particular field that compete but also cooperate. Such clusters may show good performance in terms of productivity and innovation (Porter, 1990; Porter, 1998). With such dynamics, regional clusters are assumed to be a good breeding place for start-ups and for IP-based spin-offs in particular (Mowery and Ziedonis, 2001). This thinking is also applicable to corporate spin-off firms, which often form clusters together with their parent companies and partners.

Geographical clusters were found to be of importance for the establishment of both start-ups and corporate spin-offs. Delgado, Porter and Stern (2010) found significant evidence of the positive impact of clusters on corporate entrepreneurship. They found that industries located in regions with strong clusters (i.e. a large presence of other related industries) experience higher growth in new business formation, start–up employment and formation of new establishments of existing firms (spin-offs), thereby influencing the location decisions of multi-establishment firms.

Geographical clusters were found to be important especially during the first phase of the IP based spin-offs in the ICT and life science sectors in the Netherlands and USA (Bekkers *et al.*, 2006). The IP-based spin-offs benefitted largely from geographical closeness within the cluster in the form of exchange of tacit knowledge with the PRO, firms, informal contacts with former colleagues and of possibilities to use laboratory equipment and options to attract new talents (Dahl and Pedersen, 2003). Another benefit from the regional cluster was the availability of start-up capital. Regional clusters were also important in leveraging the potential of IP based spin-offs, in view of tacit knowledge exchange and other proximity related benefits (Bekkers *et al.*, 2006). Proximity of the parent company, venture capitalists and possibly technical facilities and incubation parks formed key ingredients of such a cluster for the IP-based spin–offs in the first phase after establishment.

Summarising, proximity to a strong regional cluster is assumed to be an important explanatory factor for both establishment and early market entry success of innovation based spin-offs from KIBS firms.

Managerial institutions

Our third institutional factor is related to the parent strategy towards spin-off and the implementation of this strategy. For parent company, spinning off activities can be seen as a means to isolate new and exploratory initiatives from the core activities of the parent company (Woo *et al.*, 1989) and, thereby, represents a way to set up a play ground for riskier activities without jeopardizing the health of the company. On the other side, academics (e.g. Christensen, 1997) argue that incumbent firms with proven success in existing technologies and markets often exert strong firm-internal pressure to conform to established ways of doing things and, thereby, often ignore disruptive developments. Also, Hellmann (2007) argues that incumbent firms only invest in corporate ventures if these build upon current technologies and business, investing much less in spin-offs unrelated to their current technologies and established way of operating.

Bekkers et al (2006), comparing IPR–based spin-off processes in the Netherlands and USA, found the lack of PRO's general interest and absence of a consistent technology transfer policy were hindrances to the establishment of IP-based spin-offs in the Netherlands. The opposite was the case in the USA where most PROs had adopted policies for encouraging entrepreneurships and spin-offs, in the form of a university policy, with regard to sharing royalty rates between inventors and university, which was important for the establishment of IP-based spin-offs. Furthermore, PRO policy with regard to equity investment in the IP-based spin-off also played an important role in the establishment of IP-based spin-offs in the USA.

Spin-off companies typically lack cash, which limits their possibilities to cover patent costs, up-front license fees, cost of research facilities and marketing activities. Di Gregorio and Shane (2003) found that universities' equity investment in their spin-offs was more important than availability of formal venture capital in the early stages of a spin-off creation. It was found that universities' active pursuit of an equity programme stimulated the establishment of IP-based spin-offs (Bekkers *et al.*, 2006). Another factor of university policy affecting establishment of IP-based spin-offs was the availability of incubator parks, which allowed the spin-off to develop the university technology further in close proximity with scientist inventors. In addition, incubator parks were assumed to reduce development costs through offering subsidies and the possibility of sharing overhead costs. The communication and marketing activities of Technology Transfer Office (TTO) personnel were assumed to affect the creation of university spin-offs (Markman *et al.*, 2004). Related to commercial spin-offs, it is assumed that parent company strategy to spin-off in the form of policy related to ownerships right, equity investment, availability of co-locating and offering of subsidies, will be important for the establishment of the spin-off.

Whereas parent company strategy towards spin-off is assumed to affect the establishment of spin-offs, implementation of the spin-off strategy through organisational set-up, level and type of support, type of contractual arrangements and degree of formal distance between parent company and spin-off, will affect the spin-off's early market entry success. For a US IP-based spin-off, the transfer of Technology Transfer Offices into professional specialised teams with top experts on patent application, licensing negotiations, successful business people and spin-off entrepreneurs, had a positive effect on the subsequent chances of the spin-offs' success (Bekkers *et al.*, 2006). The TTOs functioned as professional learning organisations with professional teams of highly motivated people. This ongoing professionalizing of TTOs in the USA was assumed to positively enhance the chances of success of IP-based spin-offs (OECD, 2003).

Bekkers *et al.* (2006) argue that the importance of parent company strategy towards spin-off on establishment and subsequent success depends on whether a strong regional cluster is present. If so, such parent company support may not be needed. Research has shown that in strong regional clusters, IP-based spin-offs were supported through strong interaction with PRO staff, businesses, capital providers and entrepreneurs in comparison with areas where such regional clusters were less developed. Further research has found that an actively supporting role by the PRO (parent company) had a positive effect on the success chances of the spin-offs by offering support and facilities to overcome a lack of resources. It was found that secretarial and other facilitative support, options for housing, access to facilities such as laboratories, libraries and support in finding additional sources of funding are important for supporting spin-offs (Matkin, 2001).

Summarizing, we argue that *managerial institutions* defined as parent company strategy towards spin-offs and implementation of this strategy, are important factors for both establishment and early market entry market success of innovation based spin-offs from corporate KIBS firms.

Research strategy and data collection

In order to grasp the embedded, processual and contextual nature of the spin-off establishment process, a case study design was chosen (Denzin and Lincoln, 1994; Silverman, 2006). Following the theoretical sampling of cases, we build on the argument that multiple cases create more robust theory grounded in varied empirical evidence (Eisenhardt and Graebner, 2007). Therefore, we wanted more than one case in order to reveal the variety of entrepreneurial experience with regard to starting up a daughter enterprise. The strategy for case selection followed a homogenous sampling strategy (Patton, 1990) as we wanted similarity on issues that could interfere with the conceived challenges of the entrepreneurs. We chose to focus on innovation-based corporate spin-offs from SMEs, and that was our first and major criterion. Second, we chose three industries of major strategic importance for the development of the region in which we were operating, which are Oil and Gas, Maritime/Marine and ICT industries.

Each case includes a mother company and a daughter company. The cases varied in organisational size, although the parent company was typically a SME, defined as no more than 100 employees. Daughter companies often are smaller than mother companies. Cases differ somewhat in technology and market niche. In total, 30 extensive, semi-structured interviews were conducted as part of seven cases in 2013–14 in Norway. The data were collected through interviews with managers of the parent firm, spin-off founders and third party actors involved in the spin-off formation process. Supplementary data for the study came from websites, accounting information and press releases. The sample was reached mainly by virtue of its 'accessibility,' using a 'convenience approach' (Bryman, 2004). As data collection proceeded, we tried to 'catch' similar types of entrepreneurs so that we explored the same kinds of organisations. We applied a semi-structured interview guide, one version for mother companies and another for daughter companies, based on existing literature. Interviews lasted from 45 to 120 minutes, were audiotaped and transcribed. Following the advice of Corbin and Strauss (1990), we continually compared the data with the existing literature and also searched for new concepts or emerging links.

Analysis

Empirical setting

To address our research question we identified an empirical context where innovation has been crucial for industry development, which is an important for innovation-based spin-offs. The Norwegian oil and gas or petroleum sector and related Maritime and ICT sectors met these criteria.

The Norwegian oil and gas sector's industrial value-chain consists of activities related to oil and gas exploration and field development, petro-chemistry and oil and gas distribution. The Norwegian oil and gas industry has technological and commercial strength in large parts of the global value chain and competes in the global offshore market. The oil and gas industry is characterised by limited product differentiation and, therefore, its price is closely associated with the mechanism of supply and demand. The actors can influence supply through improved methods of production, thus, innovation within the production processes is called for. This industry is capital intensive as drilling and exploration activities are costly. In addition, drilling and exploration take place in increasingly challenging environments resulting in greater use of unmanned installations, subsea technology and drilling techniques (Fagerberg and Verspagenc, 2009). The industry is characterised by large customers with strong ties to governments yielding considerable market power. The suppliers are more fragmented in terms of company size and market power (cf. Fagerberg and Verspagenc, 2009), however, new entrants into the industry meet barriers to entry in terms of demands for capital and a high level of risk. The demand for new technology gives an incentive for the larger oil companies to invest in smaller, start-up companies. On the other hand, the high risk and cost that characterises the industry, causes it to remain conservative in actually employing the new technology. The deposits on the continental shelf created Norway's main capital industry. The growth of the oil and gas industry contributed to the maritime competency that made new applications in the offshore sector and the development of a strong Norwegian and foreign offshore environment in Stavanger.

The maritime industry in Norway is a world leading maritime cluster characterised by a unique innovation and value creation ability. The Norwegian maritime industry's development is driven by growth in world trade, energy and development of international standards. In recent decades, the Norwegian shipping industry has expanded sharply in offshore related maritime activities. Both within offshore service (e.g. supply) and oil drilling/production, Norway is at the forefront and a dominant player. This makes the Norwegian maritime industry as a whole less vulnerable to fluctuations in the global maritime transport market. Norwegian shipping companies own and operate the most modern offshore fleet in the world. For more than 40 years, shipping companies have contributed substantially to the development of the Norwegian continental shelf. Advanced missions in harsh weather and working conditions in Norwegian waters, have meant that the industry has developed the world's most modern offshore fleet, which today counts over 500 vessels. The Norwegian oil and gas industry, together with the Norwegian Maritime industry, is characterised by two out of three national clusters.

The Norwegian ICT industry is a large, profitable, highly innovative and knowledge-based sector, but is small as knowledge cluster (Andersen, 2012). It is highly centralised – mostly around Norway's capital, with the cities Trondheim and Horten as smaller centres. Few Norwegian ICT companies compete globally. Those who do are often sold out of the country when they reach a certain size and maturity, but expertise is often left in Norway. This industry's most important contribution to society is to provide a competitive arena and strategic resources to increase Norwegian innovative power, productivity and competitiveness. The low profile of the industry, due to value creation being made visible in other industries, is assumed to be a challenge, especially related to recruitment (Andersen, 2012).

Case description

We collected data from six cases, two within the oil and gas industry, two cases within the ICT industry and two cases from the maritime industry. All cases are located in one region in the south-west of Norway. The table (Table 5.1) describes the cases in relation to industry, main operations, year of establishment and number of employees.

Table 5.1 Case description

Case	*Parent company products and services*	*Year*	*Sector*	*Turn-over*	*Employees*	*No of spin-off*	*Spin-off company (case) products/ services*	*Year*	*Sector spin-off company*
A	Supplies project personnel and project information management systems to the oil and energy industry, worldwide. The system is tailor-made to large oil and gas exploration projects	1987	ICT, Oil and gas	1,3 Billion NOK in 2013	1100 (2013) 200 (2004)	10 3 Domestical and 7 Foreign	Domestically Architect service, consulting engineering construction, land- and regulation plans	2004	Public, Oil and gas (B to B)
B	ICT-service company ICT operations, web and mobile and system development	2010	ICT	26 Mill NOK (2013)	24 (2013)	0	Own developed Business Management software	Prob-ably 2016	Multiple industries (B to B)
C	Technical engineering and service company Construction, drawings, shop drawings Calculations related to strength and stability. technically services	1977	Maritime (shipping)	11,3 Mill NOK (2013)	10 (2015)	4	ICT software and service company Fully integrated solution for HSE, QA and operation support. Deliver IT and C and helpdesk services to a large number of vessels worldwide	1997	Maritime industry. Oil and gas (B to B)

Case	*Parent company products and services*	*Year*	*Sector*	*Turn-over*	*Employees*	*No of spin-off*	*Spin-off company (case) products/ services*	*Year*	*Sector spin-off company*
D	Repair company for repair and classification of vessels, rigs and barges. Deliver service, repair, survey and modification of ships, rigs and barges	1973	Maritime Oil and gas	713 Mill (2013) Consolidated 656 Mill (2011)		5	Rig service company A repair base for rigs to offer total services at harbor.	2011	Maritime, Oil and gas (B to B)
E	Originally a machining business. Today business areas are feeding and monitoring system (aquaculture) offshore valves (subsea), bridge solutions (maritime) and machining	2004 Bought as a bankrupt	Maritime/ Oil and gas and Marine (Aquaculture)	173 Mill. NOK (consolidated)	90 (2013)	6	Bridge solutions for ship	2008	Maritime, Oil and gas, Marine (Aquaculture) (B to B)
F	Offshore services Related to oil and gas exploration	2005	Oil and gas	Merged with Simens in 2010. 198 Mill NOK in 2010	NA	2	Develops and provides state of the art technology for Subsea Water Intake and Treatment (SWIT). Forms the basis for Increased Oil Recovery (IOR) and Enhanced Oil Recovery (EOR) technology	2004	Oil and gas sector

Oil and gas

Cases A and F can be classified as operating within the oil and gas industry. Case A is a bigger company providing services like project management for the industry. It has about 10 daughter companies, of which three are domestic and seven are based abroad. The spin-off we observed was providing consulting, engineering and architect services to public organisations and to the oil and gas companies, based on the side product. Case F includes the parent company, which sells offshore services related to oil and gas exploration. It has established two spin-off companies. The daughter company we approached was established to commercialise new technology for the oil and gas sector, i.e. state of the art technology for Subsea Water Intake and Treatment (SWIT) which forms the basis for Increased Oil Recovery (IOR) and Enhanced Oil Recovery (EOR) technology.

Maritime

Case D consists of a parent company that provides a repair service for vessels and has spun off four companies. The daughter company in our study is a rig-service company. Case E consists of a mother company; its main activity today is related to services and products suitable for aquaculture, subsea and maritime industries. The company developed a strategic approach towards spin-off development and has six spin-offs mostly within the maritime industry. The daughter company we contacted provides a new product for the maritime industry.

ICT

Case B includes a parent company that provides ICT-web and mobile system development. It has not yet established a spin-off but will probably do so by 2016. The spin-off company will be within ICT also. Case C consists of a parent company delivering engineering services to the maritime sector and has four spin-offs. The daughter company we approached provides ICT services for vessels worldwide (spin-off is within the ICT industry).

Results

In this article, we consider the role of institutions on the establishment and early market entry success of spin-offs from KIBS in the Oil and Gas, Maritime and ICT industries. Following the logic in our theoretical model, we first discuss the role of sectoral institutions, then regional institutions and finally we discuss managerial institutions in relation to the *establishment and early market entry* of the spin-off firms.

Sectoral institutions

In our analysis, we focus first on the effect of a strong technology regime in the form of high technological opportunity conditions and variety of technological approaches on establishment and early market entry of spin-offs. Second, we discuss the effect of patent regime on spin-off establishment and early market entry of spin-offs.

Sectoral technology regime

The interviews indicated that our cases operated in sectors or industries characterised by high technological opportunity conditions and variety of technological approaches, which promoted the establishment of spin-offs. The interviews indicated that the technology development in the sectors was fast, meaning new business opportunities for the actors:

> To solve these future problems in aqua farming, it is a long race. A lot of weird things (related to innovations) happen in the sector and many actors will try much. We (the parent company) intend to try some ideas/concepts which we believe in.
>
> CEO, parent company, Spin-off E, Maritime/Oil and gas Marine (Aquaculture) sectors

The new ideas were based on customer demands and openings in the markets, which meant opportunities for the parent companies to grow their businesses and to earn money. The interviews indicated that technological challenges could be met by a variety of technological approaches, here represented by Company E with six daughter companies (ref 3 of 6 below):

> Yes, the spin-off companies are the bridge solutions (spin-off 1), the shipcam solution (spin-off 2), which is a spinoff of from the underwater camera solutions which we further develop. We needed a generation 2, but had little capacity problem. In addition spin-off 3 is developing a new generation fully digital camera that will be basis for the next generation Ship cam. So you can say it's the bridge solutions, it is the Ship cam, and then there are valves, which are the main products in the spin-off companies. The Spin-off products are partly tailored made for the individual customer.
>
> Parent company, Spin-off E, Maritime/Oil and gas Marine (Aquaculture) sector

The reasons for establishing spin-off companies for their innovative ideas, instead of creating a new department, were many. The companies listed the following reasons for spinning off companies: to focus the technological development and the resources needed, to brand the new technology which was not associated with the parent firm and to isolate the risk of the new activity from the core activity of the parent firm. In general, the parent companies were SMEs with scarce resources and specialised capabilities.

To summarise, the innovation oriented parent companies operating in sectors with high technological opportunity conditions and a variety of technological approaches were spinning off companies for ideas lying beyond the core competence of parent company, both to focus and accelerate the technological development process and to secure the economy of core activity in the parent company. It can be suggested that:

> Proposition 1: Sectoral institutions as high technological opportunity conditions and variety of technological approaches have a positive effect on spin-off establishment.

Patent regime

Our results also indicated that a strong sectoral IP-regime affected the creation of a corporate innovation based spin-off in the oil and gas and maritime sector. With a patented idea, the immature technology was protected against infringement, i.e. protected from imitations to prevent competitors developing the same technology. The following extracts illustrate how companies implement this process:

> The background of the spin-off; I was director in a technology and development company within the oil and gas industry. It was an idea that David (employee) brought to me, after we had been in a customer meeting. He tells me that he has an idea how to treat water, sea water into the oil reservoir, so he told the idea there, and I said yes, I think we have to apply for a patent. We did, it was around Christmas 2002 and I think the patent application was in early 2003. We were then company A, a topside engineering company and this was the one idea that went on subsea technology, or it was a little of both.
>
> Spin-off company, Case F, Oil and gas sector

With a patent, the risk of imitation was reduced and the subsequent chances for getting investors to develop the idea further and for economic rent was strengthened in contrast to working with unpatented ideas/technologies:

> Regarding patent application, it takes time. And it was first after the idea was patented, then it was much easier to invest in it, than when it is just was an idea.
>
> Spin-off company, Case F, Oil and gas sector

We found that in ICT sectors, patenting was not as important:

> No, patents we do not have, because we are moving so fast. It is something about long-term versus short-term focus.
>
> Spin-off company, Case C, ICT sector

According to the theory, a strong patent regime should stimulate the potential for spin-off creation because patented ideas are defended effectively against infringement and are more attractive for potential investors. This, however, was not as evident in our cases, as illustrated in Table 5.2 below. It seems that in the ICT industry patenting is not as important, nor is it in service-related spin-offs. Another explanation might be that patents are important if a start-up wants to attract external investors. In the case of a corporate spin-off this might not be the first priority, as the parent company might provide enough capital to give it a presence in the marketplace. Thus, we suggest that:

> Proposition 2: The effect of 'a strong sectoral patent regime' on spin-off establishment is dependent on industry, type of innovation (product or service) and on available investments from the parent company.

Table 5.2 Sectoral institutions in relation to establishment and early market entry of spin-offs

	Sectoral institutions		
Case	*Tech. opportunity condition*	*Variety of technological approach*	*Patent regime*
A	Medium technological opportunities when architectural drawing (of buildings) went from manual to data drawing. Parent company had data ICT system-competence, but architectural drawing was their core competence	Medium variety of technological approaches related to architectural drawing in different data systems	Not relevant
B	Medium, - development of business management systems/online service programs can be based on new architectures/ technologies	Medium, because development of business management systems can be developed on different platforms/-technologies	NA Patent not relevant in the initial stadium of technology development
C	ISO 900 (ISM), a regulation based demand of quality systems for vessel, created a high technological opportunity condition to develop QA systems and implementation of it (which technology to be used to get it online onboard on vessels).	A variety of technological approaches for development (different ICT-platforms) and implementation (different online solutions onboard on ships) of quality-system for fishing vessels	Not Relevant. Patenting of HMSQ software not relevant because software-programs are characterised of continously developments
D	High opportunities for technological development related to projecting and delivery of large multi-discipline rigg maintenance projects.	Offering multiple	NA
E	Large possibilities for techno-logical development ship-bridge solutions,	Variation of technological approaches within ship-bridge solutions because dependent of multiple technology to function	Not relevant
F	Large possibilities for technology development within subsea water intake and treatment within oil and gas	A variety of technological approaches within Subsea Water Intake and Treatment (SWIT).	The first patent application was the first seed to what later became firm F

Regional institutions

In developing our understanding of the regional layer of our institutional model, we discuss findings related to how proximity to a strong regional cluster affects establishment and early market entry success of spin-offs from KIBS.

The results show that spin-off companies located in strong regional clusters, such as enterprises producing goods and services for the maritime and oil and gas sectors, benefitted largely from geographical closeness within the cluster in the form of (1) tacit knowledge exchange with customers, suppliers and former colleagues and (2) other proximity related benefits, including possibilities to use laboratory equipment, options to attract new talents and access to regional investors.

Knowledge about products, services, production processes, the sector and other aspects is a kind of sector specific knowledge that employees accumulate. Strong regional clusters are characterised by employer mobility and exchange of tacit knowledge.

The interviews indicated that for the spin-off companies, proximity to a strong regional cluster meant access to a specialised knowledge of mainly tacit character from specialised suppliers, customers and collaboration partners. This was of great importance in the early phase of the spin-off process, i.e. in development and testing of the technology:

> In the development of the ship bridge solution (spin-off) we collaborated much with company A (local company) and with B, i.e. the ship consultant (local company) who is upgrading ship bridge solutions.
>
> Parent company, Case E, Maritime/Oil and gas Marine (Aquaculture) sector

The interviews also indicated that geographical proximity to a cluster meant that the spin-off companies received access to a skilled and specialised workforce. For the spin-off company this meant both access to specialised manpower and reduced costs related to hiring people and training of personnel:

> Culturally I think it is important (for the daughter companies) to be located in the region. We [would] have not succeeded if we hadn't had the work culture, [or a] lot of skilled practitioners who also have a certain theoretical competence.
>
> Parent company, Case E, Maritime/Oil and gas Marine (Aquaculture) sector

Another benefit from the regional cluster was availability of start-up capital through regional cluster actors investing in the spin-off company. The regional investors operated often as demanding customers interested in the kind of technology services the spin-off company should develop:

> A local company A invited us to the Technology days. They invited the industry to come up with good ideas for new technology that could help extending the life of X (oil and gas field), or improve business to company A. We presented

the idea, and after some back and forth company A allocated 1.7 million NOK to develop our idea. On the basis that Company A (one of the largest foreign operators on the Norwegian continental shelf) allocated 1.7 million, we took the decision. It was a discussion between me and him (employee in regional collaborating company), because it would make it much easier to work with the local company, we established a joint company for the idea.

Parent company, Case D, Maritime and Oil and gas sector

He (collaborating company) had to go to Oslo (Norwegian capital) to get money, We're impatient, and when he came to us and said: listen to one thing, I cannot work with you (on the spin-off idea) anymore because I do not have budget for it. I disagree with my boss, so either I have to start my own business, or so this goes no further. We told him: "no, we have analyzed a bit, we have to get it (the funding) locally". And then we agreed on one setting, then started the spin-off company.

Parent company, Case F, Oil and gas sector

The findings are summarised in Table 5.3 below.

Table 5.3 Regional institutions in relation to establishment and early market entry success of spin-offs

	Regional institutions			
	Access to tacit knowledge through specialised suppliers, employees	*Access to demanding customers – pilot customer*	*Access to a pool of specialised employees options to attract new talents –*	*Access to industrial inventors – availability of startup capital (pre-funding)*
A	Spin-off company access to specialised competence through specialised employees	Spin-off company located in a strong regional cluster achieved clients from day one because central cluster actors needed the spin-off company services	Spin-off company located in a strong regional cluster recruited specialise employees from regional cluster	Spin-off company located in a strong regional cluster got industrial investors from regional cluster
B	Spin-off company not located in a strong ICT cluster, thereby not access to specialised suppliers	Potential spin-off company have already customers through access to parent company customers	A strong regional ICT milieu, but not good options to attract talents. Dependent of persons moving back to the city. Recruiting by stealing employees from each other and offering best people shares	Challenges with getting funding or access to industrial investors.

(*continued*)

Table 5.3 Regional institutions in relation to establishment and early market entry success of spin-offs (*continued*)

	Regional institutions			
	Access to tacit knowledge through specialised suppliers, employees	*Access to demanding customers – pilot customer*	*Access to a pool of specialised employees options to attract new talents –*	*Access to industrial inventors – availability of startup capital (pre-funding)*
C	Not strong regional ICT cluster, therefore no access to tacit knowledge through regional suppliers. But access to tacit knowledge through regional demanding customers. spin-off company supplier to a strong regional maritime cluster.	Located in a strong maritime cluster means access to demanding customer. A world leading offshore shipping company located in cluster has been an important pilot customer	Not strong ICT cluster, no options to attract new talents from the cluster	A world leading offshore shipping firm located in the region has been an important investor
D	Location in a strong cluster, with parent-company as one of the dominant cluster actors gave spin-off company access tacit knowledge through parent-company relation to suppliers and customer (through transfer of personnel from parent to spin-off).	Located in a strong regional cluster, with parent as one of the dominant cluster-actors, means access to pilot-customer through parent-company network	NA	Parent-company didn't need /wish external investor
E	Located in a strong regional cluster. It has given access to tacit knowledge through specialised suppliers for development of the technological spin-off concept	Location in a strong cluster has given access to demanding customer, shipping companies, important as demanding customers	Located in a strong cluster means access to an employee pool of skilled	N/A – Parent company not interested in other investors

	Regional institutions			
	Access to tacit knowledge through specialised suppliers, employees	*Access to demanding customers – pilot customer*	*Access to a pool of specialised employees options to attract new talents –*	*Access to industrial inventors – availability of startup capital (pre-funding)*
F	Local suppliers from the regional oil and gas cluster was important in the initial technological development, in minor degree later in development	Spin-off has during all the techno-logical developments had a close relationship with regional oil companies, and a major part of the financial funding for the technology development has also been from oil companies.		Regional cluster actors were the initial investors. Later national and international actors

The result of this study implies that proximity to a regional cluster affects early market entry success of spin-offs. However, the empirical finding did not indicate that geographical proximity to a regional cluster conditioned the establishment. Localisation in a regional cluster meant access to specialised and tacit knowledge from suppliers and the collaboration partner. For the spin-off companies, this means more efficient technology development processes because they do not have to spend time seeking out and getting to know technological suppliers or collaboration partners. Strong regional clusters are characterised by having a pool of specialised employees, which means access to specialised employees. This means reduced start-up costs in the form of reduced costs in hiring and training of employees. Proximity to regional investors also affects the early market entry success of spin-off companies in the form of reduced costs searching for potential investors.

For spin-off companies location in strong regional clusters means access to tacit knowledge and reduced start-up costs, which affect early market entry success of the spin-off. Therefore:

> Proposition 3: Proximity to a strong regional cluster is assumed to have a positive effect on the early market entry success of the spin-off.

Managerial institutions

In this paper we consider corporate spin-offs, i.e. spin-offs from KIBS-firms initiated by parent firms. For these kinds of innovation based spin-offs, managerial institutions were conceptualised as *parent company strategy towards spin-off and implementation of this strategy*, were assumed to impact both establishment and early market entry of the spin-offs.

The results from the interviews implied that the *parent company's strategy towards spin-offs* can be divided into three different, but related, dimensions: (1) strategy for sharing of IPR/royalties, (2) strategy for ownership and investments and (3) strategy for offering co-location and sharing of overhead costs. Related to *implementation of parent company strategy towards the spin-off*, i.e. how the strategy reaches the employees in the parent companies, the empirical findings indicate that the strategy's (1) organisational set up and (2) the level and type of support, affect subsequent chances of success of the spin-off. These factors affected both the establishment and early market entry success for most spin-offs. The exception was spin-off companies developing goods or services demanding minor financial investments.

Parent company strategy for spin-off

Strategy for sharing of IPR/Royalties

Clear rules related to IPR are important for the establishment of a new company to avoid conflict of interest. Among our cases, the possibility of getting shares in the company seems to have had a positive effect on the establishment of spin-offs. We found that when idea-owners and leaders of spin-offs were offered shares in the spin-off company, their motivation and dedication to work was much higher than if they were just regular employees. Parent companies apply this as part of their strategy, thus:

> We have done that in the other spin–off companies also. We have always some of the managers as co–owners, owing about 20 per cent of the shares. Then I see that they work in a completely different way. When they own it (the company) compared to only be employed, they spend more energy on it if they own it than not.
>
> Parent company, Case E, Maritime sector

Some parent companies have a strategy of 100 per cent ownership of daughter companies, to be able to fully capitalise on spin–offs later:

> We own all spin-off companies 100 per cent. It is our business model. Should not say it will last forever, but until now we have done it like that. Again, keep it simple.
>
> Parent company, Case A, Oil and gas sector

Spin-offs typically lack cash, which limits their ability to cover patent costs, up-front licenses fees, costs of research and marketing activities. From research on academic spin-off, Di Gregorio and Shane (2003) confirm that equity investment from universities in their own spin-offs is more important in the early stages of spin-off creation than the availability of formal venture capital.

Strategy for ownership and investments in spin-off companies

All the spin-off companies were owned totally or partly by their parent companies in the initial phase of development.

> In the beginning, we (the spin-off company) were owned by parent company. They contributed a lot initially.
>
> Spin-off company, Case A, Public and Oil and gas sector
>
> In our spin-off companies I am the owner and at the board of all companies. If you bring in other owners you need to operate in a so-called professional manner. Then the processes would have taken much more time and some things you would not have been allowed to do if it costs money and stuff like that.
>
> Parent company, Case E, Maritime/Oil and gas and Marine (Aquaculture) sector

This kind of support affected the spin-off's early market entry success in the form of an efficient way of getting funding and investors. A spin-off company within the oil and gas sector underlined the importance of having a parent company at its back, because the funding needed to develop the technology was huge:

> It is essential for the oil companies that you go through a qualification, a test phase, demonstrating that it works. You should qualify the technical things. It costs a sea. So, ordinary people or entrepreneurs have no chance in the ocean. We (spin-off company) have passed about 60 Million NOK, who individual can afford that? To pull off such a business like this, it's not easy to do on your own. And, it's clear that the risk for us if we'd jumped off is dramatically large related to being able to take it further To answer the question, it had not been possible, I would say without having a bigger milieu at you back.
>
> Spin-off company, Case F, Oil and gas sector

Strategy for offering co-location, sharing of overhead costs and networking to spin-offs

The parent companies also supported their spin-off companies in the initial period by offering housing and subsidies, such as covering administration and accounting costs, and subsidised prices for renting employees. These kinds of support affect the early market entry success of the spin-offs, because it means reduced start-up costs for the spin–off company.

> We support them (the daughters) with competence and capital. We are there with all the required for them to succeed. And it's a strategic decision that the Board takes.
>
> Parent company, Case E, Maritime/Oil and gas and Marine (Aquaculture*)*

> We got free housing, co–location with the mother.
>
> Spin-off company, Case C, Maritime and Oil and gas sector
>
> We were located in the conference room beside the parent company in the initial period. And we got larger offices when they enlarged the parent-company offices.
>
> Spin-off company, Case C, Maritime and Oil and gas sector
>
> It has cost me millions, the project there (one of the spin-off companies), so I hope that it finally takes off. We have spent a lot of money on it, which we have earned elsewhere. It's how we're doing with the spin-offs. To count on the project I couldn't bear, it has cost me a million.
>
> Parent company, Case E, Oil and gas, Maritime and aquaculture sector

For a spin-off company, getting access to the parent company's network was equally important as getting co-location and administrative help:

> Related to advantages of having a parent company in the back, it was not like we went to the parent to get refills economically. I do not think that has happened once. The financial risk of starting the spin-off company for parent company was related to risking their reputation – that they gave us initial "capital" in form of contacts and networks and stuff.
>
> Spin-off company, Case C, Maritime and Oil and gas sector

We further found that all parent companies offered their spin-off companies co-location and housing:

> We are offering the spin-off companies co-location and use of parent company knowledge.
>
> Parent Company, Case E, Oil and gas, Maritime and aquaculture sector

> And the spin-off company gets all the benefits of belonging to something (i.e. parent company) a little bigger, because they can use it (the resources) when they want. And those who own us – they back us up.
>
> Parent Company, Case F, Oil and gas sector

The availability of co-location with the parent company allows the spin-off to develop innovation based spin-off technology or services in close proximity with the parent in which the spin-off idea originated, using parent company equipment and personal expertise. The parent company may also reduce the costs of development through offering subsidies and the possibility of sharing overhead costs. Our findings are summarised in Table 5.4 below.

Table 5.4 Managerial institutions in relation to establishment and early market entry success of spin-offs

	Managerial institutions			
Case	*Strategy for sharing of IPR/royalties*	*Strategy for ownership and investments*	*Strategy for offering co-location and sharing of overhead costs*	*Strategy for follow-up spin-off on early stages of development*
A	Parent company offered CEO's at the spin-off company shares in the spin-off company	The parent company invested in the spin-off. 100 % ownership at start. initial period of start-up	Parent company offered spin-off company co-location and subsidised related to administration and account.	Parent company leader group (founders, interdisciplinary group) followed up spin-off company in form of weekly meetings Represents from parent company leader-group also spin-off board members
B	Parent company is going to offer CEO at spin-off companies IPR-right	Parent-company had a strategy for NOT investing in or owing the spin-off companies until parent had grown larger	Parent company strategy for offering spin-off company-co-location and sharing of overhead costs like administration is planned, also customers through own network	Parent company "innovation team" was transferred to parent company leader group, which followed up spin-off product in weekly meetings
C	Parent company offered all managers of spin-off companies shares in their companies	Parent company invested as much as possible in each spin-off company without risking the operation of the parent company	Parent company has supported the spin-off company in form of co-location, administrative tasks (account) and internal (subsidies) rates on hiring of personnel. Parent company and spin-off companies have customer-supplier relationships with each other	Parent company leader group represented with leaders from parent company and other daughter with complementary expertise is following up spin-off company by meetings once a month

(*continued*)

Table 5.4 Managerial institutions in relation to establishment and early market entry success of spin-offs (*continued*)

	Managerial institutions			
Case	*Strategy for sharing of IPR/royalties*	*Strategy for ownership and investments*	*Strategy for offering co-location and sharing of overhead costs*	*Strategy for follow-up spin-off on early stages of development*
D	Parent company offering CEO at spin-off companies IPR-rights	Parent company had a strategy for investing in and partly owing the spin-off companies.	Parent company had a strategy for offering spin-off companies co-location The spin-off company can use personnel resources from other daughter s and parent companies	The decision about spinning off the company was a strategic choice, decided by and followed up by corporate management
E	Parent company had a strategy for sharing IPR rights with founders of the spin-off companies. CEO of the spin-off companies are offered ownership in their companies, some have declined because they wish to be employees with safe income.	Parent company strategy for owning spin-off companies nearly 100%	Parent company offered spin-off companies co-location and sharing of administrative and knowledgeable resources	Parent company following up spin-off companies by weekly leader meeting, where members of parent company leader group and CEO of the spin-offs are meeting (economic, technical, branding and business development competence)
F	Parent company initially established a spin-off company to secure IPR right of the founders. founders offered shares as payment for their IPR.	Parent company invested together with one other partner 48% each	Spin-off company was offered co-location with parents the first five years	Spin-off was followed up by the board which in the initial phase was represented by leaders in the two main-owner companies. It was a close follow-up

Parent company implementation of strategy towards spin-off

Parent company organisational set-ups

Most parent companies had created internal departments or groups with dedicated managers responsible for the firm's innovation processes, spin-offs establishment and follow up. Some parent companies encouraged and followed up spin-off establishment through their leader groups. Some had even established 'innovation programmes' for all the employees, while other parent companies had created innovation groups with dedicated people working with processes encouraging innovations. These internal departments or groups were actively pursuing innovations and potential spin-off ideas. Another internal factor highlighted in the interviews with regard to encouraging innovation and spin-off processes was the existence of consistent policy on innovations and spin-off decisions that existed in the SME parent companies.

Case C had established an internal group promoting and following up parent company spin-off processes:

> We have monthly management meetings to monitor the developments in spin-off companies. In these meetings we are welcoming spin-off opportunities, if there are some good ideas.
>
> Parent company, Case C, Maritime sector

Whereas in Case A, the management team was responsible for both encouraging and follow up of parent company spin-off processes:

> Each month there is full reporting of all spin-off companies. Last night for example, we had management meeting. Then we have all the spin-off leaders at Skype. We have joint management meeting might say.
>
> Parent company, Case A, ICT and Oil and gas sector

We can conclude that parent company strategy has an important impact on the establishment of spin-offs regardless of industry and innovation type:

> Proposition 4: Parent company strategy towards spin-offs, i.e. strategy for sharing of IPR rights, strategy for equity and investments, and strategy for co-location and subsidies, affect both spin-off establishment and early market entry success.
>
> Proposition 5: Parent company implementation of strategy toward spin–off in the form of organisational set-up, affects both spin-off establishment and early market entry success.

Discussion and conclusions

The aim of this paper was to identify which institutional factors conditioned *establishment and early market entry* of innovation-based corporate spin-off from KIBS firms. To answer these questions we looked at sectoral, regional and managerial institutional layers in relation to the spin-off process. The empirical results indicate that all three institutional layers need to be taken into consideration. We further found that layers differ in their role and level of impact depending on industry and type of spin-off firm.

We observed that sectoral characteristics are especially relevant for a parent company's decision to spin-off a daughter company. Technological opportunity conditions and variety of technological approaches seem to stimulate spin-off processes. However, this was more obvious for firms operating in the oil and gas and maritime industries than the ICT industry. Furthermore, patenting was important if the daughter companies were based on an innovative product rather than service. In addition, in the ICT sector patenting does not seem as relevant as in more financially intensive sectors like the oil and gas sector. Thus, it seems that a patenting regime is less important for service based KIBS firms than KIBS firms offering services based on their own developed equipment.

We also found that regional institutions, in the form of proximity to strong regional clusters, condition KIBS spin-offs' early market entry success. Clusters provide spin-offs with access to tacit and specialised knowledge through specialised suppliers, employees and demanding customers. This is valuable for the KIBS spin-off, especially during the initial technological development, characterised by cumulative learning arising from in-depth interaction between supplier and user. Proximity to strong regional clusters also offers start-up capital through access to regional investors interested in technology development and access to specialised employees.

Parent company strategy towards spin-off and implementation of this strategy, become important in enhancing both establishment and early market entry success of corporate innovation based spin-offs from KIBS. Parent companies with a policy towards spin-offs in the form of organisational set-ups responsible for promoting and following up the parent firms' innovation and spin-off processes, condition both establishment and early market entry success. The spin-off firm's financial hardships associated with the first years of operation seem to be overcome with the help of the parent company. The only parent firm that postponed the spin-off establishment was one that could not provide financial support to its daughter company in the form of investments or equity in the spin-off. Once established, parent company implementation of strategy towards spin-off is important for further early market entry success. Innovation based spin-offs from KIBS firms in the oil and gas and maritime sectors requiring large financial investments related to technological development are strongly dependent on a parent company's equity investments, professional interdisciplinary teams providing advice in the process and secretarial support. This kind of parent-company support seem to be of minor importance when financial investment related to technological development of products and services is small, as in software development. Furthermore, co-location and administrative help from parent company are appreciated assets. Moreover, network sharing and connections to important customers and suppliers help daughter companies enter the market with greater ease.

Figure 5.2 illustrates the suggested theoretical model explaining the influence of institutional layers on spin-off establishment and early market success.

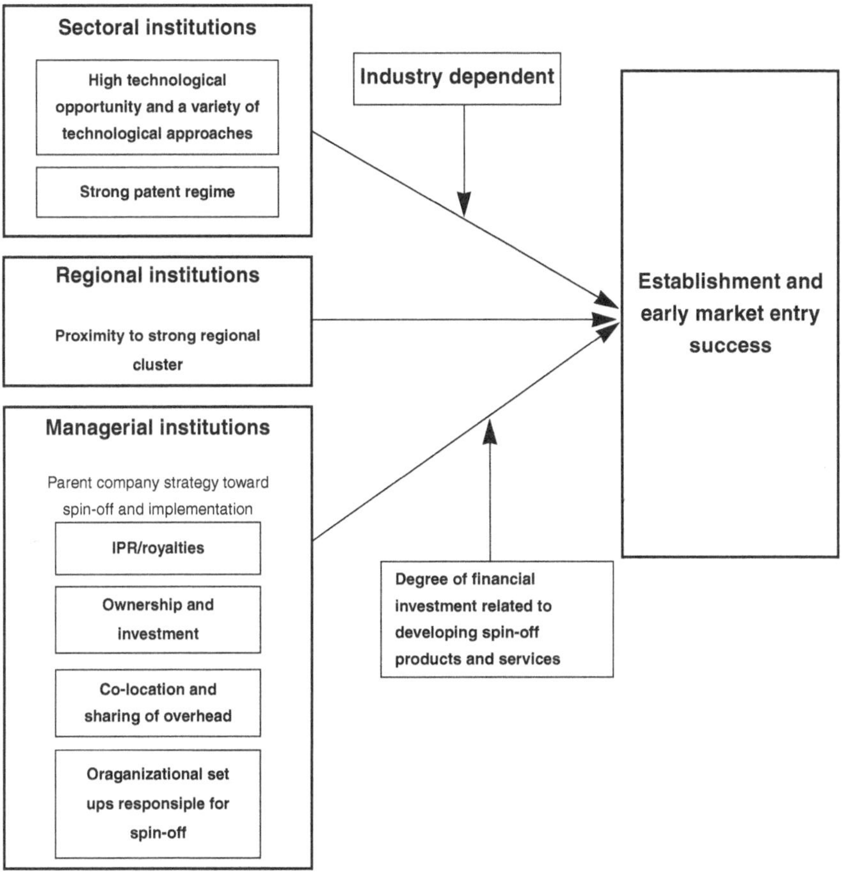

Figure 5.2 Theoretical model explaining influence of institutional layers on spin-off establishment and early market success

Implications and value

This research has several theoretical and practical implications. Prior research has, to a minor degree, focused on the effect of institutional factors on entrepreneurship in general and on the establishment and subsequent success of corporate spin-offs specifically (Gilsing, van Burg, and Romme, 2010; Kshetri, Williamson, and Schiopu, 2007). Our contribution is a better understanding of the link between the sectoral, regional and managerial institutional layers and one specific form of spin-off, i.e. corporate initiated innovation based spin-off from KIBS.

The paper provides insight to better understand the contextual conditions of the institutional aspects affecting the creation and subsequent success of KIBS-based spin-offs. In the Norwegian maritime and oil and gas sectors, characterised by high technological opportunity conditions and a variety of technological approaches, the KIBS' spin-off approach to technological innovations was a viable means of isolating a new and exploratory initiative from the core activities of the parent firm. That might be explained by the fact that oil and gas and maritime sectors are more like 'closed loop systems' with higher requirements in terms of IP rights and compliance with regulations and environmental requirements, which restrict the open flow of knowledge. On the other hand, the ICT sector is characterised by open innovation and diffusion of knowledge, which makes the patent regime and variety of approaches less of an issue for new entrants. By spinning off the technological innovations into a new company that focuses exclusively on the new product or service, the parent company could realise the potential of that new technology and capitalise on the innovation by keeping a certain extent of ownership of the spin-off company.

The study contributes to the cluster literature by providing evidence for the effect of a regional cluster on the early market entry success of spin-offs from KIBS firms. This is in accordance with the research of Delgado, Porter and Stern (2010). We find also that the regional cluster contributes to the early market entry success of the KIBS based spin-off, in the form of contribution of employment and funding to young spin-offs in regional industries. However, proximity to clusters was shown to be an important variable for oil and gas as well as the maritime sector, but of less importance for ICT sector. This can be explained by the dynamics of the industry collaboration. The maritime sector in Norway often functions as a supplier for the oil and gas industry, and proximity to a cluster supplier in these sectors provides a qualified working force and tacit knowledge transfer. At the same time, ICT, which is also often a supportive service for oil and gas, a dominant industry in Norway, still has other specific features. In the ICT sector, knowledge is often codified and easily transferable between partners, so that geographic proximity is a less important issue. Being close to oil and gas and maritime clusters would not provide ICT firms with a better workforce or tacit knowledge. At the same time, proximity to clusters in related industries might provide capital investments.

Finally, parent company strategy is of high importance both for establishment and early market entry success of spin-offs in the maritime and oil and gas sectors. The entry barriers are high as huge financial investments are connected with market entry in these sectors. Therefore, gaining financial support from the mother company in the form of ownership, investment and other support, such as co-location and sharing of overheads, is of higher importance in these sectors. In the ICT sector, entry barriers are lower and while the mother company's strategy is of importance, it is not a predominant factor in successful market entry of spin-offs in this sector.

The results also have some political implications. The formation of new enterprises is important for the development of a healthy economy. Spin-offs have a higher survival rates as ordinary start-ups, as they gain the knowledge and support both from parent companies and other actors in the regional cluster. This study provides some evidence for policy makers to move from supporting new firms in strategic activities, to design instruments or mechanisms that favour selected firms with the potential to gestate a higher number of more successful spin-offs. These specific firms could be parent firms with a strategic approach to spin-offs, evidenced by having internal policies encouraging innovations and spin-offs and organisational set-up and support activities for innovation and spin-off processes. These companies have both a motivation to create spin-offs and knowledge about how to access the different kinds of resources a spin-off need.

This study also provides some evidence with regard to what sectors policy makers should support to promote establishment of corporate spin-offs. Policy makers should design instruments supporting companies in sectors characterised by high technological opportunity conditions and a variety of technological approaches. This study shows that KIBS companies use spin-offs to realise the potential of new innovative products and services beyond the scope of the company's core business, but also to realise the potential of existent or incrementally revised products or services into new submarkets abroad.

Another finding provided by this study is the effect the regional cluster has on early market entry success and subsequent chances of success for KIBS-based spin-offs. Policy makers should design instruments or mechanisms that create meeting places, housing in the form of incubators or other forms of connections among KIBS parent firms, KIBS spin-off entrepreneurs, regional investors and suppliers of competence.

Note

1 Reference from an executive summary of an expert workshop arranged by IPTS to further understand spin-offs and with a view to the preparation of the Sixth RTD Framework Programme and the European Innovation policy.

References

Andersen, E. (2012). En kunnskapsbasert IT-næring: aktivitet og effekt. *Magma*. 1/2012.

Andersson, M. and Hellerstedt, K. (2009). Location attributes and start-ups in knowledge-intensive business services. *Industry and Innovation*, 16, 103–21.

Bekkers, R., Gilsing, V. and van der Steen, M. (2006). Determining factor of the effectiveness of IP-based spin-offs: comparing the Netherlands and the US. *Journal of Technology Transfer*, 31 (5), 545–66.

Bekkers, R. and M. van der Steen (2003). IP-based spin-offs of public research organisations in the Dutch Life Sciences and ICT sectors, in OECD, *turning science into business: patenting and licensing at public research organisations*, Paris: OECD Publishing.

Bruneel, J., Van de Velde, E. and Clarysse, B. (2013). Impact of the type of corporate spin-off on growth. *Entrepreneurship Theory and Practices*, 37, 943–59.

Bryman, A. (2004). *Social research methods*. Revised edition. Oxford: Oxford University Press.

Busenitz, L.W., Gomez, C., and Spencer, J.W. (2000). Country institutional profiles: unlocking entrepreneurial phenomena. *Academy of Management Journal*, 43 (5): 994–1003.

Chesbrough, H. (2002). Graceful exits and missed opportunities: Xerox's management of its technology spin-off organizations. *Business History Review*, 76(4), 803–37.

Chesbrough, H. and Rosenbloom, R. S. (2002). The role of the business model in capturing value form innovation: evidence from Xerox Corporation's technology spin-off companies. *Industrial and Corporate Change*, 11 (3), 529–55.

Christensen, C. M. (1997). The innovator's dilemma when new technologies cause great firms to fail. Boston: Harvard Business School Press.

Clarysse, B., Wright, M. and van de Velde, E. (2011). Entrepreneurial origin, technology endowments and the growth of spin-off companies. *Journal of Management Studies*, 48 (6), 1420–42.

Corbin, J. and Strauss, A. (1990). Grounded theory research: procedures, canons – evaluative criteria. *Qualitative Sociology*, 13 (1), 1–19.

Dahl, M.S. and Pedersen, C.O.R. (2003). Knowledge flows through informal contact in industrial clusters: myths or realities? *DRUID Working Paper* 03–01, Aalborg University.

Delgado, M., Porter, M. and Stern, S. (2010). Clusters and entrepreneurship. *Journal of Economic Geography*, 10 (4), 495.

Denzin, N.K. and Lincoln, Y.S. (1994). *Handbook of qualitative research*. Los Angeles: Sage Publications Inc.

Di Gregorio, D. and Shane, S. (2003). Why do some universities generate more start-ups than others? *Research Policy*, 32, 2009–227.

Djankov, S., McLiesh, C. and Ramalho, R. (2006). Regulation and growth. The World Bank. *Economic Letters*, 92 (3), 395–401.

Djokovic, D. and Souitaris, V. (2008). Spin-outs from academic institutions: a literature review with suggestions for further research. *Journal of Technology Transfer*, 33 (3), 225–47.

Eisenhardt, K.M. and Graebner, M.E. (2007). Theory building from cases: opportunities and challenges. *Academy of Management Journal*, 50 (1), 25–32.

European Commission (2007). Towards a European strategy in support of innovation in services: challenges and key issues for future actions. Commission staff working document. SEC (2007) 1059. Brussels: Commission of the European Communities.

Fagerberg, J. and Verspagenc, B. (2009). Innovation studies – the emerging structure of a new scientific field. *Research Policy*, 38, 218–33.

Foss, L. Iakovleva, T., Kickul, J., Oftedal, E. and Solheim, A. (2011). Taking innovations to market: the role of strategic choice and the evolution of dynamic capabilities. *International Journal of Entrepreneurship and Innovation*, 12 (2), 105–16.

Fryges, H. and Wright, M. (2014). The origin of spin-offs – A typology of corporate and academic spin-offs. *Small Business Economics*, 43 (2), 245–59.

Gallouj, F. and Weinstein, O. (1997). Innovation in services. *Research Policy*, 26, 537–56.

Gereffi, G. and Hempel, L. (1996). Latin America in the global economy: running faster to stay in place. *NACLA Report on the Americas*, 29 (4), 18–27.

Gilsing, V.A., van Burg, E. and Romme, A.G.L. (2010). Policy principles for the creation and success of corporate and academic spin-offs. *Technovation*, 30 (1), 12–23.

Goduscheit, R. C. and Brendstrup, S. (2012). Spin-offs and SME's – Potential motivations and rationales. *DRUID-paper*. 12–06. Copenhagen: CBS.

Hellmann, T. (2007). When do employees become entrepreneurs? *Management Science*, 53 (6), 919–33.

Hofstede, G. (1983). Dimensions of national cultures in fifty countries and three regions, in Deregowski, J.B., Dziurawiec, S. and Annis, R.C. (eds) *Expectations in cross-cultural psychology*. Lisse, Swets and Zeitlinger, 335–55.

Howells, J. (2001). The nature of innovation in services, in OECD (ed.) *Innovation and productivity in services*. Paris: OECD, 55–79.

Hwang, H. and Powell, W. W. (2003). Institutions and entrepreneurship, in Acs, Z.J., and Audretsch, D.B. (eds) *Handbook of entrepreneurship research. an interdisciplinary survey and introduction*. New York: Springer, 179–210.

Kshetri, N., Williamson, N.C. and Schiopu, A. (2007). Economics and politics of advertising: Evidence from the enlarging European Union. *European Journal of Marketing*, 41(3/4), 349–66.

Markman, G., Gianiodis, P., Phan, P. and Balkin, D. (2004). Entrepreneurship from the ivory tower: do incentive systems matter? *Journal of Technology Transfer*, 29, 353–64.

Matkin, G. (2001). Spinning off in the United States: Why and how?, in *STI special issue on fostering high-tech spin-offs: a public strategy for innovation*, 26. Paris, OECD.

Miles, I. (2001). Services and the knowledge-based economy, in Tidd, J. Hull, F.M. (eds) Service innovation: organizational response to technological opportunities and market imperatives. Cheltenham: Edward Elgar, 81–112.

Mowery, D.C. and Ziedonis, A. A. (2001). The geographic reach of market and non-market channels of technology transfer: comparing citations and licenses of university patents. *NBER Working Paper* No 8568. October 2001. JEL No O31, O32, R12.

Mustar, P.M., Renault, M.G., Colombo, E., Piva, M., Fontes, A., Lockett, M., Wright, B., Clarysse, B. and Moray, N. (2006). Conceptualizing the heterogeneity of research-based spin-off: a multi-dimensional taxonomy. *Research Policy*, 35, 289–308.

Nelson, R. N. (2001). Observations on the post bayh-dole rising of patenting at American universities. *Journal of Technology Transfer*, 26, 13–19.

Nås, S. O., Sandven, T. V., Eriksson, T., Andersson, J., Tegsjö, B., Lehtoranta, O., Virtaharju, M. (2003). High-tech spin-offs in the Nordic countries: summary report. ISSN: 0804–8185. STEP-gruppen. Sider: 17.

OECD, (2003). *Turning science into business: patenting and licensing at public research organizations*. Paris, OECD.

Patton, M. (1990). *Qualitative evaluation and research methods*. Beverly Hills: Sage, 169–86.

Porter, M. (1990). *The competitive advantage of nations*. London: Macmillan.

Porter, M.E. (1998). *On competition*. Boston: Harvard Business School.

Schricke, E., Zenker, A, Stahlecker, T. (2012). Knowledge-intensive (business) services in Europe. *Studies and reports European Commission. Research and innovation.*

Scott, W. R. (1995). *Institutions and organizations. ideas, interests and identities*. California: Sage.

Scott, W.R., Ruef, P.M. and Caronna, C. (2000). *Institutional change and healthcare organizations: from professional dominance to managed care*. Chicago: University of Chicago Press.

Shane, S. (2001). Technology regimes and new firm formation. *Management Science*, 47 (9), 1173–90.

Silverman, D. (2006). *Interpreting qualitative data: methods for analyzing talk, text and interaction* (Third edition). London: Sage.

Steen van der, M. (1999). *Evolutionary systems of innovations*. The Netherlands: Van Grocum

Van de Velde, E., Clarysse, B., Wright, M. and Bruneel, J. (2007). Exploring the boundary between entrepreneurship and corporate venturing: from assisted spin-outs to entrepreneurial spin-offs. Working Paper 2007/472, Ghent: Ghent University.

Woo, C., Willard, G. and Beckstead, S. (1989). Spin-offs: What are the gains? *Journal of Business Strategy*, 10 (2), 29–32.

Wood, P. (2005). A service informed approach to regional innovation – or adaption? *The Service Industries Journal*, 25, 439–45.

6 Survival of Knowledge Intensive Business Service firms

The role of agglomeration externalities

Sam Tavassoli and Viroj Jienwatcharamongkhol

Introduction

Firms (including Knowledge Intensive Business Services) are located in the regions and regions do differ with each other in terms of providing the supply of factors that enhance the innovation of firms. One of the main factors that a region may provide is externalities and knowledge spillover that occur between firms within a region. Indeed there are evidences suggesting that the more knowledge spillover in the region, the more learning opportunities, and hence the more innovation (or invention) happens in the region (Feldman 1994; Anselin *et al.*, 1997; Tavassoli and Carbonara 2014; Castaldi *et al.*, 2014). This highlights the role of geography for innovation of firms, in so-called "geography of innovation" as a field of study (Feldman and Kogler 2010). However, there is far less evidence on the role of geography on the survival of KIBS firms, especially newly start-up KIBS firms. This is an important issue to investigate because newly established KIBS firms comprises a significant share of KIBS firms in terms of number but their lives are relatively short.

Most of the studies dealing with the survival analysis of firms have neglected the role of region until recently (this will be reviewed in Section 2). Among the few survival analyses that take into account the role of region (Fotopoulos and Louri 2000; Fritch *et al.*, 2006; Falck 2007; Wennberg and Lindqvist 2010), several gaps can be identified: First, these studies rarely distinguish between various types of agglomeration externalities [an exception is Neffke *et al.*, (2012)].[1] This is however an important issues because various agglomeration externalities have various sources and consequences (Rosenthal and Strange 2004). Second, there are no studies focusing particularly on KIBS firms. This is again an important issue because KIBS firms are one of the engine of growth in the economy. Third, the lack of individual-level data has prevented most of previous studies from controlling for individual characteristics of the founder of the firms. This is important to consider, particularly for newly founded firms, because entrepreneurs are the firm's main capacity and constraint in these types of the firm (Brown and Kirchhoff 1997). The inherent characteristics of an entrepreneur, therefore, can influence the development of an entire organisation. On the other hand, looking at the literature on geography (of innovation and entrepreneurship), the firm or entrepreneur as the unit of analysis is underdeveloped (Feldman 1994; Acs and Varga 2005). When the focus is on entrepreneurial individual or firm, usually the evidences are based on case study of a particular region and systematic evidence (over all regions of a country) is still lacking (Fredin, 2014).

The aim of this paper is to analyse the role of various types of agglomeration externalities on the survival rate of the newly established KIBS firms. In order to do so, we focused on the population of the newly established Swedish KIBS firms in 1997[2]. We track the life of these firms up to 2010 and performed the survival analysis, while controlling for extensive firm-level characteristics of firms and individual-level characteristics of the founders of the firms. We find that not all types of agglomeration externalities positively affect the survival of KIBS firms. It is only Jacobs' externalities (diversity) that matters. In particular, the higher the related variety of the region in which a KIBS firm is founded, the higher will be the survival chance of it.

Overall, this paper contributes to both the literature in survival analysis of firms (by incorporating the role of region) and also it contributes to the literature in geography of innovation and entrepreneurship (by having a disaggregate unit of analysis down into the level of individual entrepreneurs and also proving systematic evidence on the role of region for KIBS firms). In particular, the contribution of the paper is as follows: (1) we distinguish between the effects of various types of agglomeration externalities on survival of firms (2) our analysis traces the life and death of the population of firms in KIBS sectors over a long period of time (3) we are able to control for extensive firm-characteristics as well as individual-characteristics of the founders of the firms.

The rest of the paper is organised as follows. Section 2 discusses how the literature explains the survival of firms and develops hypotheses concerning the role of various agglomeration externalities on the survival. Data and estimation strategy is presented in Sections 3 and 4. Section 5 presents the results of the empirical analysis. The paper concludes in Section 6 and offers suggestions for future research.

Conceptual framework and hypotheses

Survival analysis of firms has been the topic of interest in several literatures. First, the literature in industrial dynamics has extensively investigated the factors determining the survival of firms. Special attention is paid to the age and size of the firms (Audretsch and Mahmood 1995), the entry time of firms in terms of pre- or post-dominant design (Suarez and Utterback 1995; Klepper 2002), the characteristics of the industry and the stages of the industry life cycle (Fritsch *et al.*, 2006) and pre-entry experience of firms (Klepper 2002). Second, the organisational ecology literature examines the economic performance and the firm's own threshold of performance to explain the decision to exit the business entirely (Baum 1996; Gimeno et al., 1997). Third, the Resource-Based View considers profit as "the criteria of natural selection," where highly performing firms survive and poor performance leads to its eventual surmise (Penrose 1952). Here the main determinant of survival of firm is the internal characteristics of firms.

Fourth, in the same vein as the resource-based view, human capital theory points to three factors that affect the survival chances of newly found firms: (1) individual characteristics of the founder, (2) attributes and strategies of the firms; and (3) environmental conditions (Brüderl et al., 1992 among others). And finally, in evolutionary analysis, survival analysis of entities (like firms) is recently recognised as the "second moment" of the field (Essletzbichler and Rigby 2010).

Overall, there are less survival analysis studies that incorporate the effect of "region" (i.e. the business environment that a firm is born and try to survive). This area needs further investigation, as new evidence suggest that regional-level characteristics are the most important determinants of survival of firms, even more important than the conventional firm-characteristics (Falck 2007). Among the few existing studies in this vein, most of them, however, consider agglomeration externalities as equivalent as the pure size of the regions (Fotopoulos and Louri 2000; Fritch *et al.*, 2006; Falck 2007). There are very few studies that incorporate various types of agglomeration externalities in the survival analysis of firms (Neffke *et al.*, (2012) is an exception), particularly KIBS firms.

It is indeed important to incorporate the effect of the region, in which the KIBS firm is established, on the survival rate of the KIBS firms (Stephan 2011). The logic is as follows: various knowledge inputs are crucial for KIBS firms. Such knowledge inputs can be generated internally within the firm and/or it can be brought into the firm externally through knowledge spillover mechanisms. Previous studies show that such external knowledge is heavily comes from the very same region that the firm is located (Andersson and Karlsson 2007). This is because (1) innovation is a complex process which entails exploration of tacit knowledge (Polanyi 1967) and (2) such exploration of tacit knowledge is shown to be facilitated by the face-to-face interaction of economic agents in the same region (Storper and Venables 2004; Rodríguez-Pose and Crescenzi 2008). The importance of regions as the source of knowledge creation was perhaps recognised explicitly when scholars observed that innovation activities are clustered geographically in space, even more than production activities (Feldman 1994; Audretsch and Feldman 2004). Such recognition subsequently shaped a subfield in economic geography, the so-called "geography of innovation" (Feldman and Kogler 2010). One particular steam of literature in this field addresses the effect of regional characteristics on innovation of regions (Acs and Varga 2005). A conventional way to reflect the effect of regions on the performance of (innovative) firms located on those regions is through various agglomeration externalities (Rosenthal and Strange 2004).[3] The seminal study by Glaeser et al (1992) spurs the literature concerning the effect of various agglomeration externalities on the performance of regions and firms. A common way to classify agglomeration externalities are: Urbanisation, Marshallian and Jacobs' externalities.

Urbanisation externalities

Urbanisation externalities capture the overall size of the region or city, irrespective of the particular sectoral composition of the region. It can be, in principle, associated to any economic activities within the region. Larger regions provide higher access to home markets compared with smaller regions. They also provide better infrastructures. These factors may benefit the innovative behaviour of local firms. The "incubator hypothesis" states that small firms will benefit from being located initially in a high-density central metropolitan area, because they can have aces to readily available raw material, labour and other services (Hoover and Vernon 1962; Leone and Struyk 1976). However, larger regions are also characterised by substantial congestion effects, which results in high factor costs and pollution. They may also intensify the competition between local plants, which can positively affect the exit rates of the local KIBS firms (Melitz and Ottaviano 2008). Therefore, at least in theory, (pure) large region can provide economies or diseconomies for the local KIBS firms. When it comes to empirical evidence, there is indeed more evidence pointing to the congestions effects of the larger regions rather than their benefits. Mata and Portugal (1994) used the number of firms in the industry as a measure for intensity of competition and find that more competition leads to lower survival rates of newly established Portuguese manufacturing firms. Fritsch et al (2006) found the negative competition effect on the survival of newly founded German firms. Raspe and van Oort (2011) found the same negative effect for Dutch firms. Similarly, in Sweden, Neffke et al (2012) find that urbanisation externalities indeed harm plants at all ages and Lööf and Nabavi (2014) recently find that survival of newly established genuine service firms is lower in denser areas.

Marshallian externalities

Marshallian externalities refer to the specialisation of a region in a narrow set of industrial sectors. The underlying assumption in Marshallian externalities is that knowledge is predominantly industry-specific (Marshall, 1920). Therefore firms may benefit from intra-industry knowledge spillovers and this can only be supported by concentration of a particular industry in a region. Specialised cities play a larger role in exploitation of existing technologies, mostly carried out by larger firms seeking for low cost environments for mass-production (Duranton and Puga 2001). Therefore, strong specialised cities may *not* act as a desirable incubator for small and KIBS firms, operating in service sectors, for at least two reasons: first, the activities of these firms are mainly exploration of new ideas and technologies and this is in contrast with what specialised cities usually offers, according to the "nursery cities" model (Duranton and Puga 2001). Second, small and KIBS firms in business service are heavily characterised by labour-intensive rather than capital-intensive and therefore may enjoy less from the low cost environments for mass-production that a specialised cities offer. Indeed recent empirical evidence shows that specialisation negatively affect the survival rates of newly established firms in service sector of the US economy (Acs et al., 2007). Furthermore, Weterings and Marsili (2012) find that while specialisation increases the survival rate of Dutch manufacturing firms, it has the negative effect on KIBS firms.

Jacobs' externalities

Jacobs' externalities arise when firms benefit from inter-industry knowledge spillover, which occurs in diversified cities or regions (Glaeser *et al.*, 1992). An essential part of Jacobs' externalities concerns the key role of the diversified cities in fostering knowledge spillover and innovation (Jacobs 1969; Anselin *et al.*, 1997)[4]. Duranton and Puga (2001) provide the micro-foundation for such argument in their so-called "nursery cities" model. They show that an advantage of diversified cities is that firms can imitate several different processes without costly relocation. In other words, firms benefit from Jacobs' externalities because industrial diversity in the region lowers the search costs for new technologies. Therefore, firms prefer to perform their exploration activities in these cities, which in turn lead to innovation.

Accordingly, at the level of industries, Henderson et al (1995) found in the US that high levels of past industrial diversity increase the probability that a city will attract a high-tech new industry. This indicates the role of Jacobs' externalities to create a favorable environment for new high-tech industries. Similarly, Neffke et al (2011) found that the effects of Jacobs' externalities are positive only for young industries in Sweden. Acs et al., (2007) find that city diversity is an important factor for survival of new firms. This is because in the diversified and denser cities, entrepreneurs (founder of KIBS firms) can have better access to tacit knowledge, obtain social ties, and build self-confidence (Sorenson and Audia 2000). More interestingly, at the level of plant, Neffke et al (2012) find that Jacobs' externalities only contribute to the survival of Swedish plants in the first 15 years of their existence. While this finding is obtained using all manufacturing firms (plants), it makes sense to expect the similar evidence, particularly for KIBS firms, since nursery cities model is indeed about KIBS firms. In addition, it is argued that to facilitate the inter-industry knowledge spillovers, there needs to be some sort of cognitive proximity or *complementarity* between these industries (within a region), thus leading to higher learning opportunities and innovation among the firms belonging to these industries (Nooteboom 2000). The *complementarity* aspect is particularly important for innovation, as far as knowledge is drawn from a variety of sectors, as in "recombinant innovation" (Weitzman 1998). In this line of query, an important contribution has been a distinction between related and unrelated variety of technological knowledge, where related variety measures the extent to which the knowledge and skill base of two industries overlap and unrelated variety measures the otherwise in a given region (Frenken *et al.*, 2007). It is shown that the higher related variety of the region, the higher the (incremental) innovation (Tavassoli and Carbonara 2014; Castaldi *et al.*, 2014). This is because a region with higher related variety can enjoy the higher learning opportunity, higher possibility of combining different ideas through inter-industry knowledge spill-overs (Ejermo 2005; Frenken *et al.*, 2007), which eventually lead to higher innovation for the region (Feldman 1994; Audretsch and Feldman 2004). This should be particularly the case for newly established firms. This is because (1) smaller firms are more dependent on their environment and externalities in general (Fotopoulos and Louri 2000; Henderson 2003) and (2) smaller firms usually have lower internal resources, such as human capital, in order to overcome the large cognitive distance existed in a region characterised by unrelated variety of knowledge (Cohen and Levinthal 1990). Therefore, these newly established firms should have an easier time to survive in a region characterised by higher related verity type of knowledge rather than the far-fetched unrelated variety type. Among the very few existing studies, Boschma and Wenting (2007) provide evidence that the survival of British automobile producers has been positively affected by presence of local related industries such as bicycle producers.

To sum up, according to the "nursery cities" model, diversified, specialised, and large cities may co-exist. The crucial point is that diversified cities act as nurseries (incubators) for new firms engaging in exploration of new technologies, leading to innovation.[5] On the other hand, specialised cities play a larger role in exploitation of existing technologies, mostly carried out by larger firms seeking for low cost environments for mass-production.[6] Finally, as for larger and denser cities, there are more evidence on the negative effect of it on survival, because of congestions effects and competition. Therefore, we expect positive influence of Jacobs' externalities on survival rate of newly established (young) KIBS firms. On the other hand localisation externalities and urbanisation economies may not have a positive effect. Above discussions lead to the following two hypotheses:

HP1: Jacobs' externalities, particularly related variety, have a positive and significant effect on the survival rate of KIBS firms.
HP2: Localisation externalities and Urbanisation externalities do not have a positive and significant effect on the survival rate of KIBS firms.

Data

The analysis in this paper employs longitudinal individual level dataset and Firm and Establishment Dynamics database provided by Statistics Sweden.[7] The coverage is fourteen years from 1997 to 2010. The individual level dataset provides information on various characteristics of each individual in Sweden, including the age, gender, income, education, and year of immigration into Sweden. It also classifies the occupation type of each individual to be either (1) a sailor, (2) a farmer, (3) an employed (in public or private sectors) or (4) an entrepreneur (founder of a business). We focus on those individuals who are classified as "entrepreneur" and also started their firms in Knowledge Intensive Business Service (KIBS) sectors at the beginning of the period (in 1997).[8] This way we identified the population of those newly established firms that have been engaging in some degree of novelty on what they are doing. In fact, KIBS is among the top innovative sectors in Sweden with over 60 per cent of the firms engage in some forms of innovative activities (Statistic Sweden 2014).[9] Similar sectoral choice is preferred in previous Swedish survival analysis dealing with geography (Wennberg and Lindqvist 2010). This allows us to go beyond the wide definition of entrepreneurship offered by Austrian school (Kirzner 1978) and focus on the Schumpeterian innovation-based definition of entrepreneurship (Schumpeter 1934). Austrian tradition considers any business start-ups that aim to fill the gap between supply and demand in the market (regardless of the level of novelty of the business idea) while Schumpeterian tradition considers those business start-ups that have some element of creative destruction, i.e. novelty and innovativeness (Landström 2007). Therefore, the population of those newly established firms in KIBS sectors is our proxy for the newly stablished KIBS firms.[10]

We proceed as follows: as noted above, we start by identifying those individuals who were entrepreneurs in 1997. Then we match these individuals to the firms that they found and limit ourself to those firms operating in KIBS sectors. Finally, we matched these firms with the regions that they have been founded. Matching individual founders to the firm provides information on various accounting data on the level of the firms. Further matching with the region that firms are located paves the way to analyse the role of geography on survival of these KIBS firms. There are in total 4,253 individuals (firms) that were identified in 1997. We track the formation of these new innovative start-ups and their shut-downs over time up to 2010 and complement this data with extensive information on individual, firm and regional characteristics. The breakdown of firms by its survival is listed in Table 6.1 below.

Similar to previous studies of agglomeration effects on firm survival, we distinguish between various types of exit: (1) exit by closure, (2) exit by merger with others and (3) exit by acquisition by others (Globerman *et al.*, 2005; Wennberg and Lindqvist 2010). While exit by closure (termination) is generally a negative outcome, merger or acquisition is not really a sign of failure. In contrast, divesting of their equity share can be seen as the success for entrepreneurs. Therefore, we have excluded those firms who disappeared in the dataset because of merger and acquisition.

In order to measure the effect of agglomeration externalities on the survival of firms, we need to define the geographical unit in which firms are located. Some studies used municipality as the unit of analysis. However, particularly in the southern part of Sweden, sometimes municipalities are only small parts of

Table 6.1 Survival rate of KIBS firms, 1997–2010

Year	*Firms*	*Exits*	*Survival rate*
1	4,253	1,275	70.02%
2	2,978	543	57.25%
3	2,435	423	47.31%
4	2,012	285	40.61%
5	1,727	254	34.63%
6	1,473	268	28.33%
7	1,205	149	24.83%
8	1,056	116	22.10%
9	940	107	19.59%
10	833	111	16.98%
11	722	77	15.17%
12	645	76	13.38%
13	569		

KIBS firms include those firms, which are classified as codes 72-74 by the two-digit NACE (rev. 1.1) classification.
Source: *Author's calculation.*

metropolitan areas. This may cause the spatial autocorrelation problem in our analysis because, for instance, a municipality that is located at a short distance from the center of Stockholm should surely experience some of the agglomeration externalities that are generated there.[11] We instead use functional regions (*Lokal Arbetsmarknad*) because it is indeed shown in Sweden that knowledge flows transcend across municipality borders but are bound within functional regions (Andersson and Karlsson 2007). This is because knowledge production and access are found to differ across functional regions (Karlsson and Johansson 2006). This way we can internalise many spatial effects within the chosen geographical unit of analysis. There were in total 112 functional regions in 1997, which is the year of establishment of the firms and hence beginning of our survival analysis. We use the whole population of Sweden to obtain information regarding total employment in each industry, which is defined by 5-digit NACE code, within each functional region.

Estimation strategy and measurement

We use survival analysis methodology in order to examine the exit of each firm in the study. This means that we study the time duration until an event, in this case the shut-down of firms, occurs. In particular, we utilise the semi-parametric Cox proportional hazards model. The main advantage for using this model in our study is that we do not need to specify the baseline model and hence avoid the potential arbitrary and incorrect model specification. The Cox model can still be estimated by the method of partial likelihood (Cox 1972). Moreover, we can express the results conveniently as hazard ratios, so that, for instance, a hazard ratio of 2 indicates a double shut-down rate with one unit increase in the covariate, holding others constant.

A formulation of the hazard function can be written as:

$$h(t) = h_0(t)\exp(\boldsymbol{X}_{AGGL}\boldsymbol{B}_{AGGL} + \boldsymbol{Z}_1\boldsymbol{B}_{Z1} + \boldsymbol{Z}_2\boldsymbol{Z}_{Z2})$$

where $h_0(t)$ is the unspecified baseline hazard that captures the direct impact of firm's age on the survival of firm. $\boldsymbol{X}_{AGGL}$ is the vector of agglomeration externalities that are assumed to affect the survival of the KIBS firms. $\boldsymbol{Z}_1$ and $\boldsymbol{Z}_2$ are vectors of control variables, corresponding to individual – and firm-level characteristics[12], and **B**'s are the coefficients to estimate. The vector $\boldsymbol{X}_{AGGL}$ includes four agglomeration externalities variables, i.e. *Urbanisation, RelatedVariety, UnrelatedVariety*, and *Specialisation.* The vector $\boldsymbol{Z}_1$ includes *Age, Gender, Immigrant* dummy, *HighlyEducated* dummy, marital status (*Married, Divorced, Widow*), dummies indicating whether there are any changes in marital (*StatusChange*) and education (*GetDegree*), the average growth of income (*IncomeGrowth)*, and the number of firms that the founder of the KIBS firm has been associated with throughout the period (*NrFirms*). Finally, the vector $\boldsymbol{Z}_2$ includes the average growth rate of firm size, turnover, and value added

(*FirmSizeGrowth, TurnoverGrowth, ValueAddedGrowth*), dummies indicating whether a firm is founded by a division from another firm *(Spinoff)*, or through a merger and acquisition *(Merger)*.

Because *Urbanisation* is a measure of total employment within the functional region minus firm's employment and the latter takes a value of 1 for all self-employed firms, which means that the value is far bigger than all of the other variables. We then decide to use the logged value instead[13].

RelatedVariety is constructed by calculating the weighted sum of entropy within the two-digit industry codes in a given functional region. The measure follows Frenken et al., (2007) and can be formulated as

$$RelatedVariety = \sum_{g=1}^{G} P_g H_g$$

$$H_g = \sum_{i \in S_g} \frac{p_i}{P_g} \log_2 \left(\frac{1}{p_i / P_g} \right)$$

where P_g is a two-digit share of employment in a region which is calculated by the sum of five-digit shares p_i, assuming all five-digit industries i fall under a same two-digit industry $S_g (P_g = \sum_{i \in s_g} p_i)$. The idea is that all five-digit industries that fall within the same two-digit codes are technologically related and thus share a cognitive proximity, while at the same time each five-digit industry is still differentiated. Hence, the measure must be weighted by the varieties that are present within the two-digit codes.

On the other hand, *UnrelatedVariety* captures the unrelated knowledge within the region and is simply the entropy at the two-digit level:

$$UnrelatedVariety = \sum_{g=1}^{G} P_g \log_2 \left(\frac{1}{p_g} \right)$$

The last variable Marshallian externalities or *Specialisation* is an index of location quotient that captures the spatial knowledge spillover of firms in the same industry and region (Antonietti and Cainelli 2011):

$$Specialisation = \frac{I_s / I}{L_s / L}$$

where l_s is total employment within two-digit industry code s, l is total employment in a functional region, L_s is an aggregate of sectoral employment, and L is an aggregate national employment. The detail descriptions of all variables are presented in Table 6.2.

Table 6.2 Variable description

Variables	*Description*	*Source*
Agglomeration economies variables		
Urbanisation	Total employment in the functional region where the firm is located	Author's own calculation, from Firms and Establishments Dynamic database, Statistics Sweden
RelatedVariety	Weighted sum of entropy of total employment within two-digit NACE codes of the functional region	"
UnrelatedVariety	Entropy of total employment within two-digit codes of the functional region	"
Specialisation	Location quotient index of the functional region	"
Individual characteristics variables		
Age	Age of entrepreneur at time of setting up the firm (1997)	Individual database, Statistics Sweden
Gender	A dummy indicating gender of the entrepreneur, taking value of 0 for male and 1 for female	"
Immigrant	A dummy taking a value of 1 if the entrepreneur has an immigrant background by moving into Sweden from their countries of birth and 0 otherwise	"
HighlyEducated	A dummy taking a value of 1 if the entrepreneur has at least three years of university education	"
Married	A dummy taking a value of 1 if the status of the entrepreneur is married in 1997.	"
Divorced	A dummy taking a value of 1 if the status of the entrepreneur is divorced in 1997.	"
Widow	A dummy taking a value of 1 if the status of the entrepreneur is a widow/widower in 1997.	"
StatusChange	A dummy taking a value of 1 if there is a change in marital status during 1997 and 2010.	"
GetDegree	A dummy taking a value of 1 if the entrepreneur obtains a tertiary education during 1997 and 2010.	"
IncomeGrowth	The average growth rate of entrepreneur's income during 1997 and 2010.	"
NrFirms	The total number of firms that the entrepreneur is associated with during 1997 and 2010.	"
Firm characteristics variables		
FirmSizeGrowth	The average growth rate of total number of employees during 1997 and 2010.	Firms and Establishments Dynamic database, Statistics Sweden
TurnoverGrowth	The average growth rate of turnover during 1997 and 2010.	"
ValueAddedGrowth	The average growth rate of firm's value-added during 1997 and 2010.	"
Spin-off	A dummy taking a value of 1 if the firm is founded from a division from another firm in 1997.	"
Merger	A dummy taking a value of 1 if the firm is founded from a merger and acquisition in 1997.	"

It is worthy to note that an important assumption of the Cox model is that the effect of covariates is the same for firms of all ages. We think this assumption holds in our study because there is no evidence showing that individual characteristics have varying effects on the survival of firms over various ages, i.e. being time-dependent covariates. For instance, the effect of being a male entrepreneur on the survival of the firm should be the same in the year of establishment and five years later. There is, however, evidence showing that agglomeration externalities are time-dependent covariates (Neffke *et al.*, 2012). Nevertheless, it is shown in Sweden that the agglomeration externalities variables start to show varying effects after 15 years of survival of the firms (plants) (Neffke *et al.*, 2012). Since our time span is fourteen years, our study should not experience the time-dependent agglomeration externalities. Therefore, the assumption of time-independent covariates should hold in our study.

The correlation table is displayed below in Table 6.3. Generally, there is no high correlation among all pairs of variables, so multicollinearity is less likely to cause a problem in the analysis. An exception is between some of agglomeration externalities variable, which is expected. As Duranton and Puga (2001) noted, diversified, specialised, and large cities co-exist. Nevertheless, we also run the survival analysis with each of these variables separately (but not shown here for brevity reason) and can conclude that multicollinearity does not cause a problem in our interpretation of the result. Descriptive statistics are displayed in Table 6.A1 in the Appendix.

Results

We first plot the non-parametric Kaplan-Meier survival estimate of firms during the fourteen years period of the study in Figure 6.1. About 30 per cent of firms survive up to 5 years after their establishment and about 20 per cent of firms survive up to 10 years.

Next, the coefficient estimates of semi-parametric Cox proportional hazard are presented below in Table 6.4. Model (1) includes our agglomeration

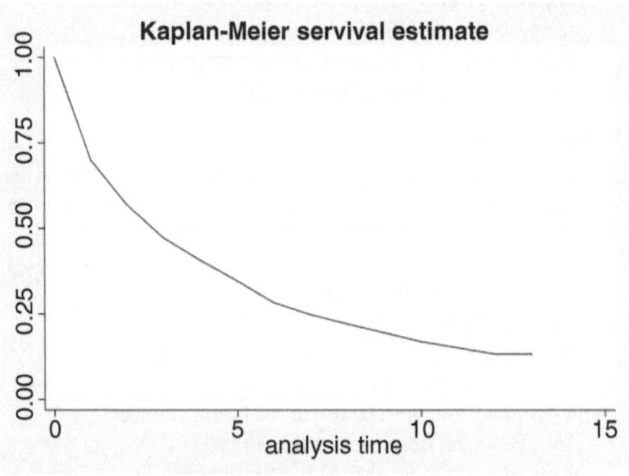

Figure 6.1 Kaplan-Meier survival estimate

Table 6.3 Correlation matrix

	URB	*RV*	*URV*	*SPC*	*AGE*	*GD*	*IM*	*HE*	*MAR*	*DIV*	*WID*	*SC*	*GD*	*IG*	*NF*	*FS*	*TG*	*VG*	*SP*	*ME*
Urbanisation	1																			
RelatedVariety	–0,002	1																		
UnrelatedVariety	0.648***	–0.219***	1																	
Specialisation	–0.149***	0.214***	–0.420***	1																
Age	–0.086***	0.156***	–0.062***	0.048**	1															
Gender	0.049**	0.131***	0,026	–0,021	–0.120***	1														
Immigrant	0.054**	0,022	0.038*	–0,011	0,006	0,021	1													
HigherEducation	0.098***	–0.083***	0.097***	–0,021	0,016	0,015	–0,007	1												
Married	–0.107***	0.054**	–0.080***	0,022	0.373***	–0,023	0.042*	0,009	1											
Divorced	0,01	0.079***	0,007	–0,019	0.139***	0.038*	0,001	–0,027	–0.437***	1										
Widow	–0,002	–0,006	0,023	0.044**	0.125***	0,022	–0,013	–0,01	–0.117***	–0.037*	1									
StatusChange	–0,002	–0,013	–0,016	–0,002	–0.038*	0.062***	0,005	–0,015	0.236***	–0.134***	–0.039*	1								
GetDegree	–0,012	–0.046**	–0,011	–0,013	–0.200***	0,03	–0,013	–0.168***	–0.082***	–0.044**	–0,016	0,011	1							
IncomeGrowth	–0,01	–0.049**	–0,02	–0,023	–0,027	–0,013	–0,005	0,02	–0,009	–0,015	0,008	0,007	–0,003	1						
NrFirms	0.034*	–0.072***	0,019	–0,024	–0.328***	0.065***	–0,004	0,025	–0.095***	–0,017	–0.047**	0.058***	0.111***	0.038*	1					
FirmSizeGrowth	–0,029	–0,027	–0,017	–0,011	–0.084***	–0.034*	0,014	–0.045**	0	–0,012	–0,004	0,01	–0,017	–0,004	0.072***	1				
TurnoverGrowth	–0.050**	–0,011	–0.065***	0,019	–0.050**	–0,011	0.050**	–0,013	0,001	–0,021	–0,002	0,013	0,009	0,017	0,011	0.062***	1			
ValueAddedG	0,019	–0.037*	0,014	–0.036*	–0.052**	0,006	0,024	–0,018	–0,015	–0,014	0,002	–0,014	0.049**	0.062***	0.033*	0.047**	0.156***	1		
Spin–off	0,001	–0,023	–0,009	–0,002	–0,002	–0,023	–0,018	–0.039*	0,008	0,001	–0,013	0,011	–0,022	–0,017	0,02	–0,009	–0,006	–0,005	1	
Merger	–0.037*	0,013	–0,018	0,011	–0,02	–0,029	–0,009	–0.039*	–0,013	0,025	–0,011	0,015	0,013	–0,014	0.046**	–0,028	–0,003	–0,001	–0,015	1

* $p<0.05$, ** $p<0.01$, *** $p< 0.001$

Table 6.4 Cox proporational hazard coefficient estimates for KIBS firms, 1997–2010

VARIABLES	*(1)*	*(2)*	*(3)*
Urbanisation	0.014	0.006	0.007
	(0.010)	(0.010)	(0.011)
RelatedVariety	–0.087***	–0.059***	–0.067***
	(0.016)	(0.015)	(0.017)
UnrelatedVariety	–0.109	–0.081	–0.086
	(0.087)	(0.083)	(0.096)
Specialisation	0.035	0.053	0.075
	(0.092)	(0.088)	(0.104)
Age		0.007***	0.011***
		(0.002)	(0.002)
Gender		–0.144***	–0.146***
		(0.031)	(0.035)
Immigrant		0.044	0.022
		(0.038)	(0.044)
HigherEducation		–0.093***	–0.107***
		(0.028)	(0.032)
Married		–0.161***	–0.190***
		(0.037)	(0.041)
Divorced		–0.196***	–0.240***
		(0.050)	(0.056)
Widow		0.096	0.037
		(0.150)	(0.171)
StatusChange		0.013	0.015
		(0.040)	(0.045)
GetDegree		0.277***	0.304***
		(0.079)	(0.093)
IncomeGrowth		0.000	–0.001
		(0.001)	(0.001)
NrFirms		0.282***	0.316***
		(0.009)	(0.011)
FirmSizeGrowth			0.036
			(0.088)
TurnoverGrowth			0.008***
			(0.002)
ValueAddedGrowth			0.006**
			(0.003)
Spin-off			0.092
			(0.111)
Merger			–0.178
			(0.155)
Observations	4,253	4,253	3,624
Loglikelihood	–28642	–28213	–22899
Chi-squared	30.91	1300	1137

The table reports the coefficient estimates. See text for the interpretation. Robust standard errors are reported in parentheses. Significance level: * $p < 0.05$, ** $p < 0.01$, *** $p < 0.001$.

externalities variables. Model (2) controls for individual-characteristics of the founders of the firms. Finally, model (3) is the full model, which controls for both individual and firm-level characteristics.

The significant coefficients of a discrete variable can be interpreted that the hazard of having the status in one group (e.g. female) is relatively different from the hazard of another group (male). The positive numbers imply a higher rate of firm exit compared to the referenced group and vice versa. The same applies for an increase in continuous variables, i.e. it suggests a positive increase in firm exits.[14] To put it differently in terms of firm's survival, we can say that a negative coefficient means that the variables affects *positively on the survival of the firms* and vice versa.

Results in model (1) shows that only Jacobs's externality (diversity) in the region favorably affects the survival of the firms, i.e. having negative coefficients. When it comes to the composition of the diversity in the region, it is indeed the related variety of knowledge (*RelatedVariety*) that affects the survival, rather than unrelated variety of knowledge (*UnrelatedVariety*). This means small KIBS firms can benefit from inter-industry knowledge spillover, but that spillover comes from other related industries and not from industries that is too far in terms of cognitive distance. The positive coefficient for urbanisation externalities (*Urbanisation*) suggests that it negatively affects the survival of firms. This is in line with previous evidence that found the negative effect of urbanisation (large and dense areas) on survival of newly established firms (Mata and Portugal 1994; Fritch *et al.*, 2006; Raspe and van Oort 2011; Neffke *et al* 2012). This is because being merely a large region can harm the survival of the firm through congestion costs and fierce competition between newly arrived and incumbent firms. However, unlike previous studies, the effect of urbanisation is not significant in our study. Marshallian externalities (*Specilisaiton*) also show a negative influence on firm survival. This result can be explained by "nursery cities" model: specialised regions may not act as a desirable incubator for small and KIBS firms, operating in business service sectors. This is because the activities of these firms are mainly exploration of new ideas and technologies and this is in contrast with what specialised cities usually offers, i.e. exploitation of existing technologies, mostly carried out by larger firms seeking for low-cost environment for a mass production base. This is in line with results of previous studies (Acs *et al.*, 2007; Weterings and Marsili 2012). However, unlike previous studies, the effect of Marshallian externalities is not significant in our study.

In model (2), we control for individual characteristics and we find that the results concerning the effects of agglomeration externalities in model (1) still hold. Considering our control variables in this model, we observe several interesting results. First, older entrepreneurs (*Age*) are associated with greater exits, albeit a small effect (from Table 6.A2, one additional year equals a 1.1 per cent increase in chance of exit, ceteris paribus). One explanation might be that older entrepreneurs are more risk averse and do not keep up with the changing business environment so their previous experiences are not applicable and result in earlier exit. Second, being a female entrepreneur (*Gender*) is associated with a lower chance of exit (13 per cent). Similarly, Giannetti and Simonov (2004) analyse entrepreneurship in Sweden and find that males tend to abandon entrepreneurial activities sooner. The results in other studies are inconclusive.[15] Third, having at least three years of university education (*HigherEducation*) equals to a relatively less exits by 9 per cent. Although a crude measure, higher education is usually perceived as a main source of human capital. Previous studies also have shown that higher education is often associated with successful start-ups (Evans and Leighton 1990; Holtz-Eakin et al., 2000). However, if the entrepreneurs set aside their time to study in order to obtain a university degree after founding the firm (*GetDegree*), it has an adverse effect, resulting to a 32 per cent increase in firm exit, probably due to the amount of workload of study needed. Also, if the founder has been involved with more than one firm, either establishing a new firm or working as an employee (*NrFirms*), the established firm in the beginning will also likely be closed down, possibly from the lack of commitment.

Finally, in our full model (3), we include both individual- and firm-level characteristics. The results concerning the effects of agglomeration externalities in previous models still hold. Turning to firm-level control variables in this model, turnover and value added growth have negative effect on survival of firms. Nevertheless, the magnitude of the effect is almost zero. This result reinforces the notion of Falck (2007) that regional-level characteristics are actually more important to firm survival than firm characteristics. All in all, the results in Table 6.4 show that both hypotheses of this paper are confirmed.

Conclusion

This paper aimed to analyse the effects of various types of agglomeration externalities on survival of KIBS firms. In order to do so, we focus on the population of Swedish firms that are founded by entrepreneurs in Knowledge Intensive Business Services (KIBS) sectors in 1997. We trace the life of these firms up to 2010. Using the semi-parametric Cox proportional hazard model, we find the followings: First, not all agglomeration externalities enhance the survival chance of KIBS firms. Second, it is only Jacobs' externalities (diversity) in the region that can positively and significantly enhance the survival rate of KIBS firms. This means being in a diverse area can benefit new firms by offering a greater exposure of new ideas and management practices that may complement firm's knowledge to innovate and help them survive. Third, not all types of diversity in the region can enhance the survival of KIBS firms. It is the related variety of knowledge in the region that matters. This is because young KIBS firms usually have lower internal resources, such as human capital, in order to overcome the large cognitive distance existed in a region characterised by unrelated variety of knowledge (Cohen and Levinthal, 1990). Therefore, these firms should have a greater chance to survive in a region characterised by higher related variety type of knowledge rather than the far-fetched unrelated variety type. Fourth, controlling for extensive sets of firm-characteristics as well as individual-characteristics of the entrepreneur, the same conclusions still hold. Therefore, for a newly – established KIBS firm, not only it matters who you are, but also *where* you are (where you establish your company).

There are several suggestions for future studies. First, we construct our agglomeration externalities variables based on predefined functional regions. An extension could be to apply a recently developed methodology using spatial micro-data to build the agglomeration externalities variable (Dubé and Brunelle 2014). Second, we construct our agglomeration externalities variables (particularly Jacobs') based on sectoral composition of the regions. An alternative way is to construct them based on composition of the skill-based or occupation of the individual entrepreneurs (founder of business) in the region (Backman and Kohlhase 2013). Third, we focus on the role of intra-regional externalities in this paper for the survival of firms, while leaving out inter-regional knowledge linkages. There are indeed good reasons to believe that newly established firms are highly dependent on their local environment and they are not large enough to establish external links (Fotopoulos and Louri 2000; Henderson 2003). Nevertheless, there are recent counter-argument pointing that knowledge intensive firms may get their knowledge through inter-regional linkages (Ponds *et al.*, 2010). Investigation of the effect of this intra- vs. inter-regional linkages on survival of KIBS firms can be an interesting topic of future studies.

Appendix 6A

Table 6A.1 Descriptive statistics

	Obs.	*Mean*	*Std.Dev.*	*Min.*	*Max.*
Urbanisation	4,253	9.788	1.849	2.944	11.612
RelatedVariety	4,253	2.365	0.964	0.000	3.490
UnrelatedVariety	4,253	0.999	0.243	–0.029	1.363
Specialisation	4,253	1.014	0.191	0.021	5.680
Age	4,253	44.115	12.332	16	83
Gender	4,253	1.311	-	0	1
Immigrant	4,253	0.167	-	0	1
HigherEducation	4,253	0.514	-	0	1
Married	4,253	0.559	-	0	1
Divorced	4,253	0.121	-	0	1
Widow	4,253	0.010	-	0	1
StatusChange	4,253	0.136	-	0	1
GetDegree	4,253	0.034	-	0	1
IncomeGrowth	4,253	1.207	9.738	–1.000	373.000
NrFirms	4,253	3.011	2.009	1.000	14.000
FirmSizeGrowth	4,251	0.036	0.206	–0.800	5.000
TurnoverGrowth	3,625	0.545	5.078	–1.000	249.500
ValueAddedG	3,644	0.455	4.478	–78.000	85.500
Spin-off	4,253	0.018	-	0	1
Merger	4,253	0.017	-	0	1

Table 6A.2 Hazard ratios estimation results of KIBS firms, 1997-2010

VARIABLES	*(1)*	*(2)*	*(3)*
Urbanisation	1.014	1.006	1.007
	(0.011)	(0.010)	(0.011)
RelatedVariety	0.916***	0.943***	0.935***
	(0.015)	(0.014)	(0.016)
UnrelatedVariety	0.897	0.922	0.918
	(0.078)	(0.077)	(0.088)
Specialisation	1.035	1.054	1.078
	(0.095)	(0.093)	(0.112)
Age		1.007***	1.011***
		(0.002)	(0.002)
Gender		0.866***	0.864***
		(0.027)	(0.030)
Immigrant		1.045	1.022
		(0.040)	(0.045)
HigherEducation		0.911***	0.898***
		(0.026)	(0.028)
Married		0.851***	0.827***
		(0.031)	(0.034)
Divorced		0.822***	0.787***
		(0.041)	(0.044)
Widow		1.101	1.038
		(0.165)	(0.178)
StatusChange		1.013	1.015
		(0.040)	(0.045)
GetDegree		1.319***	1.356***
		(0.105)	(0.126)
IncomeGrowth		1.000	0.999
		(0.001)	(0.001)
NrFirms		1.325***	1.372***
		(0.012)	(0.015)
FirmSizeGrowth			1.037
			(0.092)
TurnoverGrowth			1.008***
			(0.002)
ValueAddedGrowth			1.006**
			(0.003)
Spin-off			1.096
			(0.122)
Merger			0.837
			(0.130)
Observations	4,253	4,253	3,624
Loglikelihood	–28642	–28213	–22899
Chi-squared	30.91	1300	1137

The table reports the exponential of the coefficients, which is the hazard ratios. A hazard ratio >1 decreases the likelihood of survival. A hazard ratio <1 increases the likelihood of survival. Robust standard errors are reported in parentheses. Significance level: * $p < 0.05$, ** $p < 0.01$, *** $p < 0.001$.

Notes

1 Neffke *et al.*, (2012) does not study the very small firms, although majority of firms in the economy are actually very small. We precisely focused on such sample size in this paper.
2 1997 is the earliest year that we have our full dataset ready in three levels of individual, firm, and region. Therefore, choosing 1997 provide us with the longest survival analysis possible. Changing the starting point of the survival analysis from 1997 to later years reveal similar results, albeit shorter time span.
3 Agglomeration externalities are costs or benefits that firms gain by being located in geographical proximity of other economic agents.
4 As Jane Jacob argued, many innovations are the result of "adding new work to old ones".
5 This is in line with entrepreneurial regime (Winter, 1984).
6 This is in line with routine regime (Winter, 1984).
7 Företagens och Arbetställenas Dynamik (FAD) in Swedish.
8 KIBS sectors are classified as two-digit European NACE codes 72–74.
9 These innovating activities are: introducing product, process, marketing, and organization innovations as well as engaging in ongoing/abandoned product or process innovation.
10 Although there are other datasets to be used for identifying anentrepreneurial firm, such as Community Innovation Survey (CIS) or patent data, none of these dataset offers a possibility of tracing *newly established* entrepreneurial firms over an extended period of time. Therefore we cannot use them in our survival analysis.
11 Those studies which used municipality as the geographic unit of analysis constructed their agglomeration externality indicators based on a number of geographical potentials (Neffke *et al.*, 2012).
12 The vectors of control variables can be attributed to internal factors and resource-based view (Penrose 1952).
13 We also used population density as an alternative measure of Urbanization externalities. Main results are very similar.
14 We also display the results as hazard ratios in Table 6.A2 in the Appendix. This aids for a convenient interpretation of the magnitude of each covariate. For example, the hazard ratio for *Age* in column 5 of 1.007 can be interpreted as a 1.007 – 1 = 0.7 per cent increase in firm exit for one additional year of enterpreneur's age. Whereas the hazard ratio of 0.866 for *Gender* means 1 – 0.866 = 13.4 per cent decrease in firm exit when the entrepreneur is female.
15 Various studies find that male entrepreneurs lead firms that stay in business longer than female counterpart (Holmes and Schmitz Jr 1996; Taylor 1999, among others). On the other hand, Watson (2012) finds that there is little difference in networking between male- and female-controlled Australian SMEs after controlling for size, education, industry and size.

References

Acs, Z. J., Armington, C., and Zhang, T. (2007). The determinants of new-firm survival across regional economies: the role of human capital stock and knowledge spillover. *Papers in Regional Science*, *86*(3), 367–91.

Acs, Z. J., and Varga, A. (2005). Entrepreneurship, agglomeration and technological change. *Small Business Economics*, *24*(3), 323–34.

Andersson, M., and Karlsson, C. (2007). Knowledge in regional economic growth-the role of knowledge accessibility. *Industry & Innovation*, *14*(2), 129–49.

Anselin, L., Varga, A., and Acs, Z. (1997). Local geographic spillovers between university research and high technology innovations. *Journal of Urban Economics*, *42*, 422–48.

Antonietti, R., and Cainelli, G. (2011). The role of spatial agglomeration in a structural model of innovation, productivity and export. *Annals of Regional Science, 46*, 577–600.

Audretsch, D., and Feldman, M. (2004). Knowledge spillovers and the geography of innovation, in J. V. Henderson, and J. Thisse, *Handbook of Urban and Regional Economics: Cities and Geography, Volume 4* (pp. 2713–39). Amsterdam: North Holland Publishing.

Audretsch, D. B., and Mahmood, T. (1995). New firm survival: new results using a hazard function. *The Review of Economics and Statistics*, 97–103.

Backman, M., and Kohlhase, J. (2013). The influence of diversity on the formation, survival and growth of new firms. *Working Paper Series in Economics and Institutions of Innovation, No 337.*

Baum, J. A. C. (1996). Organizational ecology. In S. R. Clegg, and C. Hardy, *Studying organization: theory and method,* 71–108.

Boschma, R. A., and Wenting, R. (2007). The spatial evolution of the British automobile industry: Does location matter? *Industrial and Corporate Change, 16*, 213–38.

Brown, T., and Kirchhoff, B. (1997). Resource acquisition self-efficacy: measuring entrepreneur's growth ambitions. *Frontiers in Entrepreneurship Research*, 59–60.

Brüderl, J., Preisendörfer, P., and Ziegler, R. (1992). Survival chances of newly founded business organizations. *American Sociological Review*, 227–42.

Castaldi, C., Frenken, K., and Los, B. (2014). Related variety, unrelated variety and technological breakthroughs: an analysis of US state-level patenting. *Regional Studies*. doi:10.1080/00343404.2014.940305.

Cohen, W., and Levinthal, D. (1990). Absorptive capacity: a new perspective on learning and innovation. *Administrative Science Quarterly, 35*(1), 128–52.

Cox, D. R. (1972). Regression models and life tables (with discussion). *Journal of the Royal Statistical.*

Dubé, J., and Brunelle, C. (2014). Dots to dots: a general methodology to build local indicators using spatial micro-data. *The Annals of Regional Science, 53*(1), 245–72.

Duranton, G., and Puga, D. (2001). Nursery cities: urban diversity, process innovation, and the life cycle of products. *American Economic Review*, 1454–77.

Ejermo, O. (2005). Technological diversity and Jacobs' externality hypothesis revisited. *Growth and Change, 36*(2), 167–95.

Essletzbichler, J., and Rigby, D. L. (2010). Generalized Darwinism and evolutionary economic geography. In R. Boschma, and R. Martin, eds *The handbook of evolutionary economic geography* (pp. 43–61). Cheltenham: Edward Elgar.

Evans, D. S., and Leighton, L. S. (1990). Small business formation by unemployed and employed workers. *Small Business Economics, 2*(4), 319–30.

Falck, O. (2007). Survival chances of new businesses: do regional conditions matter? *Applied Economics, 39*(16), 2039–48.

Feldman, M. P. (1994). *The geography of innovation.* Dordrecht: Kluwer Academic.

Feldman, M. P., and Kogler, D. F. (2010). Stylized facts in the geography of innovation. In B. Hall, and N. Rosenberg, *Handbook of economics of technical change.* Oxford: Elsevier.

Fotopoulos, G., and Louri, H. (2000). Location and survival of new entry. *Small Business Economics, 14*(4), 311–21.

Fritsch, M., Brixy, U., and Falck, O. (2006). The effect of industry, region, and time on new business survival- a multi-dimensional analysis. *Review of Industrial Organization, 28*(3), 285–306.

Giannetti, M., and Simonov, A. (2004). On the determinants of entrepreneurial activity: Social norms, economic environment and individual characteristics. *Swedish Economic Policy Review, 11*(2), 269–313.

Gimeno, J., Folta, T. B., Cooper, A. C., and Woo, C. Y. (1997). Survival of the fittest? Entrepreneurial human capital and the persistence of underperforming firms. *Administrative Science Quarterly*, 750–83.

Glaeser, E., Kallal, H., Scheinkman, J., and Shleifer, A. (1992). Growth of cities. *Journal of Political Economy*, *100*, 1126–52.

Globerman, S., Shapiro, D., and Vining, A. (2005). Clusters and intercluster spillovers: their influence on the growth and survival of Canadian information technology firms. *Industrial and Corporate Change*, *14*, 27–60.

Henderson, J. V., Kuncoro, A., and Turner, M. (1995). Industrial development in cities. *Journal of Political Economy*, *103*, 1067–90.

Henderson, V. J. (2003). Marshall's scale economics. *Journal of Urban Economies*, *53*, 1–28.

Holmes, T. J., and Schmitz Jr, J. A. (1996). Managerial tenure, business age, and small business turnover. *Journal of Labor Economics*, *14*(1), 79–99.

Holtz-Eakin, D., Rosen, H. S., and Weathers, R. (2000). Horatio Alger meets the mobility tables. *Small Business Economics*, *14*(4), 243–74.

Hoover, E. M., and Vernon, R. (1962). *Anatomy of a metropolis*. New York: Anchor Books.

Jacobs, J. (1969). *The economy of cities*. New York: Random House, Inc.

Karlsson, C., and Johansson, B. (2006). Dynamic and entrepreneurship in a knowledge-based economy. In C. Karlsson, B. Johansson, and R. Stough, *Entrepreneurship and dynamics in the knowledge economy* (pp. 12–46). New York: Routledge.

Kirzner, I. M. (1978). *Competition and entrepreneurship*. Chicago: University of Chicago press.

Klepper, S. (2002). Firm survival and the evolution of oligopoly. *RAND Journal of Economics*, 37–61.

Landström, H. (2007). *Pioneers in entrepreneurship and small business research*, Vol. 8. New York: Springer.

Leone, R. A., and Struyk, R. (1976). The incubator hypothesis: evidence from five SMSAs. *Urban Studies*, *13*(3), 325–31.

Lööf, H., and Nabavi, P. (2014). Survival, productivity and growth of new ventures across locations. *Small Business Economics*, *43*(2), 477–91.

Marshall, A. (1920). *The principles of economics*. London: MacMillan.

Mata, J., and Portugal, P. (1994). Life duration of new firms. *The Journal of Industrial Economics*, 227–45.

Melitz, M. J., and Ottaviano, G. I. (2008). Market size, trade, and productivity. *The review of economic studies*, *75*(1), 295–316.

Neffke, F. M., Henning, M., and Boschma, R. (2012). The impact of aging and technological relatedness on agglomeration externalities: a survival analysis. *Journal of Economic Geography*, *12*(2), 485–517.

Neffke, F., Svensson Henning, M., Boschma, R., Lundquist, K., and Olander, L. (2011). The dynamics of agglomeration externalities along the life cycle of industries. *Regional Studies*, *45*(1), 49–65.

Nooteboom, B. (2000). *Learning and innovation in organizations and economies*. Oxford: Oxford University Press.

Penrose, E. (1956). *The theory of the growth of the firm*. Oxford: Oxford University Press.

Polanyi, M. (1967). *The tacit dimension*. London: Routledge and Kegan Paul.

Ponds, R., van Oort, F., and Frenken, K. (2010). Innovation, spillovers and university–industry collaboration: an extended knowledge production function approach. *Journal of Economic Geography*, *10*, 231–55.

Raspe, O., and van Oort, F. (2011). Growth of new firms and spatially bounded knowledge externalities. *The Annals of Regional Science*, *46*(3), 495–518.

Rodríguez-Pose, A., and Crescenzi, R. (2008). R&D, spillovers, innovation systems and the genesis of regional growth in Europe. *Regional Studies*, *42*(1), 51–67.

Rosenthal, S. S., and Strange, W. C. (2004). Evidence on the nature and sources of agglomeration economies. *Handbook of Regional and Urban Economics*, *4*, 2119–71.

Schumpeter, J. A. (1934). *The theory of economic development*. Cambridge, MA: Harvard University Press (1st edn. 1911).

Sorenson, O., and Audia, P. G. (2000). The social structure of entrepreneurial activity: geographic concentration of footwear production in the United States 1940–89. *American Journal of Sociology*, *106*(2), 424–62.

Statistic Sweden. (2014). *Innovation activity in Swedish enterprises 2010–12*. Stockholm: Statistiska centralbyrån.

Stephan, A. (2011). Locational conditions and firm performance: introduction to the special issue. *The Annals of Regional Science*, *46*(3), 487–94.

Storper, M., and Venables, A. (2004). Buzz: face-to-face contact and the urban economy. *Journal of Economic Geography*, *4*, 351–70.

Suarez, F. F., and Utterback, J. M. (1995). Dominant designs and the survival of firms. *Strategic Management Journal*, *16*(6), 415–30.

Tavassoli, S., and Carbonara, N. (2014). The role of knowledge variety and intensity for regional innovative capability. *Small Business Economics*, *43*(2), 493–509.

Taylor, M. P. (1999). Survival of the fittest? An analysis of self–employment duration in Britain. *The Economic Journal*, *109*(454), 140–55.

Watson, J. (2012). Networking: gender differences and the association with firm performance. *International Small Business Journal*, *30*(5), 536–58.

Weitzman, M. L. (1998). Recombinant growth. *Quarterly Journal of Economics*, 331–60.

Wennberg, K., and Lindqvist, G. (2010). The effect of clusters on the survival and performance of new firms. *Small Business Economics*, *34*(3), 221–41.

Weterings, A., and Marsili, O. (2012). Spatial concentration of industries and new firm exits: Does this relationship differ between exits by closure and by M&A? *Regional Studies*, 1–15.

Winter, S. G. (1984). Schumpeterian competition in alternative technological regime. *Journal of Economic Behavior and Organization*, *5*, 287–320.

Part III

KIBS and their contribution to regional competitiveness and economic development

7 Entrepreneurship and KIBS

Key factors in the growth of territories

Joaquín Alcazar, Norat Roig-Tierno, Alicia Mas-Tur and Belén Ribeiro-Navarrete

Introduction

There are some studies that have shown a positive relationship between entrepreneurial activity and the growth of territories such as cities, regions (Reynolds *et al.*, 2005, 2002; Carree, 2001; Reynolds, 1999; Fritsch, 1997), and countries (Carree *et al.*, 2000; Carree and Thurik, 1999; Thurik, 1999). In addition to presenting theoretical arguments that entrepreneurial activity positively affects the economic growth of territories, researchers have empirically tested and corroborated this theory in several studies.

One of the first authors to highlight the effect of new firm creation and expansion on economic growth and employment was David Audretsch in the eighties. Drawing upon a detailed analysis of all US firms for the period 1969–76, Audretsch discovered that small, recently formed firms were responsible for 81 per cent of net job creation. Studies by the GEM (Reynolds *et al.*, 2000) have confirmed this positive relationship between business creation and growth.

Subsequent research (Audretsch *et al.*, 2007, 2002; Hölzl, 2006; Storey, 2005) has confirmed entrepreneurship's contribution to higher levels of economic development. Studies have shown that entrepreneurship's contribution depends not only on the number of new firms, but also on new firms' quality and growth potential.

Thus, business initiatives in various forms (newly created firms, young innovative companies, university spin-offs, technology-based firms, etc.) act as vehicles that not only allow knowledge created within an organisation to be commercialised in other firms, but also allow knowledge spillovers to arise in the form of competencies, products, and innovations (Audretsch, 2004; Lindholm and Dahlstrand, 1997). In this sense, business initiatives are becoming a key component of innovative activity.

Nevertheless, not all newly created firms affect economic growth equally. Scholars have typically distinguished between firms with lower growth expectancy (Cooper *et al.*, 2004) and firms with greater growth potential (Ács and Naudé, 2013; Acs, 2008). Thus, to maximise the effect of public spending on economic growth (Shane, 2009), entrepreneurship policy should focus on firms whose features (technology-based firms, innovative companies, etc.) afford them greater growth potential (Heirman and Clarysse, 2004).

Scholars have identified a series of factors that may indicate high growth (Pickernell *et al.*, 2013). These factors include the existence of explicit growth targets (Dutta and Thornhill, 2008; Wiklund and Shepherd, 2003), aspects such as capability, necessity, and opportunity (Davidsson, 1991), and characteristics of the entrepreneur like education and experience (Cooper *et al.*, 1994), technology, degree of innovation (Allen and Stearns, 2004), and business background (Wiklund and Shepherd, 2003).

In particular, newly created firms face financing barriers as well as other types of barriers (e.g. limited internal resources, imperfect information) (Mas-Tur and Ribeiro Soriano, 2014). Public financial aids must therefore be supplemented by indirect support such as advisory or consulting services that encourage connections and knowledge transfer services (Lundvall *et al.*, 2000).

Knowledge Intensive Business Services (KIBS) act as sources of external knowledge, while contributing to innovation and hence the growth of territories. Miles (2005) defined KIBS as services targeting private firms and public institutions that undertake complex operations to resolve problems and that view the role of human capital as essential. KIBS perform a range of tasks that are essential for innovation by integrating knowledge intensive activities into other firms' production processes (García-Quevedo *et al.*, 2012).

In this research, we will observe how KIBS, as tools that enhance entrepreneurship, can encourage the growth of territories. To do so, we thoroughly reviewed the literature on both entrepreneurship and KIBS. We then drew conclusions and made recommendations that can be implemented by economic policy makers.

Entrepreneurship and Knowledge Intensive Business Services

Entrepreneurship

The number of research articles that examine entrepreneurship and its influence on economic development is growing (Lundström and Stevenson, 2005). In economic theory, entrepreneurship has been modelled as an occupation lying somewhere between self–employment and paid work (Murphy *et al.*, 1989; Evans and Jovanovic 1989). People try to become entrepreneurs if earnings and non-monetary benefits of self-employment exceed the salary plus perks of employment. Thus, entrepreneurial spirit is often synonymous with self-employment. In many cases, self-employment is not a choice but rather a necessity, so scholars sometimes distinguish between necessity and opportunity entrepreneurs –for example, in the Global Entrepreneurship Monitor (GEM, see Reynolds *et al.*, 2005).

In any case, entrepreneurship encompasses personal characteristics that fall into either of two categories: (1) a system of values (beliefs about the outcomes of entrepreneurship, wealth creation, self-interest, self-realisation through business, etc.) or (2) a set of skills and aptitudes that serve as the instrument to achieve such a system of values.

One of the longest standing definitions of the entrepreneur is that of Schumpeter (1950, 1961), who defined the entrepreneur as the coordinator of production and the agent of change. As such, the 'Schumpeterian' entrepreneur is somehow above the rest: he or she is an innovator. Researchers who share this view of entrepreneurship believe that entrepreneurship is of little importance during the early stages of economic development. Conversely, they believe that the entrepreneurship is especially important during the later stages of development, when economic growth is driven by knowledge and skill. During the early stages of development, entrepreneurship may play a less prominent role because growth is largely driven by the factor of accumulation (Ács and Naudé, 2013).

Gries and Naudé (2011) adopt a definition of synthesis that combines different points of view and behaviour. This definition has two features. First, it reflects changes in academic thinking on entrepreneurship depicting entrepreneurship as a process whereby people exploit market opportunities through the creation and growth of new enterprises. Second, this definition emphasises the value of the process and opportunities of entrepreneurship. Other definitions that emphasise behaviour assign entrepreneurial spirit a key role in developing countries. Kirzner (1973) describes the entrepreneur as someone who enables adaptation to change by detecting opportunities.

There are other definitions based on entrepreneurial behaviour. These definitions emphasise the dimension of risk-taking entrepreneurship. Kanbur (1979, p.773) described the entrepreneur as someone who 'manages the production function' by paying wages, which are more secure than profits, and by embracing risks and uncertainties in production.

Whereas early studies restricted entrepreneurship's scope to innovation and business creation, recent studies depict entrepreneurship as a social phenomenon that reflects the broader institutional features of society. Entrepreneurship cares not only for the success of the business (profits), but also for subjective and non-economic well-being. Thus, the entrepreneurial spirit is a catalyst for structural change and the institutional development of territories. Entrepreneurship as a social phenomenon broadly refers to new economic activity that leads to changes in the market (Davidsson, 2004).

Policy makers in the European Union have sought to develop a framework to promote entrepreneurship across several countries within a single program, namely the program of entrepreneurship and innovation (EIP). This program requires systematic political coordination at and across all levels. The EIP is an entrepreneurship promotion project that offers financial assistance for enterprises, promotes the Enterprise Europe Network, and supports eco-innovation. The program is an important form of support for entrepreneurship.

Although a recent evaluation of the Centre for Assessment and Strategy Services, perform in 2011, analysed the programme's relevance, efficiency, and effectiveness, no such assessment has been made of the programme's real impact in terms of business growth, job creation, or contribution to welfare. Nevertheless, after monitoring participants and beneficiaries, the review yielded the following conclusions about operational performance and principles inputs:

- general objectives are sound and execution processes are measures taken by participating Member States;
- the programme is on track to achieve the expected effects, and stakeholders believe that the available budget and resources are appropriate.

Politicians have recognised the importance of entrepreneurship's role in promoting economic development. In many countries, business education is a key action to enhance technical knowledge (the basis of knowledge intensive services) and stimulate entrepreneurial activity.

KIBS and territorial development

During the 1980s, a debate arose among policy makers and academics regarding approaches to promoting local economies (Eisinger 1989). Instead of trying to attract foreign investment, regional governments attempted to capitalise on existing potential within their local territories (endogenous development) to drive job creation and industrial development. Creating knowledge intensive services as instruments to facilitate the emergence, creation and development of companies is a natural approach within this strategic context.

KIBS include professional services (strategic advisory services, design, engineering, consulting, etc.) and technical services (IT services, advertising, and marketing, etc.) (Nielsen and Lassen, 2012; Simmie and Strambach, 2006). Thus, KIBS generate and disseminate knowledge, which is crucial for innovation. Together with access to knowledge intensive services, innovation policies have acquired a fundamental role within public policy because of innovation's key role in the success of firms and territories. Ultimately, KIBS are a fundamental element for knowledge transfer, information exchange, and the innovation process (Mas-Verdú *et al.*, 2009).

Along these lines, KIBS act as a source of external knowledge while contributing to entrepreneurship and innovation. Miles (2005) defined KIBS as a set of services for private companies and public institutions where complex problem-solving operations and human capital are essential. KIBS offer consulting on a range of tasks that are essential for entrepreneurship by integrating knowledge intensive activities into other companies' production process (Garcia-Quevedo *et al.*, 2012).

One factor that may affect a company's development is the existence of an entrepreneurship policy that provides external resources in the territory where the company operates (Brooksbank, 2008; Massey, 2006). These resources are defined, in many cases, as advisory services (Curran and Storey, 2002;

Mole, 2002; Turok and Raco, 2000). The following list by Bennett and Robson (2003) highlights some of the advisory services offered through entrepreneurship policy:

- business strategy
- organisational structure
- marketing
- market research
- advertising and PR
- product and service design
- new technologies
- information technology (IT) services
- Human Resources (HR)
- tax
- finance
- etc.

Mas-Verdú *et al.*, (2011) reported that KIBS cover a variety of services ranging from advertising to legal services provided through consulting, engineering, and technical analysis. Bettencourt *et al.*, (2002) defined the companies that provide such services as those whose main activity is the creation, stockpiling, and dissemination of knowledge with the aim of developing a service or product to meet customers' needs.

KIBS have the same features as any other service (intangibility, inseparability, heterogeneity, and perishable nature), but in addition, high *personalisation*, risk, uncertainty, and credibility also characterise KIBS. Hence, this multi-faceted nature makes the evaluation of KIBS complex and difficult for consumers (Javalgi *et al.*, 2011; Zeithaml, 1981). The quality of KIBS, like the quality of industrial design or software development, is difficult to measure. Therefore, the quality and credibility of these services is indirectly observed from the quality and credibility of the KIBS provider (Javalgi *et al.*, 2011). This is why countries that specialise in knowledge economies enjoy competitive advantage as suppliers of KIBS.

Leading KIBS providers are usually private companies with high levels of knowledge or experience in a particular discipline or domain such as technical service or customer support (Drongelen and Bilderbeek, 1999). Miles and Darroch (2006) established two types of KIBS:

1. Traditional professional services such as consulting, management, advertising, accounting, and legal services based on social and institutional knowledge.
2. Services derived from technical knowledge and the transfer of this knowledge. Technology-based services include IT-related services, architecture and engineering services, medical and pharmaceutical research services, design, and R&D.

KIBS and entrepreneurship policies

The economic rationale for public intervention in business is based on market distortions and market failure. These are the results of information asymmetries. These asymmetries are of three types: lack of sensitivity to entrepreneurship, access to information about funding, and external advice (NESTA, 2013; Audretsch *et al.*, 2007; Storey, 2005).

First, there is a general lack of information and lack of knowledge regarding the possible benefits of starting a business (Storey, 2005). Educational and cultural policies aimed at students, unemployed people, and even workers can counteract this first market failure. The right kind of political education can raise awareness and train young people in the basics of business management. The overall aim of these policies is to promote a change of attitude so that people and society as a whole adopt a more open vision of entrepreneurship.

A second are where information is commonly lacking relates to access to financing. The financial market is particularly sensitive to information asymmetries. Because start-ups have little or no history and offer investors little guarantee of recovering their investment, financial institutions are unable to fully assess risks. Intervention in the form of systems of credit guarantees responds to this type of market failure.

The third, market failure derived from information asymmetries is linked to ignorance (or even scepticism) of entrepreneurs regarding the benefits of receiving external expert advice on strategic business issues. This information asymmetry can justify public subsidies aimed at facilitating expert advice or management training to SMEs or start-ups. Public policies can help to mitigate for the entrepreneur's ignorance regarding problems and solutions that advisory services can provide. This third type of public policy relates to entrepreneurs' access to knowledge intensive services.

Thus, from an economic and social point of view, entrepreneurship policies seek to stimulate the emergence of new productive activities by individuals (Henrekson and Stenkula, 2010). Nevertheless, policy makers must establish measures and specific actions that allow the effective promotion of entrepreneurship because some controversy remains over which programs are most appropriate to support and promote innovation-oriented entrepreneurial spirit (Audretsch, 2004).

From a general perspective, Storey (2005) distinguished between entrepreneurship policies with a 'hard' component and those with a 'soft' component. Hard policies consist of financial support (grants and loans). Soft policies, in contrast, consist of consulting services for entrepreneurs at different stages of the business project (before the launch, during the start-up phase, etc.) whereby entrepreneurs receive technological assistance, access to infrastructures, and so forth. Therefore, KIBS are a kind of 'soft' policy. KIBS are thus a key factor for the creation and distribution of knowledge at the regional, national, and international levels (Javalgi *et al.*, 2011).

Notably, depending on the nature of entrepreneurship, design and management of 'soft' entrepreneurship policies should vary. Lundström and Stevenson (2005) argue that entrepreneurship policy is influenced by its context. Accordingly, the design and implementation of entrepreneurship policy must reflect the economic and social situation as well as the conditions and nature of business activities within each territory.

Because of their characteristics, SMEs have special capabilities to innovate in cooperation with other companies. This allows SMEs to optimise their use of internal knowledge resources and combine these resources with those of partners (Muller *et al.*, 2001). KIBS act as a source of external knowledge and contribute to innovation by customers. KIBS are thus co-innovators (García-Quevedo *et al.*, 2013), together with the young innovative companies to which they provide their services. But this relationship goes much further, and KIBS themselves also benefit from their interaction with other companies by accessing skills and the capacity for innovation. This creates a virtuous 'win-win' circle in which all agents benefit (Muller *et al.*, 2001).

Conclusions

Even in countries with a tradition of public interventionism in economic activity, state intervention has seldom been part of the business environment. Public policy for business has mainly been concerned with preventing large firms from abusing their market power. To protect against such actions, governments have used three instruments: regulation, the defence of competition, and government property. The 1980s, however, heralded the reduction or closure of many regulatory bodies and marked the beginning of a privatisation process. For some, this was a sign that government intervention was receding, but an alternative explanation exists. In reality, a new policy agenda to promote business activity emerged. Unlike traditional policy instruments seeking to curb activities of large corporations, current policy instruments focus on SMEs. In addition, whereas traditional policy instruments were generally implemented nationwide, entrepreneurship policies are implemented at all government levels.

Nevertheless, the growing trend towards globalisation and the knowledge economy has led certain traditional policy instruments to become less relevant (Gilbert *et al.*, 2004) in their promotion of economic growth and employment. Conversely, policies linked to innovation and driving knowledge intensive entrepreneurship have become increasingly prominent among both academics and the political agenda of policy makers at all administrative levels (local, regional, national, and supranational). Innovation, entrepreneurship, and knowledge, especially when properly combined, play a fundamental role in economic well-being.

Despite the complex relationship between entrepreneurship and economic growth, there is a strong belief that governments can improve economic growth by affecting conditions for entrepreneurship (Hart, 2003). In general, entrepreneurship policy aims to create an environment and a support system that promotes the emergence of new entrepreneurs and new companies capable of overcoming problems that may arise early on in the venture (Lundström and Stevenson, 2005;

Stevenson and Lundström, 2002). As previously discussed, one factor that can affect a company's development is the existence of an entrepreneurship policy – based on the provision of knowledge intensive resources – in the territory where the company operates. KIBS are thus not only a source of external knowledge for the company, but also a resource that contributes to entrepreneurship and innovation in the territory where the company operates.

In this sense, efforts by regional governments to promote entrepreneurship could gain in efficiency if the actions implemented were more specific. This could be achieved through selective and segmented policies with programmes adapted to the specific needs of the entrepreneurial ventures as a function of their characteristics. Regardless of the nature of these policies, greater emphasis should be placed on supplementing financial actions like subsidies and tax incentives – the most common government actions – with indirect actions (Lerner, 2009), in particular those giving start-ups access to knowledge intensive services.

References

Acs, Z. (2008), Foundations of high impact entrepreneurship, *Foundations and Trends in Entrepreneurship*, 4 (6), pp. 535–620.

Ács, Z.J. and Naudé, W.A. (2013), Entrepreneurship, stages of development, and industrialization, in Szirmai, A., Naudé, W.A. and Alcorta, L. (eds) *Pathways to industrialization in the 21st century*. Oxford: Oxford University Press, Chapter 14.

Allen, K. and Stearns, T. (2004), Technology entrepreneurs, in Gartner, W.B., Shaver, K.G., Carter, N.M. and Reynolds, P.D. (eds), *Handbook of entrepreneurial dynamics: the process of business creation*, Thousand Oaks, CA: Sage Publications.

Audretsch, D. (2004), Sustaining innovation and growth: public policy support for entrepreneurship, *Industry and Innovation*, 11 (3), pp. 167–91.

Audretsch, D. B., I. Grilo, and A. R. Thurik (2007), Explaining entrepreneurship and the role of policy: a framework, in D. B. Audretsch, I. Grilo, and A. R. Thurik (eds), *Handbook of research on entrepreneurship policy*. Cheltanham: Edward Elgar, pp. 1–17.

Audretsch, David B., Roy Thurik, Ingrid Verheul and Sander Wennekers (eds) (2002), *Entrepreneurship: determinants and policy in a European-US comparison*, Boston/Dordrecht/London: Kluwer Academic Publishers.

Bennett, R., and Robson, P. (2003), Changing use of external business advice and government supports by SMEs in the 1990s, *Regional Studies*, 37 (8), pp. 795–811.

Bettencourt, B. R., Kim, I., Hoffmann, A. A., and Feder, M. E. (2002), Response to natural and laboratory selection at the Drosophila hsp70 genes, *Evolution*, 56 (9), pp. 1796–1801.

Brooksbank, P. A. (2008), Fast constructive recognition of black box symplectic groups, *Journal of Algebra*, 320 (2), pp. 885–909.

Carree, M., (2001), Does unemployment affect the number of establishments? A regional analysis for US, *Regional Studies*, 36 (4), pp. 389–98.

Carree. M. A. and Thurik. A. R., (1999), Industrial Structure and Economic Growth, in D. B. Audretsch and A. R. Thurik (eds), *Innovation, industry evolution and employment*, Cambridge: Cambridge University Press, pp. 86–110.

Carree, M. A., A. van Stel, A. R. Thurik and S. Wennekers, (2000), Economic development and business ownership: an analysis using data of 23 OECD countries in the period 1976–96. Institute for Development Strategies Discussion Paper 00–6.

Cooper, A. C., Gimeno-Gascon, F. J., and Woo, C. Y. (1994), Initial human and financial capital as predictors of new venture performance, *Journal of Business Venturing*, 9 (5), pp. 371–95.

Cooper, R.G., Edgett, S.J. and Kleinschmidt, E.J. (2004), Benchmarking best NPD practices II, *Research Technology Management*, 47 (3), pp. 50–60.

Curran, J., and Storey, D. J. (2002), Small business policy in the United Kingdom: the inheritance of the small business service and implications for its future effectiveness, *Environment and Planning C*, 20 (2), pp. 163–78.

Davidsson, P. (1991), Continued entrepreneurship: ability, need, and opportunity as determinants of small firm growth, *Journal of Business Venturing*, 6 (6), pp. 405–29.

Davidsson, P. (2004), *Structure-acoustic analysis; finite element modelling and reduction methods*. Lund: Lund University.

Davidsson, P. (2005), Methodological approaches to entrepreneurship: past research and suggestions for the future. *Small Enterprise Research*, 13 (1), pp. 1–21.

Drongelen, I. and Bilderbeek, J. (1999), R & D performance measurement: more than choosing a set of metrics, *R&D Management*, 29 (1), pp. 35–46.

Dutta, D. and Thornhill, S. (2008), The evolution of growth intentions: toward a cognition-based model, *Journal of Business Venturing*, 23 (3), pp. 307–32.

Eisinger, N. (1989), A note on the completeness of resolution without self-resolution, *information processing letters*, 31 (6), pp. 323–26.

Evans, D. S., and Jovanovic, B. (1989), An estimated model of entrepreneurial choice under liquidity constraints, *The Journal of Political Economy*, 97 (4), pp. 808–27.

Fritsch, M., (1997), New firms and regional employment change, *Small Business Economics*, 9, pp. 437–48.

García-Quevedo J, Mas-Verdú F. and Montolio D. (2013), What types of firms acquire knowledge intensive services and from which suppliers? *Technology Analysis and Strategic Management*, 25 (4), pp. 473–86.

Garcia-Quevedo, J., Mas-Verdú, F., and Polo-Otero, J. (2012), Which firms want PhDs? An analysis of the determinants of the demand, *Higher Education*, 63 (5), pp. 607–20.

Gilbert, N., Boyle, S., Fiegler, H., Woodfine, K., Carter, N. P., and Bickmore, W. A. (2004), Chromatin architecture of the human genome: gene-rich domains are enriched in open chromatin fibers, *Cell*, 118 (5), pp. 555–66.

Gries, T., and Naudé, W. (2011), Entrepreneurship and human development: a capability approach, *Journal of Public Economics*, 95 (3), pp. 216–24.

Hart, O. (2003), Incomplete contracts and public ownership: Remarks, and an application to public-private partnerships, *The Economic Journal*, 113 (486), pp. C69–C76.

Heirman, A. and Clarysse, B. (2004), How and why do research-based start-ups differ at founding? A resource-based configurational perspective, *Journal of Technology Transfer*, 29 (3/4), pp. 247–68.

Henrekson, M., and Stenkula, M. (2010), Entrepreneurship and public policy, in Acs, Z. J., Audretsch, D. B. (eds), *Handbook of entrepreneurship research*, pp. 595–637.

Hölzl, W. (2006), Gazelles, innovation Watch, Europe INNOVA, scoping paper 31.05.2006, EU: Brussels.

Javalgi, R. R. G., Gross, A. C., Benoy Joseph, W., and Granot, E. (2011), Assessing competitive advantage of emerging markets in knowledge intensive business services, *Journal of Business and Industrial Marketing*, 26 (3), pp. 171–80.

Kanbur, S. M. (1979), Of risk taking and the personal distribution of income. *The Journal of Political Economy*, pp. 769–97.

Kirzner, I. M. (1973), *Competition and entrepreneurship*. Chicago: University of Chicago Press.

Lerner, J. (2009), *Boulevard of broken dreams: why public efforts to boost entrepreneurship and venture capital have failed – and what to do about it*. Princeton, NJ: Princeton University Press.

Lindholm Dahlstrand, Å. (1997), Growth and innovativeness in technology-based spin-off firms, *Research Policy*, 26 (3).

Lundstrom, A. and Stevenson, L. (2005), *Entrepreneurial policy: theory and practice*. New York: Kluwer Academic Publishers.

Lundvall, A., and Zetterström, C. (2000), Cataract extraction and intraocular lens implantation in children with uveitis, *British Journal of Ophthalmology*, 84 (7), 791–93.

Mas-Tur, A., and Soriano, D. R. (2014), The level of innovation among young innovative companies: the impacts of knowledge-intensive services use, firm characteristics and the entrepreneur attributes. *Service Business*, 8 (1), pp. 51–63.

Mas-Verdú, F., Baviera-Puig, A., and Martínez-Gómez, V. (2009), Entrepreneurship policy and targets: the case of a low absorptive capacity region, *International Entrepreneurship and Management Journal*, 5 (3), pp. 243–58.

Mas-Verdú, F., Wensley, A., Alba, M., and Álvarez-Coque, J. M. G. (2011), How much does KIBS contribute to the generation and diffusion of innovation? *Service Business*, 5 (3), pp. 195–212.

Miles, I. (2005), Knowledge intensive business services: prospects and policies, *Foresight*, 7 (6), pp. 39–63.

Miles, M. P. and Darroch, J. (2006), Large firms, entrepreneurial marketing processes, and the cycle of competitive advantage, *European journal of marketing*, 40 (5/6), pp. 485–501.

Mole, K. (2002), Business advisers' impact on SMEs an agency theory approach, *International Small Business Journal*, 20 (2), pp. 139–62.

Murphy, T. H., Miyamoto, M., Sastre, A., Schnaar, R. L., and Coyle, J. T. (1989), Glutamate toxicity in a neuronal cell line involves inhibition of cystine transport leading to oxidative stress, *Neuron*, 2 (6), pp. 1547–58.

Müller, P., Li, X. P., and Niyogi, K. K. (2001), Non-photochemical quenching. A response to excess light energy, *Plant Physiology*, 125 (4), pp.1558–66.

Nielsen, S. L., and Lassen, A. H. (2012), Images of entrepreneurship: towards a new categorization of entrepreneurship. *International Entrepreneurship and Management Journal*, 8 (1), pp. 35–53.

Pickernell, D., Senyard, J., Jones, P., Packham, G., and Ramsey, E. (2013), New and young firms entrepreneurship policy and the role of government – evidence from the Federation of Small Businesses survey, *Journal of Small Business and Enterprise Development*, 20 (2), pp. 358–82.

Reynolds, A. (1999), *Electoral systems and democratization in Southern Africa*. New York: Oxford University Press.

Reynolds, D. A., Quatieri, T. F., and Dunn, R. B. (2000), Speaker verification using adapted Gaussian mixture models, *Digital Signal Processing*, 10 (1), pp. 19–41.

Reynolds, P., Bygrave W., Autio, E., Cox, L. and Hay, M. (2002), Global entrepreneurship monitor – 2002 executive report. Babson College, Kauffman Foundation, London Business School. Available at: www.gemconsortium.org

Reynolds, P.D., Bosma, N. Autio, E. Hunt, S. DeBono, N. Servais, I. Lopez-Garcia P. and Chin N. (2005), Global entrepreneurship monitor: data collection design and implementation 1998–2003, *Small Business Economics*, 24, pp. 205–31.

Schumpeter, J. A. (1950), The march into socialism, *The American Economic Review*, pp. 446–56.

Schumpeter, J. A. (1961), *The theory of economic development*. New York: Oxford University Press.

Shane, S. (2009), Why encouraging more people to become entrepreneurs is bad public policy, *Small Business Policy*, 33 (2), pp. 14–49.

Simmie J., Strambach S. (2006), The contribution of knowledge-intensive business services (KIBS) to innovation in cities: an evolutionary and institutional perspective. *Journal of Knowledge Management*, 10 (5), pp. 26–40.

Stevenson, L., and Lundström, A. (2002), Beyond the rhetoric: defining entrepreneurship policy and its best practice components. *Swedish Foundation for Small Business Research.*

Storey, D. J. (2005), Entrepreneurship, small and medium sized enterprises and public policies, in Z.J. Acs and D.B. Audretsch (eds), *Handbook of entrepreneurship research*. New York: Springer, pp. 473–511.

Thurik, A. R., (1999), Entrepreneurship, industrial transformation and growth, in G. D. Libecap (ed.), *The sources of entrepreneurial activity, in advances in the study of entrepreneurship, innovation and economic growth*, Vol. 11, Greenwich, CT: JAI Press, pp. 29–66.

Turok, I., and Raco, M. (2000), Developing expertise in small and medium-sized enterprises: an evaluation of consultancy support, *Environment and Planning C*, 18 (4), pp. 409–28.

Wiklund, J. and Shepherd, D. (2003), Aspiring for, and achieving growth: the moderating role of resources and opportunities, *Journal of Management Studies*, 40 (8), pp. 1911–41.

Zeithaml, V. A. (1981), How consumer evaluation processes differ between goods and services, Marketing of services, 9 (1), pp. 25–32.

8 Contribution of Knowledge Intensive activities to regional competitiveness

Production function approach

Anna-Leena Asikainen and Giovanni Mangiarotti

Introduction

In the 27 Member States of the European Union, there are relatively stark differences between NUTS2 (NUTS stands for Nomenclature of Territorial Units for Statistics) regions in terms of both research and development (R&D) and innovation inputs and outputs. In 2010, 30 regions with the highest R&D intensity accounted for almost 40 per cent of the total R&D expenditure in EU27. Moreover, the statistics indicate that the lower the national level of R&D intensity, the narrower the variation among the country's regions. In addition, researchers measured by headcount tend to be highly concentrated in certain regions. Similarly, innovation output measured as patents stems from regions with industrial strongholds. (Eurostat 2013) These differences translate into gaps in Gross Domestic Product (GDP) per capita, providing a rationale for regional policies aiming at evening out the situation.

In narrowing the gaps in GDP per capita in the era of knowledge-based production, the regions' capability to generate, absorb and exploit new knowledge to create jobs and growth plays a significant role. Observation regarding the importance of local knowledge production originates in empirical work, finding that academic research conducted in a region's universities improves the private sector's innovation outcomes (Jaffe, 1989). The identified impact is mediated via knowledge spillovers, which occur more often and are larger if the firms are located within geographical, technological or horizontal proximity of each other (Jaffe, 1989). Hence, knowledge spillovers may affect regional convergence, and they are also the focus of and underlying rationale for many policy interventions (Lychagin *et al.*, 2010), such as clusters (Porter, 1990; Beaudry and Breschi, 2003; Iammarino and McCann, 2006).

The role of private knowledge production infrastructures has been highlighted in research on the activities, practices and impact of Knowledge Intensive Business Services. Empirical research shows that Knowledge Intensive Business Services (KIBS) are crucial for production, diffusion and use of technologies, and consequently also for economic growth (Mas-Verdú *et al.*, 2011; Hauknes and Knell, 2009; Rodriguez and Camacho, 2008). In addition, the small European countries seem to have KIBS in their top three R&D specialisations (Asikainen, 2014).

Via their knowledge creation and diffusion activities the KIBS firms have a positive effect on the regional competitiveness, development (Danek *et al.*, 2014;

Muller and Zenker, 2001) and innovation performance of young innovative firms (Mas-Tur and Ribeiro Soriano, 2014), which are in a prime position with respect to enlarging existing and creating new businesses. The KIBS firms largely serve the local market (Asikainen, 2015), and they are less likely to collaborate with international partners (Camacho and Rodríguez, 2005; Sundbo and Gallouj, 2000; Tether, 2005). This means that their networks are restricted to local innovation networks, and thus their impact should mainly be observed locally as an increase in overall productivity and competitiveness. An additional reason for the relevance of analysing the KIBS and regions arises from the fact that scale economies are non-existent in services with highly-tailored output and hence analyses focusing on small geographical areas should be able to capture all features of the phenomena (Rubalcaba *et al.*, 2010). Moreover, many of the current economic and societal challenges (globalisation, social and environmental issues) are faced locally; combatting them also emphasises the role of regional policies and knowledge production structures (OECD, 2011).

Regional differences in economic performance, innovation inputs and outputs, and labour force qualification levels are significant and lead to disparities in competitiveness. The objective of this study is to assess the relationship between a subsector of KIBS, namely high-tech knowledge intensive services (HKIS), and regional competitiveness. Several studies have shown the non-negligible role of KIBS in economic growth at country-level (Gallaher and Petrusa, 2006), however, the regional impact of HKIS has not been analysed in empirical research. Due to the diverging levels of regional competitiveness, it is deemed essential to analyse whether the selected factors have a differing effect in highly competitive regions rather than in weakly competitive ones. Building on the much-tested Cobb-Douglas production function approach, the aim is to capture the impact of HKIS along with other factors advancing regional competitiveness in 139 NUTS2 regions. The other factors included in the analysis encompass physical and intangible capital and labour quality. Following both the previous empirical research in the field of innovation analysis and recent trends in policy-oriented work (Coad and Rao 2008; Coad *et al.*, 2014), the following analysis is run using quantile regression.

This chapter continues with a short description of the regional innovation system which forms the theoretical basis for the analysis; and it continues with a review of the characteristics and role of KIBS in the economy. After introducing the regional knowledge production function, data and results are discussed and the chapter is finished with conclusions and recommendations for further research in the field.

Regional competitiveness

The innovation system-concept offers a theoretical rationale for the regional approach to competitiveness and KIBS. Research on innovation systems started between the late 1980s and early 1990s (Freeman, 1987; Nelson, 1993; Lundvall, 1992). National innovation system is defined by Nelson (1993) as "a set of institutions whose interactions determine the innovative performance of national firms." Hence, the system identifies and characterises the structure, functioning and roles of every actor. In the frame of novelty creation process, the system describes all the activities and relations occurring between the actors. Due to the definition focusing on structure, functioning and behaviour, an innovation system can be generally applied to national, regional, sectoral and technological contexts, which constitute the backbone of competitiveness of the entity under the loop at any level.

Actors in the system are members from the business sector, research and higher education institutions, public authorities involved with science, technology, innovation and industrial policies, and private and public financiers. Institutions contain framework conditions (intellectual property rights and tax laws, health, safety and environmental regulations), technical and managerial standards, rules, norms, habits and practices internal to the business enterprises, etc. Activities refer to knowledge production for and in the innovation process, steering of the demand for side activities, and activities supporting function, evolution and dynamics of the system (Edquist, 2011).

According to Lundvall *et al.*, (2002) learning capabilities of individuals, organisations and regions are the success factors for national innovation systems. Just like national innovation systems, regional innovation systems are defined by their knowledge base and learning processes, inputs and demand, institutions, basic technologies, type and structure of interactions between organisations, and processes of variety and selection generation (Malerba, 2002). Further characterisations of a regional innovation system emphasise the role of knowledge production without making a difference between public and private producers. First, Doloreux (2002) classifies the RIS components into firms, institutions, knowledge infrastructure, and innovation policy. Second, Autio (1998) divides the actors into knowledge generators, diffusers and exploiters. The two classifications suggest an integral part for KIBS in RIS.

According to Porter (1990), regional innovation capability, i.e. a region's ability to generate new knowledge, constitutes a significant source of competiveness. Basically, regional competitiveness improves when a region's relative share in the economy relative to other regions increases (Boschma, 2004). Regions compete against each other in attracting and retaining a highly educated labour force, direct investment, and high-growth innovative enterprises (Florida, 2002).

Regional competitiveness is composed of, and bases its evaluation on, the same elements as national competitiveness. Public and private knowledge infrastructures as part of regional innovation systems construct the backbone of the region's competitiveness, and when these are aggregated, national competitiveness also depends on them. Competitiveness sets requirements for the quantity and quality of regional resources, institutions and infrastructure, the level of economic dynamism and the interaction across sectors of the economy.

In recent years, the role of interactions between actors has received more attention, both in research and policy spheres. The increasing importance of external cooperation and knowledge sourcing supports technology transfer and interactive learning in the system (Carlsson *et al.*, 2002; Edquist, 2011). Difficulties related to transferring tacit knowledge over longer distances support the idea of regional knowledge production represented by knowledge and technology intensive activities as a source of competitiveness (Griliches, 1979; Pakes and Griliches, 1984). Research-intensive universities are the main source of tacit knowledge (Jaffe, 1989; Jaffe *et al.*, 1993) because they produce new human capital in terms of highly qualified persons and new knowledge that benefits the economy via spillovers (Arrow, 1962).

Overall, the central factors contributing towards the success of RIS refer to traditional public and private infrastructure, labour force quality and the spillovers from public and private knowledge infrastructure. In the following, there is further characterisation of the role private knowledge infrastructure has in terms of KIBS and its subsector.

Role of knowledge-intensive activities

Characterizing knowledge intensive activities

Despite the central role of knowledge in its production process, KIBS share certain features and functionalities with other services. Among the similarities are the simultaneous production and consumption of tailored outputs and due to that, an intensive cooperation between the service provider and customer. Similarities are visible also on the side of outputs. KIBS outputs have high information content (Sirilli and Evangelista, 1998), and as in other services they are intangible, often non–replicable and their production has high labour intensity. Furthermore, the output is not necessarily a separate, single item but may consist of a set of goods and/or services (Gallaher and Petrusa, 2006). Differentiating the final output and its production process in services is difficult. Overall, innovation in services takes many different forms and due to this, the simultaneous presence of product, process and organisational innovations is more frequently found in services than in manufacturing (Cainelli *et al.*, 2006). Typically, innovations arise as solutions to daily problems and as ad hoc developments rather than as outcomes of highly resourced project-based systematic exploration (Gallouj *et al.*, 1997; Crevani *et al.*, 2011; Crevani *et al.*, 2011). As the outputs are developed in tight collaboration with clients, the risks are more likely technical than market related. Technical risks are accentuated by the fact that it is impossible to produce a test piece, whereas market risks arise from the potentially low-level of novelty linked to the extensive involvement of clients in the process (Gallaher and Petrusa, 2006).

Recent decades have witnessed a remarkable growth in KIBS due to the tendency of manufacturing service firms to outsource their non-core activities and functions, and the strengthened role of knowledge and the highly skilled workforce in production and organisations (Hipp and *et al.*, 2013). In KIBS, the role of knowledge and clients is more important than in other services. In a broad sense, Knowledge Intensive Business Services (KIBS) use knowledge as their main input and produce knowledge as their main output (Gallouj, 2002).

This knowledge is accumulated in the employees and organisational routines (Rubalcaba *et al.*, 2008). Due to the dual role of knowledge in production, the KIBS firms contribute strongly via their core activities, i.e. accumulation, creation, or dissemination of knowledge, to spillovers and transfer of knowledge between firms and industry (Hertog, 2000; Miozzo and Grimshaw, 2005; Gallouj and Windrum, 2008).

The KIBS are activities which are typically used by their clients as intermediary inputs leading to improved quality and efficiency in their production as they either complement or substitute internal services (Muller and Doloreux, 2009; Rubalcaba and Kox, 2007). Specialised, but not necessarily science-based knowledge embedded in organisational routines and highly educated labour force characterises KIBS (Hertog, 2000; Rubalcaba *et al.*, 2008). Work is conducted in projects and aims at solving a client's very specific problems using advanced technology, strategic input and high-level expertise (Miozzo and Grimshaw, 2005; Miles *et al.*, 1995; Muller and Zenker, 2001). In addition, KIBS maintain the business processes outsourced to them (Miozzo and Grimshaw, 2005). Due to these features, the KIBS firms compete on service quality instead of price (Tether, 2002; Tether and Hipp, 2002). Moreover, KIBS is able to create new markets and business opportunities by recombining previously acquired knowledge.

Overall, the added value offered by KIBS stems from specialised expertise allowing for the efficient division of labour in the client firm, and for innovation and knowledge transfer. KIBS offers either substitutes for or complements of in-house skills, knowledge, expertise and resources. By being suppliers of high-level professional knowledge and expertise in narrowly defined functional or technical fields (Windrum and Tomlinson, 1999), KIBS become integrated into the client firm's production process facilitating and co-producing innovations (Muller and Doloreux, 2009) as well as advancing a commercialisation of novelties (García-Quevedo and Mas-Verdú, 2008). Knowledge exploitation by KIBS' clients encompasses a knowledge transmission process and a re-engineering process conducted in cooperation by a KIBS firm and its client (Muller and Zenker, 2001) aiming at translating the potentially generic knowledge to match with the client's cognitive context. Moreover, KIBS firms provide strategic technical and organisational knowledge whose incorporation or exploitation might not take place without the support of KIBS (Wood, 2002).

Innovation in KIBS is based on combining existing and new knowledge in a novel way (Miles, 2005). Overall, the KIBS innovation strategies resemble innovation strategies in industry because KIBS produce non-technological innovations which are based on R&D to a larger extent than is commonly believed (Hipp *et al.*, 2013). In comparison to other sectors, KIBS are more likely to cooperate with firms from other economic activities and the cooperation is more extensive; hence the client relationship is central to the innovation process (Miles, 1999). Overall, external cooperation, knowledge sourcing and outsourcing make up a significant part of R&D in services. In KIBS, the clients are strongly integrated into R&D, innovation and production process (Hertog, 2000). Moreover, strong and very strong ties between KIBS and client firms are beneficial for most innovation types (Amara *et al.*, 2009). KIBS not only serve both public and private clients but also amend the public knowledge production infrastructure (universities, research and technology organisations, etc) (Miozzo and Grimshaw, 2006; Hertog, 2000; Gallouj and Windrum, 2008).

Knowledge-intensive activities and the regional innovation system

The role of KIBS in RIS is manifold; they have a central role in knowledge creation, collection and diffusion process, connect actors and via these activities generate and help others to generate new knowledge in the system. Wood (2005) represents KIBS as supporters of "regional adaptability" as they are in the position to adapt the generic technologies and commercialisation processes to the needs of the regional economy, independent of the economic activity.

KIBS' exchanges with other actors not only remain as one-on-one interactions, but they act as knowledge brokers in wider collaborative settings and networks (Hipp, 1999). Due to KIBS' dependency on regional actors' business needs, KIBS strengthen regional specificities and possibly contribute to maintenance and growth of regional diversities/differences because of the cumulative nature of knowledge acquired from KIBS (Wood, 2005).

Overall, the importance of KIBS in and for RIS is non-negligible. Their role is strengthened when the focus of analysis moves from actors to activities and interactions in RIS. The inter-organisational interactions for innovation mainly take place within regional borders between local actors (Asheim, 1996; Braczyk *et al.*, 1998; Cooke, 2005). The benefits of geographical proximity and agglomeration are due to knowledge externalities (Glaeser *et al.*, 1992; Jaffe *et al.*, 1993; Feldman, 1994; Audretsch and Feldman, 1996). In their activities, the KIBS firms create intended and unintended knowledge externalities, which build ground for further knowledge generation, and which then contribute towards technological progress and economic growth (Audretsch and Feldman, 1996; Jaffe *et al.*, 1993; Dumais *et al.*, 2002; Thompson, 2006). There are three types of externalities depending on the concentration of firms in a given industry. Marshall-Arrow-Romer (MAR) externalities originate from the specialisation of a region in a single industry; Jacobs externalities stem from the complementarity of knowledge provided by a diversified pool of firms and industries within a given region; and Porter externalities are generated by competition among local firms concentrated in the same industry.

The externalities can be further divided into pure knowledge, sometimes termed as technological, and into pecuniary externalities (Scitovsky, 1954; Griliches, 1982). The type of interdependence between firms (sectors), whether direct or indirect, separates these externalities from each other. Direct interdependence takes place in the case of knowledge externalities, as there is no need for any third parties or mechanisms to intervene in order to generate the benefits from externalities. Indirect interdependence occurs when the impact of the externalities is mediated via the price mechanism (Scitovsky, 1954). Pure knowledge externalities, known also as pure knowledge spillovers (Griliches, 1982) or disembodied knowledge, take place when firm (sector) A benefits from R&D conducted in firm (sector) B. In other words, knowledge is accessible for other enterprises without a cost, and the output of a firm (sector) depends not only on its own inputs but also on the inputs of other firms (sectors). Pecuniary externalities complicate pricing of outputs (that are to be used by other sectors as intermediate inputs) with high knowledge content. Due to pure knowledge externalities, the price of a product does not fully reflect the quality of goods or services. Moreover, the price tends to vary depending on the market structure.

KIBS generate both pure knowledge and pecuniary externalities. The very business of KIBS in creating, acquiring, exploiting, diffusing and selling knowledge, emphasises the role of pecuniary externalities linked to the price mechanism (Antonelli, 2008a; 2008b). Pecuniary externalities touch KIBS to a greater extent than other activities, as they concern the price of both input and output leading to variation in profits. When the price of knowledge as an input is lower than the equilibrium price, the KIBS gain from the situation in terms of higher profits. However, a decrease in the price of KIBS output has a negative effect on profits. Changes in the price are linked to the structure of the sector, mainly to the number of active firms in the region. The effect of input price dominates when a higher density of KIBS firms creates positive agglomeration effects. But at a certain point, when the market produces an oversupply of knowledge, the price of knowledge begins to fall and appropriability losses increase. Hence, these price-altering effects are behind the diminishing returns on agglomeration of KIBS firms. (Antonelli *et al.*, 2010) The dynamic impact on productivity is derived in a similar vein: higher input prices first improve productivity as they compensate for the cumulating appropriability losses; after the oversupply threshold is reached, prices start to decline and lead to lower productivity. Empirical evidence supports the existence of pecuniary externalities in the relationship between KIBS and manufacturing (Ciriaci *et al.*, 2013).

Knowledge-intensive activities and productivity

KIBS contribute to the regional innovation pool and economic growth in different ways: innovating and creating growth in its own right, facilitating innovations and acting as sources of knowledge for firms carrying out other economic activities, and transferring knowledge (Mas-Verdú *et al.*, 2011; Muller and Zenker, 2001; Gallaher and Petrusa, 2006).

Empirical research on the connections between services and productivity is relatively common. The first wave of studies concluded that the higher the share of services, the lower the productivity and its growth (Baumol, 1967). Later research has acknowledged the need to account for the diversity of services and the importance of knowledge and innovation for service evolution (Baumol, 2002). Besides the development within the service sector, innovation and knowledge generated by, diffused and spilled over from services are of benefit to other related economic activities, both in services and manufacturing.

Overall, economic activities and diffusion of knowledge are generally accepted to have a significant role in the economy. However, empirical analysis of their role for regions is still meagre. A recent empirical study focusing on the 1980–2008 period and 17 European countries suggests that at regional level, the size of the service sector measured in terms of employment improves productivity growth (Maroto-Sanchez and Cuadrado-Roura, 2013). Moreover, in the previously mentioned study both physical capital and qualified labour force are found to drive productivity development.

Overall, services have increasingly replaced industry as a source of labour productivity (Desmarchelier *et al.*, 2013; Corrado *et al.*, 2009). In a simulation study, KIBS have been identified as a source of economic growth. Industry retains its role as a source of economic growth through its demand for KIBS (Desmarchelier *et al.*, 2013). Moreover, recent results indicate that R&D acquired from KIBS improves innovation performance in manufacturing (Ciriaci *et al.*, 2013) and innovation likelihood in SMEs (Muller and Zenker, 2001). KIBS play a significant role in the more advanced and innovative economies and regions (Hipp *et al.*, 2013). In these regions, the competitiveness of manufacturing (also services) is dependent both on KIBS and the manufacturing firms' ability to interact with KIBS and absorb the acquired knowledge (Chadwick *et al.*, 2008; Cooke and Piccaluga, 2006; Simmie and Strambach, 2006).

The focus of this work is a subsector of KIA, namely the high-tech knowledge intensive services, which for example entail information and communications technologies (ICT). There are two main reasons for the choice. First, the impact of ICT services exceeds the boundaries of the activity. Second, the ICT industry is the main driver of growth in European economies. As a general purpose technology, ICT acts as an enabler of the novelty creation process, a catalyst for complementary co-inventions, and a facilitator of new modes of operation in other businesses. Typically, its impact is seen as increasing efficiency and productivity (Jalava and Pohjola, 2007).

In particular, diffusion and exploitation of ICT has improved productivity in services (van Ark *et al.*, 2003). Innovations in HKIS will not be left in the margin, but have an extensive impact as the novel goods and services developed in the sector contribute to finding solutions for environmental, technological and societal challenges of our time. HKIS produce applications and services that support global cooperation, networking, knowledge transfer, labour mobility, strategic agenda development and policy coordination in all fields of economy (Vickery and Wunsch-Vincent, 2009). Furthermore, innovations in the field of intelligent infrastructures, green technologies, nano- and biotechnologies are to a large extent based on a smart use of ICT. Moreover, many advances in the private and public sector are built on novelties in software development. The existence of high quality, local ICT services supports the productivity and competitiveness of the whole economy by being the first stop for advices and support in the adoption and use of new applications in other sectors (Hanna, 2010).

Based on both theoretical considerations and empirical work in the field, it seems justified to assume that HKIS has a positive impact on regional competitiveness.

Regional knowledge production function

To analyse the role of HKIS for regional productivity, a knowledge production function is estimated. One of the most commonly applied measures for competitiveness is (labour) productivity. The selected specification is limited in the number of explanatory variables partly due to data availability, and partly due to the aim of obtaining results that are both robust and straightforward to interpret. The output variable is regional productivity, while the input variables are capital, intangible capital, labour and HKIS. Following Griliches (1979) and Jaffe (1989) the relationship between inputs and output is assumed to take the form of the Cobb-Douglas production function.

$$Y_{it} = \alpha K_{it}^{\beta_1} L_{it}^{\beta_2} HKIS_{it}^{\beta_3} IK_{it}^{\beta_4} Z_{it}^{\beta_5} e^{\varepsilon_{it}}$$

Log-linearizing this yields:

$$y_{it} = \alpha + \beta_1 k_{it} + \beta_2 l_{it} + \beta_3 hkis_{it} + \beta_4 ik_{it} + \beta_5 z_{it} + \varepsilon_{it}$$

Regional productivity (y) is operationalised as labour productivity and measured by gross value added at basic prices per total employment. Among the explanatory variables capital (k) is measured as gross fixed capital formation, quality of the labour force (l) as a share of the population aged 15–64 with tertiary education, HKIS ($hkis$) as a share of HKIS in total employment, and intangible capital (ik) is measured by patents per million of the active population.

Unfortunately, capital investments are flow variables. However, given the high correlation between stock and flow measures in cross-section, results obtained using a similar production function estimation with a flow measure have been reported to have little impact on the results (Crepon *et al.*, 1998). Finally, due to the skewed distribution of countries in the different quantiles, it was necessary to include country controls (z) in the regression equation. All the variables are expressed as logarithmic transformations.

Due to the impossibility of credibly setting pre-defined threshold values splitting regions into categories of high, medium and low competitiveness, relative measures to indicate the level of competitiveness are applied. As the aim is to analyse whether the highly competitive regions differ from less competitive regions in their ability to exploit knowledge and technologies supplied by HKIS, the chosen estimation method is quantile regression (QR). Instead of calculating the regression coefficients at the mean value, the quantile regression estimates the coefficients at selected quantiles along the competitiveness distribution. The QRs are estimated at the 25th, 50th and 75th percentile (henceforth, the quantiles will be referred to as Q1, Q2, Q3).

In addition to the estimation method, there are two important decisions to be made in the empirical productivity analysis. The first decision is related to the definition of the dependent variable, and whether it is expressed in levels or as growth rates. Considering the time period under the loop, the main reason for relying on the productivity level lies in its lower exposure to cyclicality than growth rate.

The second central issue to be addressed is the endogeneity of the explanatory variables. Typically, endogeneity can be controlled using an instrument variable approach. However, it might be very difficult to identify suitable instruments, especially in the analysis of regional productivity where the selection of variables is relatively limited. Alternatively, the explanatory variables can be inserted into the regression either as lagged values or moving averages. While not perfectly addressing the endogeneity, this approach helps to tackle simultaneity bias and accounts for lagged effects.

Data and results

All the data used in the analysis comes from Eurostat's Regional Statistics database (http://ec.europa.eu/eurostat/web/regions/overview). The definition for the geographical unit follows Eurostat's Nomenclature of Territorial Units for Statistics (NUTS). The focus will be on the NUTS2-level since these regions possess a sufficient level of autonomous decision-making power over policies affecting regional competitiveness.

Due to the time frame of the analysis the industry classification is Nace Rev. 2, in which the high-tech knowledge intensive services (HKIS) contain the following activities (NACE codes in parenthesis):

- Motion picture, video and television programme production, sound recording and music publishing activities (59)
- Programming and broadcasting activities (60)
- Telecommunications (61)
- Computer programming, consultancy and related activities (62)
- Information service activities (63)
- Scientific research and development (72)

The analysis focuses on the year 2012 due to reasons related to data and general economic development. By and large, the Eurostat database contains regional figures from the year 2000 onwards. As the goal is to produce evidence using the most recent data and as the EU countries have suffered from an economic downturn since 2008, it was decided to focus on the year 2012. This is the latest year for which a sufficient number of observations and their lagged values (i.e. year 2011) are available. In addition, turbulence in the economy is still observable in some countries, although in most countries development in terms of growth and economic outlook in general are turning positive. Due to the still imbalanced macroeconomic situation across countries, it was considered important that the sample also contains those large countries where the economy is doing well, such as Germany and the United Kingdom. This consideration also had an impact on the selection of the explanatory variables.

The number of regions included in the estimations varies depending on variable availability. However, it seems that a minimum of 400 regions should be covered by the most common statistics. Even dividing the observations into quantiles retains enough observations to produce reliable and representative results.

To illustrate the regional divergence with respect to the variables of interest, mean, maximum and minimum values by NUTS2-region are displayed in Table 8.1. Gross value added (GVA) is presented at basic price per total employment, intangible capital per million of active population, tertiary education as a share of total employment and high-tech knowledge intensive services as a share of total employment. Every variable confirms the stark differences across regions, the biggest difference indicated by the intangible capital for which the maximum value is a thousand times higher than the lowest value.

A glimpse at the variation in the dependent variable at country-level is presented in Figure 8.1. Countries with less than 3 regions are excluded from Figure 8.1. The lowest and highest adjacent values, which are not necessarily the highest and lowest values of the entire distribution, are represented by the highest and lowest bars in the figure. As for the boxes, the lowest and highest bars represent the 25th and 75th percentiles, whereas the bar inside the box is the median value. Based on Figure 8.1, competitiveness as measured by the gross value added per employment is low in Bulgaria, the Czech Republic, Greece, Hungary, Poland, Romania and Slovakia. High competitiveness is observed in Sweden, Denmark, Finland, France and Austria. It indeed seems that variation in competitiveness across regions is smaller in countries with overall low competitiveness.

Table 8.1 Mean, maximum and minimum for main variables

	Mean	*Max*	*Min*
GVA	50.1	184.6	7.6
Intangible capital	82.9	612.2	0.6
3rd Education	28.2	71.7	11.8
HKIS	2.5	7.6	0.7

Source: Regional Statistics Database, Eurostat, Authors' calculations

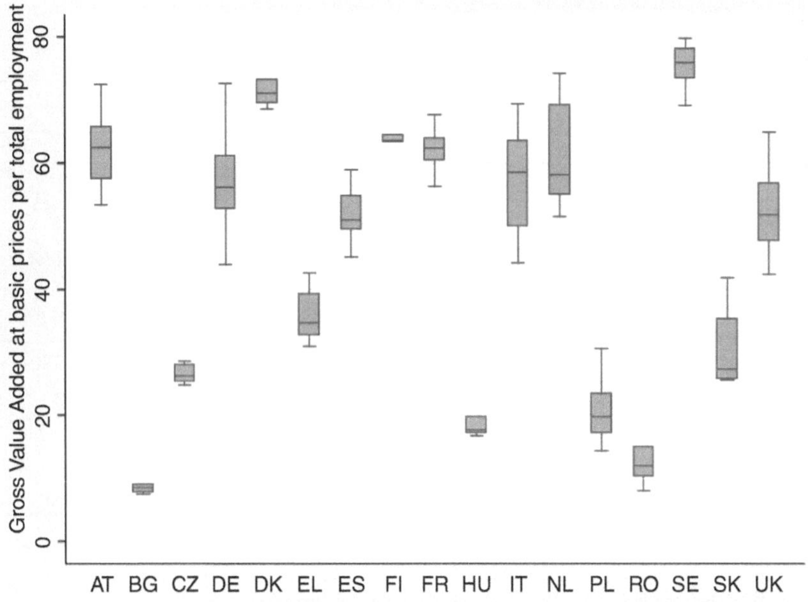

Figure 8.1 25th, median and 75th percentiles of regional gross value added by country

Source: Regional Statistics Database, Eurostat, Authors' calculations

The econometric analysis contains an estimation of two specifications for the regional knowledge production function. The first specification relies on capital, quality of labour force and HKIS as explanatory variables (Model 1), whereas the second specification adds to also intangible capital (Model 2). Tables 8.2 and 8.3 display the quantile regression results for the two specifications. Comparisons are provided both across specifications as well as quantiles. All the explanatory variables are lagged by one year, i.e. referring to the year 2011.

In Model 1, all coefficients for HKIS and capital are positive and statistically significant. Labour force quality has a positive effect in all quantiles, but is statistically significant only in Q2 and Q3. Coefficient sizes remain largely the same across all quantiles for capital, whereas they seem to vary for both HKIS and quality of labour force. The HKIS coefficient has its highest value in Q1, after which it gradually loses its size in higher quantiles. The effect of labour force quality increases from Q2 to Q3. Hence, it seems that the role of HKIS is higher in less competitive regions whereas in highly competitive regions the quality of labour force has a bigger impact.

In Model 2, all the coefficients have positive signs and apart from HKIS in Q1 all are statistically significant. The impact of capital reflects the results of Model 1 in terms of coefficient size; only in Q2 in Model 2 the coefficient size is somewhat lower. The labour force variable peaks in Q3 with a clearly higher coefficient than

Table 8.2 Regional productivity and HKIS in 2012

Model 1	*Q1*		*Q2*		*Q3*	
	Coef.	*Std.*	*Coef.*	*Std.*	*Coef.*	*Std. err.*
HKIS	.154***	.041	.140***	.027	.096*	.054
Capital	.301***	.073	.300***	.044	.291***	.053
3rd education	.134	.086	.184***	.067	.349**	.158
Country effect			Included			
Constant	4.163***	.356	4.266***	.227	4.516***	.320
Number of observations	139		139		139	
Pseudo-R^2	0.869		0.832		0.775	

Note: * p<0.1; ** p<0.05; *** p<0.001, standard errors in parentheses, explanatory variables lagged by one year

Source: Regional Statistics Database, Eurostat, Authors' calculations

Table 8.3 Regional productivity, HKIS and intangible capital in 2012

Model 2	*Q1*		*Q2*		*Q3*	
	Coef.	*Std.*	*Coef.*	*Std.*	*Coef.*	*Std. err.*
HKIS	.040	.030	.112***	.026	.077**	.037
Intangible capital	.078***	.010	.048***	.014	.039*	.023
Capital	.300***	.045	.269***	.042	.291***	.060
3rd education	.234***	.051	.195***	.066	.328**	.139
Country effect			Included			
Constant	3.532***	.290	4.013***	.241	4.161***	.481
Number of observations	137		137		137	
Pseudo-R^2	0.888		0.841		0.781	

Note: * p<0.1; ** p<0.05; *** p<0.001, standard errors in parentheses, explanatory variables lagged by one year

Source: Regional Statistics Database, Eurostat, Authors' calculations

in Q2 or Q1, whereas HKIS has a smaller coefficient in Q3 than in Q2. When the labour force variables are significant in the same quantiles for both models, their coefficient sizes are roughly of the same magnitude and they reflect the same pattern, i.e. higher effect in Q3 than in Q2. When the coefficient is statistically significant, HKIS has the lowest value in Q3 for both models. Intangible capital has a positive and statistically significant effect across all quantiles. However, the coefficient size declines towards the higher end of the competitiveness distribution. Overall, comparison across the models indicates no drastic changes in coefficient significances and sizes; hence, the base specification seems stable and the results are robust. The most interesting result, however, is that in both specifications in Q1 the most important factor contributing towards regional competitiveness is gross fixed capital formation.

Overall, the results suggest that among the three variables aiming to capture features considered typical for a highly competitive region in the era of the knowledge-based economy, the most traditional factor of production has the highest impact. The impact of labour force quality is highest in the most competitive regions, and in those regions it has a stronger effect than any other variable. Nevertheless, HKIS and intangible capital also improve regional competitiveness in all regions but their effect is stronger in regions with lower competitiveness. These results restate the divergence in production structure and the factor of competitiveness among regions. Regions with lowest competitiveness rely on capital as the distinctively most important contributor to competitiveness, hence, indicating strong industrial basis in those regions. From the EU policy perspective, these results call for fine-tuned approaches depending on the regions' positions in the competitiveness distribution. "One size fits all regional policies" would only harm the aggregate development. At country level in regions with low competitiveness and small differences across regions the policies can be more similar.

Conclusions

The objective of the analysis is to look into differences in the impact of high-tech knowledge intensive services, capital, intangible capital and quality of labour force on regional competitiveness in three different competitiveness categories. Regions are divided into classes based on their location and the gross value added per employment distribution, which is then used as the definition for competitiveness. It is assumed that due to the most competitive regions' higher absorptive capacity measured by quality of labour force they would be in the best position to exploit the knowledge (Cantwell and Janne, 1999) and technologies offered by HKIS.

The results indicate clear differences between the three competitiveness classes. In the highest class, the share of employment with tertiary education clearly drives the results. Also, HKIS and intangible contribute to competitiveness but with a smaller effect. In the lowest class, while intangible capital and quality of labour force improve competitiveness, investments in gross fixed capital have the biggest impact.

Due to the strong role of a highly educated labour force in the most competitive regions, it is crucial to maintain and even improve the current level of education. As all the included factors have a positive effect on competitiveness, it is important to focus on the interplay of different factors – i.e. HKIS, innovative enterprises and education providers. As these regions already have a strong knowledge base, regional policies should concentrate on an efficient and effective transfer of knowledge both within and outside of the region. In very practical terms, these observations mean further strengthening of public-private partnerships, closer industry-science collaboration, measures to attract talent from outside the EU and continuous investments in higher education institutions.

In the less competitive regions, the quantity and quality of R&D and innovation is not sufficient to boost competitiveness, thus, it is essential to guarantee that the framework conditions support innovation in enterprises, such as sufficient and easily attainable financing for private R&D. In addition, these regions would benefit from demand side policies, such as public procurement for innovative goods and services and R&D vouchers. The least competitive regions may face competition from emerging economies which will call for a (re)profiling of production. It may require identification of new niche markets or upgrading production in terms of novelty of offering and level of productivity via R&D and innovation. Finally, the less competitive regions might benefit from policy comparison and learning exercises with highly competitive regions that have a similar production structure.

Overall, these results question the empirical analyses conducted on pooled data for EU NUTS2-regions. Production structure across the regions varies so drastically that producing estimated results at mean is likely to give a misleading picture of the reality. However, as the descriptive analysis indicates, there is more variety in competitiveness across regions in higher competitiveness countries than in lower. Hence, at national level the less competitive countries may rely on similar industrial and innovation policies across regions, whereas these results suggest that due to larger variation in competitiveness in highly competitive countries there is a need for more varied policy measures.

Acknowledgements

This study has been carried out within Open Innovation Strategies in IT Service Firms (OISIS)–project funded by Fonds National de la Recherche Luxembourg through its CORE Programme (Contract: C12/IS/3982385).

References

Amara, N., Landry, R., Doloreux, D. 2009. Patterns of innovation in knowledge-intensive business services. *Service Industries Journal*, 29(4), pp. 407–30. Available at: www.tandfonline.com/doi/abs/10.1080/02642060802307847.

Antonelli, C., Patrucco, P.P., Quatraro, F. 2010. Productivity growth and knowledge externalities? An empirical analysis of agglomeration economies in European regions. *Economic Geography*, 87(1), pp. 23–50. Available at: www.mendeley.com/catalog/productivity–growth–pecuniary–knowledge–externalities–empirical–analysis–agglomeration–economies–eur

Ark, van B., Inklaar, R., McGuckin, R.H. 2003. ICT and productivity in Europe and the United States. Where do the differences come from? *CESifo Economic Studies*, 49(3), pp. 295–318. Available at: www.eco.rug.nl/~inklaar/papers/ictdecompositionrev2.pdf.

Arrow, K. 1962. Economic welfare and the allocation of resources for invention. In Nelson, R. (ed.) *The rate and direction of inventive activity*. Princeton: Princeton University Press, pp. 609–26.

Asheim, B. T. 1996. Industrial districts as 'learning regions': a condition for prosperity. *European Planning Studies*, 4, pp. 379–400.

Asikainen, A–L. 2014. Small country strategies in complementing national innovation systems. Forthcoming in *International Journal of Business Innovation and Research*. Available at: http://www.inderscience.com/info/ingeneral/forthcoming.php?jcode=ijbir

Asikainen, A–L. 2015. Innovation modes and strategies in knowledge intensive business services. *Service Business*, 9(1), pp. 77–95. Available at: http://link.springer.com/article/10.1007%2Fs11628-013-0219-5

Audretsch, D. B., Feldman, M. P. 1996. R&D spillovers and the geography of innovation and production. *American Economic Review*, 86, pp. 630–40.

Autio, E. 1998. Evaluation of RTD in regional systems of innovation. *European Planning Studies*, 6(2), 131–40.

Baumol, W.J. 1967. Macroeconomics of unbalanced growth: the anatomy of an urban crisis. *American Economic Review*, 57, pp. 415–26.

Baumol, W.J. 2002. Services as leaders and the leader of the services. In Gadrey, J., Gallouj, F. (eds) *Productivity, innovation and knowledge in services*. Edward Elgar, pp. 147–63.

Beaudry, C., Breschi, S. 2003. Are firms in clusters really more innovative? *Economics of Innovation and New Technology*, 12, pp. 325–42.

Boschma, R.A. 2004. Competitiveness of regions from an evolutionary perspective. *Regional Studies*, 38, pp. 1001–14.

Braczyk, H. J., Cooke, P., Heidenreich, M. 1998. *Regional innovation systems: the role of governance in a globalised world*. London: Routledge.

Cainelli, G., Evangelista, R. and Savona, M. 2006. Innovation and economic performance in services: a firm-level analysis. *Cambridge Journal of Economics*, 30, pp. 435–58.

Camacho, J., Rodríguez, M. 2005. How innovative are services? An empirical analysis for Spain. *Service Industries Journal*, 25(2), pp. 253–71. Available at: www.tandfonline.com/doi/abs/10.1080/0264206042000305448.

Cantwell, J., Janne, O. 1999. Technological globalisation and innovative centres: the role of corporate technological leadership and locational hierarchy. *Research Policy*, 28(2–3), pp. 119–44.

Carlsson, B., Jacobsson, S., Holmén, M., Rickne, A. 2002. Innovation systems: analytical and methodological issues. *Research Policy*, 31(2), pp. 233–45.

Chadwick, A., Glasson, J., Lawton Smith, H. 2008. Employment growth in knowledge intensive business services in Great Britain during the 1990s – Variations at the regional and sub-regional level. *Local Economy*, 23(1), pp. 6–18.

Ciriaci, D., Montresor, S. Palma, D. 2013. Do KIBS make manufacturing more innovative? An empirical investigation for four European countries. JRC Technical Reports. IPTS Working Papers on Corporate R&D and Innovation, 04/2013. Luxembourg: Publications Office of the European Union. doi:10.2791/58386.

Coad, A., Daunfeldt, O., Holzl, W., Johansson, D., Nightingale, P. 2014. High-growth firms: introduction to the special section. *Industrial and Corporate Change*, 23(1), pp. 91–112. Available at: http://icc.oxfordjournals.org.proxy.bnl.lu/content/23/1/91.full?sid=58bb118c–d749–4483–9695–1b8797760520

Coad, A., Rao, R. 2008. Innovation and firm growth in high-tech sectors: a quantile regression approach. *Research Policy*, 37(4), pp. 633–48. Available at: http://linkinghub.elsevier.com/retrieve/pii/S0048733308000152.

Cooke, P. 2005. Regionally asymmetric knowledge capabilities and open innovation: exploring 'globalisation 2'– A new model of industry organisation. *Research Policy*, 34, pp. 1128–49.

Cooke, P., Piccaluga, A. 2006. *Regional development in the knowledge economy*. London: Routledge.

Corrado, C., Hulten, C., Sichel, D. 2009. Intangible capital and US economic growth. *Review of Income and Wealth*, 55(3), pp. 661–85.

Crepon, B., Duguet, E., Mairesse, J. 1998. Research, innovation, and productivity: an econometric analysis at the firm level. *Economics of Innovation and New Technology*, 7, pp. 115–58.

Crevani, L., Palm, K., Schilling, A. 2011. Innovation management in service firms: a research agenda. *Service Business*, 5(2), pp. 177–93. Available at: http://link.springer.com/10.1007/s11628-011-0109-7

Danek, A. H., Fraps, T., von Müller, A., Grothe, B., Öllinger, M. 2014. Working wonders? Investigating insight with magic tricks. *Cognition*, 130(2), pp. 174–85.

Desmarchelier, B., Djellal, F., Gallouj, F. 2013. Knowledge intensive business services and long term growth. *Structural Change and Economic Dynamics*, 25, pp. 188–205. Available at: http://dx.doi.org/10.1016/j.strueco.2012.07.003

Doloreux, D. 2002. What we should know about regional systems of innovation. *Technology in Society*, 24(3), pp. 243–63.

Dumais, G., Ellison, G., and Glaeser, E. L. 2002. Geographic concentration as a dynamic process. *Review of Economics and Statistics*, 84, pp. 193–204.

Edquist, C. 2011. Design of innovation policy through diagnostic analysis: identification of systemic problems (or failures). *Industrial and Corporate Change*, 20(6), pp. 1725–53.

Feldman, M. P. 1994. Knowledge complementarity and innovation. *Small Business Economics*, 6, pp. 363–72.

Florida, R. 2002. *The rise of the creative class*. New York: Basic Books

Freeman, C. 1987. *Technology policy and economic performance: lessons from Japan*. London: Pinter.

Gallaher, M.P., Petrusa, J.E. 2006. Innovation in the U.S. service sector. *Journal of Technology Transfer*, 31(6), pp. 611–28. Available at: http://link.springer.com/10.1007/s10961–006–0018–4

Gallouj, F. 2002. Innovation in services and the attendant myths. *Journal of Socio–Econonomics*, 31, pp. 137–54.

Gallouj, F., Weinstein, O. 1997. Innovation in services. *Research Policy*, 26, pp. 537–56.

Gallouj, F., Windrum, P. 2008. Services and services innovation. *Journal of Evolutionary Economics*, 19(2), pp. 141–48. Available at: http://link.springer.com/10.1007/s00191-008-0123-7

García-Quevedo, J., Mas-Verdú, F. 2008. Does only size matter in the use of knowledge intensive services? *Small Business Economics*, 31(2), pp. 137–46. Available at: http://link.springer.com/10.1007/s11187-007-9090-x

Griliches, Z. 1979. Issues in assessing the contribution of of Research and Development in productivity growth. *Bell Journal of Economics*, 10, pp. 92–116.

Griliches, Z. 1982. The search for R&D spillovers. National Bureau of Economic Research. NBER Working Paper No. 3768.

Hanna, N.K. 2010. *Enabling enterprise transformation: business and grassroots innovation for the knowledge economy*. New York: Springer Science + Business Media. DOI 10.1007/978-1-4419-1508-5_6.

Hauknes, J., Knell, M. 2009. Embodied knowledge and sectoral linkages: An input–ouput approach to the interaction of high- and low-tech industries. *Research Policy*, 38(3), pp. 459–69. Available at: http://linkinghub.elsevier.com/retrieve/pii/S0048733308002217.

Hertog, P. 2000. Knowledge-intensive business services as co-producers of innovation. *International Journal of Innovation Management*, 4(4), pp. 491–528. Available at: www.worldscientific.com/doi/abs/10.1142/S136391960000024X.

Hipp, C., Gallego, J., Rubalcaba, L. 2013. Shaping innovation in European knowledge–intensive business services. *Service Business*, 9(1), pp. 41–55. Available at: http://link.springer.com/10.1007/s11628–013–0217–7.

Iammarino, S., McCann, P. 2006. The structure and evolution of industrial clusters: transactions, technology and knowledge spillovers. *Research Policy*, 35: 1018–36.

Jaffe, A.B. 1989. Real effects of academic research. *American Economic Review*, 79(5), pp. 957–70.

Jaffe, A., Trajtenberg, M., Henderson, R. 1993. Geographic localization of knowledge spillover as evidenced by patent citations. *Quarterly Journal of Economics*, 108, pp. 577–98.

Jalava, J., Pohjola, M. 2007. ICT as a source of output and productivity growth in Finland. *Telecommunications Policy*, 17(52). Available at: www.sciencedirect.com/science/article/pii/S0308596107000535.

Lundvall, B. A. 1992. *National systems of innovation: towards a theory of innovation and interactive learning*. London: Pinter.

Malerba, F. 2002. Sectoral systems of innovation and production. *Research Policy*, 31(2), pp. 247–64.

Maroto-Sanchez, A., Cuadrado-Roura, J.R. 2013. Do services play a role in regional productivity growth across Europe? In Cuadrado-Roura, J.R.(ed.) *Service industries and regions. Growth, location and regional effects*. Berlin: Springer-Verlag.

Mas-Tur, A., Ribeiro Soriano, D. 2014. The level of innovation among young innovative companies: the impacts of knowledge-intensive services use, firm characteristics and the entrepreneur attributes. *Service Business*, 8(1), pp. 51–63.

Mas-Verdú, F., Wensley, A., Alba, M., García Alvarez-Coque, J.M. 2011. How much does KIBS contribute to the generation and diffusion of innovation? *Service Business*, 5, pp. 195–212.

Miles, I., 1999. Services in national innovation systems: from traditional services to knowledge intensive business services. In Schienstock, G., Kuusi, O. (eds), *Transformation towards a learning economy*. Helsinki: The Finnish National Fund for Research and Development, Sitra Report No 213.

Miles, I. 2005. Innovation in Services. In Fagerberg, J., Mowery, D., Nelson, R. (eds) *The Oxford handbook of innovation*. Oxford: Oxford University Press.

Miles, I., Kastrinos, K., Flanagan, K., Bilderbeek, R., Den Hertog, P., Huntink, W., 1995. Knowledge intensive business services: users, carriers and sources of innovation. Luxembourg: *Report to the EC DG XIII Sprint EIMS Programme*.

Miozzo, M., Grimshaw, D. 2005. *Knowledge intensive business services: organizational forms and national institutions*. Cheltenham and Northampton: Edward Elgar.

Muller, E., Doloreux, D. 2009. What we should know about knowledge-intensive business services. *Technology in Society*, 31(1), pp. 64–72. Available at: http://linkinghub.elsevier.com/retrieve/pii/S0160791X08000705.

Muller, E., Zenker, A. 2001. Business services as actors of knowledge transformation: the role of KIBS in regional and national innovation systems. *Research Policy*, 30(9), pp. 1501–16.

Nelson, R. 1993. *National innovation systems: a comparative study*. NewYork: Oxford University Press.

OECD. 2011. *Regions and innovation policy*. Paris: OECD.

Pakes, A., Griliches, Z. 1984. Patents and R&D at the firm level: a first look. In Griliches, Z. (ed.) *R&D, patents, and productivity*. Chicago: University of Chicago Press, pp. 55–72.

Porter, M.E. 1990. *The competitive advantage of nations*. Free Press: London.

Rodriguez, M., Camacho, J.A. 2008. Are KIBS more than intermediate inputs? An examination into their R&D diffuser role in Europe. *International Journal of Services Technology and Management*, 10, pp. 254–72.

Rubalcaba, L., Gago, D., Gallego, J. 2010. On the differences between goods and services innovation. *Journal of Innovation Economics*, 1(5), pp. 17–40.

Rubalcaba, L., Gallego, J., Hipp, C., Gallouj, C., Savona, M., Djellal, F., Fornahl, D. 2008. Towards a European strategy in support of innovation in services. A review of key evidence and policy issues. Final Report Brussels. INNOVA Innovation Watch.

Rubalcaba, L., Kox, H. 2007. *Business services in European economic growth*. Basingstoke and New York: Palgrave Macmillan.

Scitovsky, T. 1954. Two concepts of external economies. *Journal of Political Economy*, 62, pp. 143–51.

Simmie, J., Strambach, S. 2006. The contribution of KIBS to innovation in cities: an evolutionary and institutional perspective. *Journal of Knowledge Management*, 10(5), pp. 26–40. Available at: http://dx.doi.org/10.1108/13673270610691152.

Sirilli, G. Evangelista, R. 1998. Technological innovation in services and manufacturing: results from Italian surveys. *Research Policy*, 27(9), pp. 881–99.

Sundbo, J., Gallouj, F. 2000. Innovation as a loosely coupled system in services. *International Journal of Services Technology and Management* 1(1), pp. 15–35.

Tether, B. 2002. Who co-operates for innovation, and why: an empirical analysis. *Research Policy*, 31, pp. 947–67. Available at: www.sciencedirect.com/science/article/pii/S004873330100172X.

Tether, B.S. 2005. Do services innovate (differently)? Insights from the European Innobarometer Survey. *Industry and Innovation*, 12(2), pp. 153–84. Available at: www.tandfonline.com/doi/abs/10.1080/13662710500087891.

Tether, B.S., Hipp, C. 2002. Knowledge intensive, technical and other services: patterns of competitiveness and innovation compared. *Technology Analysis & Strategic Management*, 14(2), pp. 163–82.

Thompson, P. 2006. Patent citations and the geography of knowledge spillovers: evidence from inventor and examiner added citations. *Review of Economics and Statistics*, 82, pp. 383–89.

Vickery, G., Wunsch-Vincent, S. 2009. R&D and innovation in the ICT Sector: toward globalization and collaboration. The global information technology report 2008–2009: mobility in a networked world. World Economic Forum.

Windrum, P., Tomlinson, M. 1999. Knowledge-intensive services and international competitiveness: a four country comparison. *Technology Analysis and Strategic Management*, 3, pp. 391–408.

Wood, O. 2005. A service-informed approach to regional innovation – or adaptation? *The Service Industries Journal*, 25(4), pp. 429–445.

Wood, P. 2002. European consultancy growth: nature, causes and consequences. In Wood, P. (ed.) *Consultancy and innovation: the business service revolution in Europe*. London and New York: Routledge.

9 KIBS as a factor in meetings industry competitiveness creation in Krakow, Poland

Krzysztof Borodako, Jadwiga Berbeka and Michał Rudnicki

Introduction

Contemporary markets are characterised by complexity and changeability. They result from many factors like globalisation and its impact on the nature of competition; the emergence and growth of e-commerce; over-capacity in most industries; currency fluctuations; volatile global monetary flows, modification of customer preferences; the emergence and growth of new services; a revolution in retail; the emergence of new technologies; new business models; numerous alliances and joint ventures (Rastogi, 2000), just to mention some of the most important elements. These circumstances make competition on many markets quite fierce and the meetings industry is not an exception.

In such a complex and competitive environment, enterprises have to use their resources as effectively as possible. Human resources are crucial in reaching business targets. Due to organisational capabilities, innovativeness, expertise, and creative approaches to business, people are one of the most significant factors in firms' competitive advantage. Employees in an enterprise are expected to identify new opportunities, think and act innovatively, discover new paths to growth, develop the necessary capabilities and deploy them rapidly to be ready to cope with challenges. Many of these tasks and activities are knowledge intensive. As creators, owners, and users of knowledge, employees constitute the most valuable resource in the organisation. Certainly, other resources are also important. However, all resources of an enterprise are limited – the number of employees, material resources, capital. This imposes the necessity of cooperation, also in the form of external outsourcing of some services, including Knowledge Intensive Business Services (KIBS).

The literature stresses a close relation between competitiveness and the use of KIBS at a micro, mezzo and macro level (Huggins, 2011). At a micro-level, an increase of KIBS outsourcing is connected with the growing pressures of competitiveness on firms in most sectors, as customer expectations are growing (Huggins and Izushi, 2007, Borodako *et al.*, 2014b). Within this environment, firms focus on their core activities, i.e. those they anticipate will provide them with a competitive edge, and increasingly search for sources of external knowledge as part of their innovation management strategies (Abramovsky *et al.*, 2004; Sako, 2006; Huggins and Johnston, 2010). Similarly, at a mezzo and macro-level, regional and national competitiveness is considered to be sustained by facilitating innovation among respective business communities (Porter, 1990; Huggins and Izushi, 2011). This can also be observed within cities. In summary, an increased demand for knowledge as a result of changing competitiveness conditions causes a rise in KIBS vendors and the scope of cooperation. Firms require KIBS suppliers to undertake activities that are not carried out in–house and to access knowledge that is unavailable in-house (Huggins, 2011).

This highlights the fact that nowadays knowledge constitutes the quintessential competitive resource of enterprises (Rastogi, 1999, p. 34). Knowledge can be acquired through learning: permanent learning of staff, learning-by-doing and in the process of cooperation. Within the process of cooperation, knowledge sharing takes place, especially tacit knowledge. This, in turn, provides the basis for developing skills, capabilities, expertise, and innovation in the firm.

The mezzo scale – a regional or city perspective – assumes a "knowledge-based" economic growth path, resulting from new combinations of commercial and technological expertise (Simmie *et al.*, 2004). This approach includes the conventional emphasis on technological innovation and investment, including spin–off from regional universities, and the encouragement of entrepreneurial small-medium enterprises. KIBS are also credited with being urban-based bearers of economic competitiveness and growth (Wood, 2006). The meetings industry profits from many different types of services, including KIBS.

The meetings industry itself, if operating well, is able to make a significant contribution to the national and urban economies of host destinations. Moreover, it also enhances the overall growth of travel and tourism sectors (Sangpikul and Kim, 2009). This results in intensified competition among host destinations that are keen to attract more meetings and events.

The overall impact of the meetings industry is assessed to be larger than the spending associated with attending meetings and events (Braun and Rungeling, 1992; Kim, Chon, and Chung, 2003; Borodako *et al.*, 2014a). The amount of direct spending initiates a broad set of economic interactions that produce additional spending in other sectors of a region or city's economy (Braun and Rungeling, 1992), which constitutes an indirect impact (Borodako *et al.*, 2014a). Thus, the economic impact of meetings and events can be doubled or tripled due to the extensive indirect impact on the host's economies (Lee and Back, 2005), and meetings and events can also provide year-round demand, especially during the off or shoulder seasons (Oppermann, 1996). In addition to economic contributions, intangible benefits generated from the meetings industry cannot be ignored. These may include associated social and cultural benefits to the host destinations, the exchange of ideas and knowledge, the cultivation of business contacts, the provision of forums for continuing education and training, and the facilitation of technology transfers (Dwyer and Forsyth, 1997; Sangpikul and Kim, 2009).

The meetings industry is very complex. There is a wide range of businesses involved in the industry e.g. congress and convention venue suppliers, hotels, airlines, travel companies, catering services, professional conference organisations, event agencies, creative industries, urban transport services, Knowledge Intensive Business Services – to mention the most significant (Borodako *et al.*, 2014d, 2014c). The management of meetings and events requires many subcontractors and deep knowledge in this area. Some kinds of meetings (e.g. huge congresses or incentive travel) are prepared in close cooperation with different types of partners that deliver their experience and knowledge in the process of organisation. Service suppliers involved with events (meetings) possess very similar features as those of KIBS vendors.

The comparative quality of the expertise offered by KIBS firms, including event agencies, raises the quality of final products: meetings and events. It is stressed in the literature that the contribution of KIBS depends on their size, which impacts resources. Large KIBS firms have the potential due to their international human resources and market reach. Small-medium firms usually depend on key staff, including the experience of founders, the recruitment of partners and employees, the quality of interaction with major clients and larger KIBS, continuous project-based learning, formal and informal collaborative arrangements, and other forms of inter-firm networking (Wood, 2006).

To make cooperation more effective, all aspects of the meetings industry should be profoundly research in order to better understand how the various elements are interrelated and to examine their relationships within the industry and city (Smith and Garngam, 2006; Yoo and Weber, 2005).

The meetings industry is a very competitive market. The issue of competitiveness can be discussed at a micro, mezzo and macro level, which involves focusing on the competitiveness of an individual enterprise involved in meeting organisation, the competitiveness of the meetings industry in a city and the competitiveness of a city as a destination on the tourism market. In the paper we would like to concentrate on the micro and mezzo levels.

The aim of this paper is to assess if event-oriented services can be considered KIBS and how they support the increased competitiveness of Krakow's meetings industry.

The text will be structured as follows. In the first part, a literature review is presented concerning: Knowledge Intensive Business Services, with a special focus on event services, the meetings industry and competiveness. In the second part, the research methodology is described; in the third research results are provided. The final section consists of conclusions and further research recommendations and managerial implications.

Literature review

KIBS – the scope of the term

The literature contains many definitions of KIBS (Knowledge Intensive Business Services) formulated by a number of researchers who draw attention to various characteristics that are inherent to this service sector. These definitions can be found in works written by Amara, Landry, and Doloreux (2009), Den Hertog (2000), Leiponen (2006), Muller and Doloreux (2009), Muller and Zenker (2001), and Starbuck (1992). One of the first and most popular was proposed by Miles: these are service companies which, through the creation, accumulation and dissemination of knowledge, contribute to the business processes of other enterprises (Miles *et al.*, 1995). KIBS are identified primarily with their knowledge intensive contribution to other enterprises, taking into account both enterprises in the private and public sectors (Muller and Doloreux, 2007), thereby making a strategic contribution to their business activities (Miozzo and Grimshaw, 2005). These services combine various kinds of highly specialised knowledge in formulating concrete solutions to the problems facing an enterprise (Koschatzky and Stahlecker, 2006). On the basis of the analysis carried out by Miles identifying the properties that are characteristic of these services, as well as the solutions proposed by Muller and Doloreux (2009), we can enumerate the following characteristic traits of KIBS:

1 such services are based on professional knowledge, where knowledge-intensity is also construed within the scope of the exchange conditions between the service provider, recipient or purchaser of services (Miles, 2005),
2 the enterprises offering them take on complicated (complex) activities of an intellectual character, where human capital is the dominating factor (Alvesson, 1995),
3 companies offering highly specialised business services are themselves a basic source of information and knowledge or use knowledge to create intermediate services for the production processes of their clients,
4 enterprises offering KIBS play a large role in the competitiveness of enterprises and provide their services primarily for business.

The above-mentioned characteristics demonstrate that these services are not intended for private consumption, but are rather applied exclusively to the B2B market. Recipients of such services, along with the knowledge they transmit, are other enterprises or public organisations (Strambach, 2001). In turn, enterprises offering KIBS take advantage of the following sources of knowledge: the government, public enterprises and institutions (Yang and Yan, 2010), as well as private enterprises and organisations (den Hertog, 2000).

KIBS are also described as "four high" (Yang and Yan, 2010): a high degree of knowledge, a high degree of technology, a high degree of interaction and a high degree of innovation. The last aspect in particular has attracted a great deal of interest from researchers since it was observed that in the course of providing services for the client, KIBS providers are also required to constantly introduce innovation, acquire knowledge, learn new technologies and create new knowledge that is appropriate for the technological and production requirements of the client, only to implement it, thereby contributing at the same time to its innovative development. Therefore KIBS are highly innovative and act as a driver of innovation in other companies (Hartshorn and Wheeler, 2002), as well as facilitate innovations in other economic sectors (Hipp and Grupp, 2005). KIBS are also called "bridges for innovation" (Czarnitzki and Spielkamp 2003), because they play a leading role in processes and act as facilitators, carriers and sources of innovation (Den Hertog, 2000; Den Hertog and Bilderbeek, 2000; Miles *et al.*, 1995). This issue is of particular significance now because creating innovation is one method of increasing the competitiveness of enterprises that cooperate with KIBS.

Examples of KIBS sectors include computer services, research and development (R&D) services, legal, accountancy, management services, architecture, engineering, technical services, advertising, and market research (Miles, 2005). The literature also includes many examples of KIBS classifications (Miles *et al.*, 1995; Wong and He, 2005; National Science Board, 1995; Javalgi *et al.*, 2011). Despite the large number of research studies conducted with respect to KIBS (Miles *et al.*, 1995; Bilderbeek *et al.*, 1998; Muller, 2001; Kam and Singh, 2004; Toivonen, 2007 and many others), there is no identification of new services (compared to the first KIBS classification) that could be assigned to their group. Research conducted by a team of authors has shown that event services, in particular those provided for industry meetings, have identical characteristics to those of other types of services provided by KIBS (Borodako, Berbeka and Rudnicki, 2014c), hence the proposal of a new KIBS classification that would take event services into account.

Event services as KIBS

The KIBS classification elaborated by the authors for event services includes: organisation, promotion and/or management of events, such as business and trade shows, conventions, conferences and meetings, activities connected with artistic presentations, creation of stands and other display structures and sites as well as technical event maintenance as part of other professional, scientific and technical activities not otherwise classified. These services are provided by specialised companies and organisations (among others, professional conference organisers (PCOs) and destination management companies (DMCs), professional event agencies) on the B2B market that also meet one of the traits characteristic of KIBS (except for those mentioned above).

An event can be defined as a one-time or infrequently occurring event outside normal programmes or activities of the sponsoring or organising body (Getz, 2005) or as "that which is different from a normal day of living" (Goldblatt, 2005). These definitions characterise the same events without revealing the essence and enormous role they play on the B2B market. The significance and role of this issue represents a growing part of event management literature [Tassiopoulos (2000); Shone and Parry (2001); Allen O'Toole, McDonnell and Harris (2002); Goldblatt (2007); Silvers (2004); Van der Wagen (2004); Getz (2005), and Bowdin, Allen, O'Toole, Harris, and McDonnell (2006)]. A number of books are devoted to managing specific types of events, including those by Morrow (1997) on exhibitions (trade and consumer show), Rogers (2003) on conventions, Supovitz and Goldblatt (2004) on sports events.

An event service or more precisely an event organisation service can be understood as a cycle of activities consisting of design, planning, organisation, production and evaluation connected with the organisation of various types of events that serve the realisation of business objectives with respect to the external and internal environment: clients, decision-makers, partners, opinion-making groups and employees. Leaving aside other issues connected with definition of this issue, we would like to emphasise that people are a key element of production, similarly as within KIBS. Their participation in various activities is their main contribution. Of these activities, human capital/social capital is a major input factor (Capik and Drahokoupil, 2011; Jansen, Curşeu, Vermeulen, Geurts, and Gibcus 2011), enabling other firms and organisations to utilise the knowledge and skills or talents of KIBS (Miles, 2008) including event services. Some researchers (McCabe *et al.*, 2000; Montgomery, Rutherford, 1994; Rogers, 1998) underline that organisational, negotiation, communication and creative skills as well as flexibility are even more important than having the specific knowledge needed to provide event services. However, a review of the literature (Fenich, 2014, 2015; Rogers, 2013; Ferdinand, Kitchin, 2012) indicates a wide range of various areas in which people working within this industry should be knowledgeable. These areas also include KIBS services, which thanks to, among others, a unique combination of knowledge in various areas, KIBS create specialists in the field of event services. Moreover, event services, much like other KIBS, are characterised by a high degree of tacit knowledge (Hu, Chang, Lin, and Chien, 2006; Segal-Horn, 2006; Tether, 2005; Windrum and Tomlinson, 1999). Since they are no longer viewed as the conveyors of specific information, KIBS (including event services) are regarded as knowledge intermediaries in interactive processes with their clients as related to tacit knowledge and problem–solving (den Hertog, 2000; Miles, 2005). Muller and Doloreux (2009) acknowledge highly specialised business services as a unique subsector of services because the services provided by KIBS are complicated, unstructured and characterised by their adaptation to the unique needs of the client (Bettencourt *et al.*, 2002, p. 101). These same traits describe event services, which support and create innovation and "customisation, in other words unique solutions linked to "tacit knowledge", which remains an aspect of the human factor. Event design, a human invention, is a critical element that will determine the success of the whole event (Fenich, 2014). Only knowledge, experience and imagination, or limitless imagination and use of the newest technology and innovation allow for the creation of an unforgettable event that achieves the client's objectives. Moreover, personal contact with the ordering party/client plays a significant role during the design phase of an event. Despite greater remote access to such services (thanks to new communication technology), they demonstrate a high degree of interaction between service suppliers and clients, much like other KIBS (Camacho and Rodriguez, 2005; Muller and Doloreux, 2009).

The relationships KIBS enjoy with their clients or users are highly important for the achievement of innovations (Gadrey and Gallouj, 1998; He and Wong, 2009; Landry, Amara and Doloreux, 2012; Sundbo and Gallouj, 2000). The literature stresses that KIBS are often involved in an interactive learning process with their clients and other firms (Den Hertog, 2000). Both types of interaction – with clients and firms – generate new knowledge (Muller and Zenker, 2001) that may result in the development of service innovations. Knowledge–based service firms also benefit from close collaboration with suppliers. Many times technological knowledge and support from suppliers has resulted in enhanced service innovation performance (Hsueh, Lin and Li, 2010). The research results obtained by groups of authors (Borodako, Berbeka and Rudnicki, 2015) reach identical conclusions, which confirm the justified inclusion of event services within KIBS.

The meetings industry

The meetings industry plays a significant role in the functioning of all types of business entities: associations, corporations, institutions, non-profit organisations, national and local government entities and even certain social groups. In the literature, the concept of the meetings industry is used with increasing frequency, although until recently it was synonymous with concepts such as business, MICE or MEEC tourism. Leaving aside the small discrepancies in definition, the meetings industry is defined as an industry that "designs, coordinates and ensures realisation of group experiences, which makes it possible for communities, organisations and enterprises to develop in order to achieve their planned objectives" (MPI, 2014).

The meetings and conventions market is a complex and fragmented industry with many stakeholders: suppliers, agencies/professional intermediaries and industry organisations. The combination of such a wide spectrum of interested parties means that certain locations adopt an active policy of attracting events, thereby ensuring an income for many economic entities (Cheung, Law, 2002; O'Brien, Shaw, 2002). Numerous studies (Lee, Back, 2005; Grado, Strauss, Lord, 1998) demonstrate that certain players in the meetings industry obtain considerable profits by serving meeting participants, which leads to a perception of this industry as a very profitable activity for the entire region. The economic significance of this sector is evident in the numbers: a business traveller has from 300 per cent to 600 per cent more expenses than a typical tourist at the destination (Business Events Council of Australia, 2013). Calculations made by the International Congress and Convention Association (ICCA) indicate that international conferences generate a daily expenditure of $678 per participant (ICCA, 2013). Moreover, each dollar spent in the meetings industry triggers a wide range of economic interactions, which bring additional economic effects (secondary expenses, employment, income and taxes) in other sectors of the economy of a given region (Hu, Hiemstra, 1996). The Global Association of the Exhibition Industry (UFI) estimates that one euro paid to the organiser of an industry event for participation leads to an expenditure of 7 to 10 euros in other sectors of a given region (UFI, 2010).

Competiveness

Competitiveness is generally defined as the ability of entrepreneurs to design, produce, and market goods and services, the prices and non-price qualities of which form a more attractive package of benefits than those of competitors (International Institute for Management Development, 1994).

The OECD definition states that competitiveness means the ability of enterprises, industries, regions or national and international groups to face international competition and assure a relatively high rate of return on the resources used and relatively high and sustained employment (Stankiewicz, 2005, p. 36)

So, competitiveness can be discussed at a micro, mezzo and macro level. At a microeconomic level, enterprises that achieve a greater than average improvement in the quality of offered products and/or a reduction in their relative costs, enabling them to increase their profits and /or market share (Dunford, Louri, Rosenstock 2001, p. 109), are competitive.

Porter (1985, p. 102) defines competitiveness as implementation of a value creating strategy by a company and not simultaneously implemented by a competitor and these strategies are not easily duplicated. Many authors stress that business networks and relationships with business partners are a source of SME competitiveness (Gracia *et al.*, 2011; Karaev *et al.*, 2007; Bek *et al.*, 2013).

The literature demonstrates that competition between enterprises has also been extended to competition between territories such as cities, local units, regions and countries (Gorzelak, Jałowiecki, 2000). At a mezzo level, the competitiveness of an industry can be discussed, including the meetings industry.

In tourism the competitiveness of a destination is key. It is defined as the ability of a destination to attract tourists and satisfy their needs and wants (Enrigth and Newton, 2004). Often added to this definition is the ability of a destination to sustain its market share and power, protecting and developing it over time (d'Hauteserre, 2000).

Analogous to the competitiveness of a destination on the tourism market is the competitiveness of a destination, a city in particular, on the meetings industry market. Unlike a specific manufacturing product, competition among cities has a different structure in terms of integrating all the characteristics of the destination visited (Vanhove, 2006), which include the integration of 41 different sectors consisting of both small and large branches of business which supply individual products and services (e.g. convention centres, hotels, airlines, seaways, railways, rental car companies, travel agencies, PCOs, DMCs, intermediaries of many different types, including: event agencies, creative industry representatives; restaurants, cultural venues, universities (Lundberg, Stavenga, and Krishnamoorthy, 1995).

Collaboration among small- and medium-sized enterprises (SMEs) is an emerging approach to industrial competitiveness. Since first observed in Western Europe, government agencies and private foundations have experimented with policies to stimulate and accelerate forms of inter-firm collaboration — commonly called manufacturing networks. It is assumed that cooperative behaviour helps SMEs firms compete (Setyawan, 2015).

There are many factors influencing the competitiveness of an industry. There are research results concerning the competiveness of some industries and its determinants. For example, seven factors were determined as critical in the pharmaceutical industry: a benchmarking strategy and knowledge structure; organisational culture; information technology; employee involvement and training; leadership and commitment from senior management; a learning environment and resource control; and evaluation of professional training and teamwork. (Hung, Huang, Lin, Tsai, 2005).

It is stressed however, that competitiveness and the resulting rewards can be obtained by taking advantage of knowledge management and intensive learning both by employees and organisations (Liua, Chenb, Tsaia, 2004).

Lee and Park (2002) argue that competition on the meeting market demands not only very attractive convention facilities, but also **high-quality services** provided during the meeting programmes. Leask and Spiller (2002) stress that the provision of "a quality service" with supporting arrangements and assistance during the meetings may encourage repeat business and customer satisfaction. According to Lee and Back (2005), as the competition among convention destinations intensifies, a positive destination image is also an important factor.

Data and methodology

This research was conducted in the city of Krakow, the most popular destination in Poland and one of the key tourist cities in Central and Eastern Europe. At the same time, in the global rankings of Tholon's Top 100, in recent years Krakow also ranked very high as an outsourcing destination (e.g. Tholons, 2014), also surpassing Dublin as the best European destination for many years.

The study incorporated two methods – an online questionnaire and individual in–depth interviews – IDI (semi-structured). In the case of the questionnaire, the database consisted of four groups of firms from Krakow selected for the study: (a) recommended by the Convention Bureau (CB) Krakow PCOs and DMCs, and (b) business restaurants, (c) convention centres (business hotels with conference rooms and private convention venues – other than hotels) and (d) event companies (mostly event agencies and technical support agencies). A total of 90 replies were received by the research team (a return ratio of 37.34 per cent), but for some questions almost 20 respondents did not respond, which makes the sample slightly smaller. All key players on the market were invited to participate in the study and many accepted. This makes the results reliable, but does not allow for generalisations. After collecting information from the questionnaire, interviews were conducted with key stakeholders of business tourism in Krakow. The authors had prepared a scenario for the interview but the interviewer had some scope of freedom in leading the conversation. In the research sample, over one quarter of respondents were in a top management position and one fifth in a mid-level position. One in three companies was a micro-company and the same sample percentage was represented by young companies (with up to nine years on the market). Also, more than one third of the entities (34.4 per cent) were lodging venues (with conference infrastructure), followed in second by event companies (PCOs, DMCs, tourism operators) (24.4 per cent).

The aim of the interviews was to examine cooperation between the organiser, customer and other subcontractors with a special focus on knowledge transfer, innovation, competitiveness and product quality (events). Eighteen interviews were conducted: six with PCOs, eight with event agencies and four with technical support firms. The largest group of companies (eleven entities) were firms with considerable experience on the market (between 10 and 20 years on the market). Five firms have operated on the market for up to ten years and three for more than 20 years. An analysis of the material collected during interviews was carried out with GABEK software – GABEK® (Ganzheitliche Bewältigung von Komplexität – Holistic Processing of Linguistic Complexity). The total recorded time was 15 hours and 14 minutes, which allowed the authors to identify 527 sentences, 315 expressions and 43,937 generated connections. The study was carried out under the Research Project financed by the National Science Centre pursuant to decision DEC-2011/01/D/HS4/03983.

Results

Survey

The development of the meetings and events industry as mentioned before requires different sources of competitiveness. One is the knowledge and expertise delivered by the vendors of such kinds of services. The results of the questionnaire conducted among key players on the Krakow market demonstrates how the importance of particular services differs. The most important services for business tourism (tourists who participate in meetings and events) were those connected with event organisation, promotion and management (40 per cent). It is obvious that these services are ranked first among others. But this market is strongly based on cooperation among many stakeholders (private and sometimes public) and the position of these services proves how crucial managing and organising a meeting or event can be. Advertising services was ranked second – one in five respondents declared that this kind of service is crucial for the industry. In third were again services directly connected with the event and meeting industry – event services (artistic agencies). Next, were types of services that are typical outsourcing services – IT and programming services and Accounting and tax advisory services. These five types of services cover more than 80 per cent of all indications in the survey (Table 9.1).

Within the analysis of the meetings industry and cooperation among different partners, the results of the second question concerning the scope of collaboration with different types of KIBS vendors were very interesting. The first four types of services used most often by the studied companies were so called firm-KIBS, i.e. advertising, IT, accounting and legal services. More than 80 per cent of the respondents indicated the first two types of services – advertising and IT. This might mean that services strictly connected with events for a particular company are less important because they cooperate with temporary partners in area of event organisation and management. More than two thirds of the companies use legal and accounting services. Firm–KIBS were followed in the ranking by event-oriented KIBS – Table 9.2. They are used by more than the half of the respondents' firms (respectively 60 per cent, 56.3 per cent and 53.1 per cent).

Table 9.1 The most important types of KIBS services for business tourism (events)

Type of KIBS services	*Per cent (N=85)*
Services connected with event organisation, promotion and/or management	40.00
Advertising services	20.00
Event services	8.24
IT and programming services	7.06
Accounting and tax advisory services	5.88
Management advisory and PR services	4.71
Market research	4.71
Legal services	2.35
Research and experimental development	2.35
Technical support for events	2.35
Specialist design services	1.18
Employment agencies	1.18

Source: Own elaboration.

Table 9.2 The share of respondents cooperating with KIBS vendors

No	*Type of KIBS services*	*Number of records*	*Cooperation with KIBS* *Yes* *[%]*
1	Advertising	N=82	87.80
2	IT and programming services	N=82	84.15
3	Accounting and tax advisory services	N=84	70.24
4	Legal services	N=83	69.88
5	Services connected with event organisation, promotion and/or management	N=82	59.76
6	Technical support for events	N=80	56.25
7	Event services	N=81	53.09
8	Market research	N=81	50.62
9	Management advisory and PR services	N=80	46.25
10	Specialist design services	N=79	41.77
11	Employment agencies	N=80	40.00
12	Architectural activities	N=81	39.51
13	Technical analysis	N=81	30.86
14	Research and experimental development	N=79	25.32

Source: Own elaboration.

Some factors were mentioned to explain why and how companies in the meetings industry are motivated to use KIBS vendors. Respondents were asked to evaluate the significance of these parameters: to what degree they influence the decision to use or not to use KIBS. The first two are related to quality and refer to the final product delivered to the customer and the relationship during previous cooperation. More than three quarters of respondents declared that the aspiration to improve the quality of the final product is important (42 per cent) or very important (35 per cent) in the decision to cooperate with KIBS. On a similar level, the strong influence (40 per cent) of the quality of relations during cooperation on the decision was expressed. The aim of increasing the competitiveness of the company was treated as less important than the quality of the product, but it was still very highly ranked. It should be stressed that similar values were noted with respect to quality of relations (in the case of moderate and strong influence – respectively 40 per cent and 20 per cent). Respondents evaluated the knowledge of employees with a KIBS company very high – strong influence (47 per cent) and very strong influence (31 per cent). For these four factors, weak and very weak influence was mentioned seldom (between 1.4 per cent to 8.6 per cent of replies). The last, but by no means the least important parameter might suggest that companies tend to focus on high-quality and professional services (products) that for this reason may be expensive. The very low rates for reducing costs with respect to strong and very strong influence and high rates in weak and very weak influence can be treated as proof that Krakow's industry tends to offer a first-class product for international and demanding clients (Table 9.3).

In this part of the results, we can claim that many companies active on the event and meetings industry market use different types of KIBS services. The dominant type of services are related closely with business operation (so called Firm-KIBS), but according to representatives of Krakow's companies, event-oriented service vendors also play a very important role in the development of

Table 9.3 Reasons for cooperating with KIBS vendors

Influence of the following factors on a decision to cooperate with KIBS	*Quality of the final product*	*Quality of relations during cooperation*	*Increasing competitiveness*	*Knowledge of personnel*	*Reducing costs*
N	*69* *Per cent*	*70* *Per cent*	*70* *Per cent*	*71* *Per cent*	*69* *Per cent*
Very strong influence	34.78	14.29	24.29	30.99	15.94
Strong influence	42.03	40.00	40.00	46.48	28.99
Moderate influence	7.25	20.00	20.00	11.27	18.84
Weak influence	2.90	8.57	4.29	2.82	13.04
Very weak influence	1.45	4.29	1.43	1.41	8.70
No influence	11.59	12.86	10.00	7.04	14.49

Source: Own elaboration.

the market. A majority of questionnaire respondents declared that maintaining the high quality of the final product (event, meeting) as well as increasing competitiveness and acquiring knowledge from employees of KIBS companies have a very strong influence on the decision to outsource to KIBS.

Interviews

Delivery of a competitive product on the event and meetings industry market requires meeting many conditions. To learn more about relations among the stakeholders participating in the entire process of event or meeting preparation and management, it is worth examining the results of interviews with key players on Krakow's market. The first expression studied in the research and related to competitiveness is "service of the highest quality". Innovations were the focus in the network of the most connected terms expressed during the interviews. The adopted minimum number of connections among the terms was five – which means there were at least five sentences in all interviews where connected terms occurred. Other important terms closely related to high quality were: technical KIBS, event design, professionalism, specialisation and technical event services. As we can see on the graph the terms related to *service of the highest quality* are those that somehow support firms to achieve this high level – professionalism, specialisation. Both are strongly based on the knowledge and experience of contractors – so the examined contractor can afford high-quality services and improve the competitiveness of the firm (Figure 9.1).

Knowledge transfer is one of the key features of KIBS. It can be accomplished via different means but should always lead to increased competitiveness of the company. The study respondents – leading meetings industry managers in

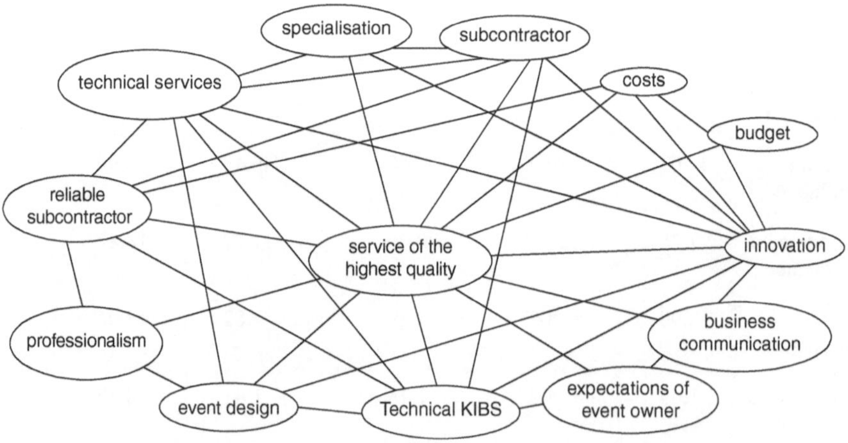

Figure 9.1 Network connections as an expression of "service of the highest quality"
Source: Own elaboration.

Krakow – often referred to the pull and push of ideas and knowledge. They discussed attempts to collect detailed expectations from an event owner or listen to subcontractors who are professionals in AVL technics or other staging and non-staging activities. In both directions the flow of knowledge leads to its distribution among many project partners (meeting or event organisation). The transfer of knowledge (Figure 9.2) results from the expertise of PCOs as well as event agencies, and as mentioned before some aspects of knowledge come back to them (the PCO or event agency) through learning by doing with cooperation partners. All these aspects stress three conclusions from this part of the analysis. Companies deliver expert knowledge to their clients through meetings and industry events. Second, the companies that tend to be more competitive outsource services related to knowledge. And third, when event agencies (or PCOs) act as intermediaries between the event or meeting owner and their clients (participants) it leads to increased innovation in the firm and support for more dynamic development.

When we think about event companies, we are usually referring to an actor who tries to manage all aspects of an event or meeting and in the meantime to generate original solutions fulfilling the expectations of the event owner. As mentioned in the previous part of the analysis based on the questionnaire, the companies confirmed that they mostly buy services from firm-KIBS vendors (i.e. accounting, advertising, IT or legal services) to lead their business. In the case of the interview, we noted that event-oriented firms (called Event-KIBS) mostly connect this activity with market-KIBS or technical-KIBS. Both kinds of services are more external to companies and at the same time more connected to the client. This means they are part of the process of creating innovation driven by participants and partly by the owner (the user of this product). An increasingly important role in this industry is played by technology (in the preparation, organisation, commu-

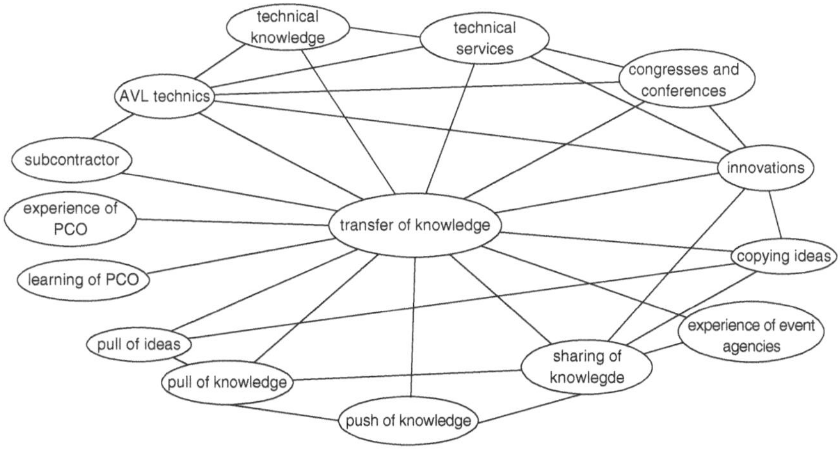

Figure 9.2 Network connections of the expression "transfer of knowledge"

Source: Own elaboration.

nication and evaluation phase). This observation is somewhat confirmed by the results of the study – Event-KIBS are related to Technical-KIBS, as well as technical event services and AVL techniques (Figure 9.3). Connections on the graph with the expression "need to cooperate" show how important Technical-–KIBS and AVL techniques are. This need is mostly motivated by a lack of resources – equipment and knowledge of how to use it efficiently. It was already mentioned that clients (owners of an event or meeting) require high professionalism and at the same time a company specialised in a specific area.

The key term in the context of Event-KIBS and innovation is competitiveness. In the study this term was not defined explicitly because it could be understood differently by respondents during the interviews. But an analysis of the interviews reveals that managers often refer to competition on the market. These terms are not the same, but very connected, because competition inspires hard work to create a potential competitive advantage on the market. How to gain such an advantage was mentioned by many respondents and is reflected on the graph (Figure 9.4). This can be connected with marketing and operational activities. Some managers suggest that competition pressures them to build good relations or greater trust with partners ("regular contacts"). This trust is based on the experience that the partner is reliable and delivers only high-quality services (mostly based on expertise). Some respondents suggested that one way of acquiring new knowledge is networking by participating in industry contests and applying for innovation or industrial awards. Some operational activities undertaken by companies can refer to the use of a leading technical event services vendor or building the creative programme of the event. Both can support companies in increasing competitiveness and gaining the above-mentioned competitiveness advantage.

Managers of PCOs and event agencies referred very often in the interviews to the new knowledge shared by members of the whole project team (consisting of

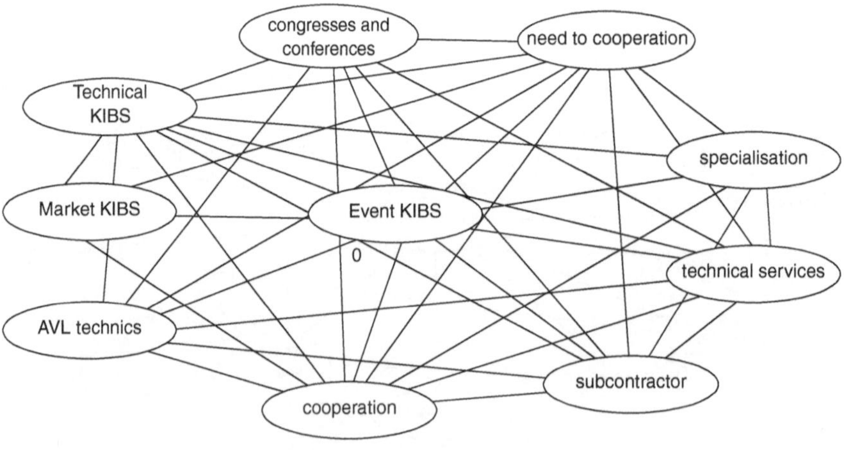

Figure 9.3 Network connections of the expression "event KIBS"
Source: Own elaboration.

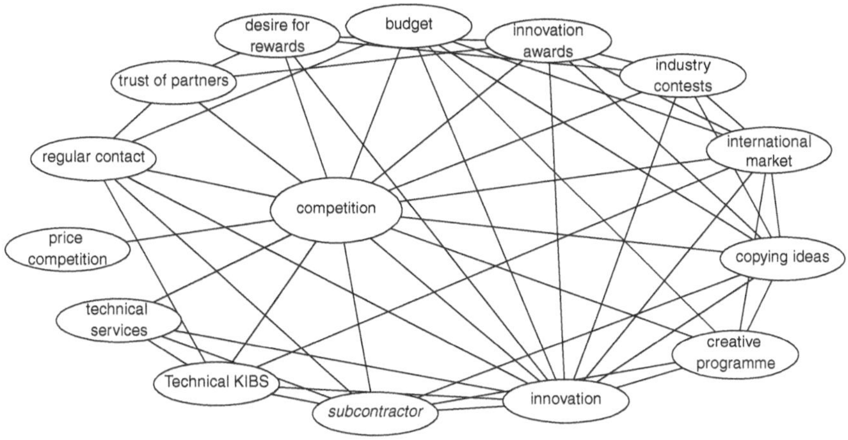

Figure 9.4 Network connections of the expression "competition"
Source: Own elaboration.

many firms). Event companies (agencies and PCOs) offer their clients many years of experience and mainly tacit knowledge. This expertise can be divided into two segments: organisational and technical. Both are very important for the success of an event and both provide value-added to the competitiveness of their clients and on a mezzo level – to regional competitiveness.

Discussion

Careful desk research based on the most common databases such as Ebsco, Scopcus or ScienceDirect does not deliver many results on the topic of cooperation between KIBS and event companies. The topic of the research in the context of events or meetings mostly concerns event visitors or participants and sometimes relations between the event and destination – city or regional (Ziakas, Costa, 2010; Jago et al., 2010; Kostopoulou, Vagionis, Kourkouridis, 2013). From the point of view of our study, we identified papers concerning festivals and exhibitions (as events and meetings) with connection to the terms knowledge, innovation, outsourcing and quality. Unfortunately, review of the above-mentioned database did not identify any relevant documents focused directly on the competitiveness of event companies – they refer mainly to the image of the destination and public funding of different events (mostly festivals).

Knowledge

One of the most important keywords in the study is knowledge – this refers to KIBS but also (as the results of the interviews reveal) to the event industry. Our results demonstrate that knowledge sharing takes place during the organisation of an event or meeting. This result is in line with the research conducted by Ragsdell and Jepson (2014), who studied knowledge sharing between festival volunteers. As they argue during the event, the members of the team (in this case – volunteers) exchange knowledge in the context of individual, organisational and technological barriers. Our study looked deeper into the relations between different stakeholders of an event – especially KIBS vendors. Another study (Abfalter, Stadler, Müller, 2012) focused on knowledge sharing during events, in particular the problem of key attributes of the festival. They are seasonal and organised by permanent and seasonal staff (that has to obtain and use knowledge fast and efficiently). The results of this study indicate that informal and flexible project structures, depending on the level of participation of team members based on their commitment, involvement, responsibility and seniority, are important to the successful sharing of knowledge with newcomers and seasonal members. The results presented in this chapter are in line with those already presented because they show that the issue of knowledge sharing during an event is a fundamental issue of the management and team as a whole. Cooperation with KIBS vendors reduces the risk of failure with respect to responsible positions and functions during the event and requires professional execution. Romiti and Sarti (2012) argued that a network member's cooperation within the scope of event organisation can improve the creation and sharing of knowledge – which confirms our results from PCOs and event agencies.

Innovation

The organisation of an event requires proper resources tangible and non-tangible. To attract visitors or participants, organisers need to implement some kind of innovation. According to our study, the most popular ways of implementing new solutions at an event is to hire a KIBS vendor who delivers unique solutions or copy solutions from another market (from abroad). These results are in line with those formulated by Carlsen *et al.*, (2010). They claimed that many forms of agenda, market, service, organisational and financial innovation are available to festival managers.

Outsourcing

This area of study identified in the literature focuses on outsourcing both kinds of services – knowledge and non-knowledge based. From our point of view, the work of Lamminmaki (2008), who focused on accounting and outsourcing management in the hotel industry, is of great interest. Although her studies verified the dependence of accounting activities in a hotel on the decisions to outsource, her findings indicate that the quality of the hotel was a key factor in deciding on the scope of an external services purchase.

Quality

Our research is in line with other studies that referred to quality and satisfaction. A complementary study was conducted by Bruwer (2014, 2015), who examined festival visitors in relation to their satisfaction with generic festival and logistic features, comfort amenities, the festival venue and service staff as indicators of quality perceptions. Logistic features refer to transportation services, but usually this kind of service is managed by the PCO or event agency – which requires specific knowledge of how to coordinate it successfully. Our results concerning the high significance (strong influence) of achieving a high-quality final product on the decision to outsource KIBS (including Event-KIBS) can be confirmed by conclusions from a study carried out in Greece (Papadimitriou 2013). According to this study, service quality dimensions were stronger predictors of visitor behavioural intentions than satisfaction from participation in the festival. Wong, Wu and Cheng (2014) also claimed in their study that the high quality of the event as well as the image is important factors in the success of the festival. Our respondents in Krakow similarly reveal that the main reason to outsource KIBS is to achieve high service quality, which seems key to the success of an event. Event quality was also addressed by Hiller (2015), who focused more on cooperation between event management and music bands as a way of ensuring the high quality of an event. From the point of view of event management, these bands are subcontractors that contribute emotions and creative input to an event, which can also be considered a kind of tacit knowledge.

In conclusion it should be stated that all the results generated for Krakow represent a case study and require other research in other parts of the world for comparison. However, the data gathered during the study prove that services delivered by contractors (partners) during the organisation of an event can be included in the category of Knowledge Intensive Business Services. Additionally, it would be beneficial for industry associations to intensify the network among KIBS-vendors – especially among Event-KIBS and other types of KIBS firms to better use the knowledge they possess. Another suggestion would be to improve communication among all stakeholders in the city to draw their attention to the potential of Event-KIBS and increase the innovation and competitiveness of the industry and destination

Acknowledgement

Project financed with a National Science Centre grant pursuant to decision DEC–2011/01/D/HS4/03983.

References

Abfalter, D., Stadler, R. and Müller, J. (2012). The organization of knowledge sharing at the Colorado Music Festival. *International Journal of Arts Management*, 14(3), pp. 4–15.

Abramovsky, L., Griffith, R. and Sako, M. (2004). *Offshoring of business services and its impact on the UK economy*. London: Advanced Institute of Management.

Allen J., O'Toole W., McDonnell I. and Harris R. (2002). *Festival and special event management* (2nd edn). Milton: Wiley.

Alvesson, M. (1995). *Management of knowledge-intensive companies*, Berlin and New York: de Gruyter.

Amara, N., Landry, R. and Doloreux, D. (2009). Patterns of innovation in knowledge-intensive business services, *The Service Industries Journal*, 29(4), pp.407–30.

Bek, M. A., Bek, N. N., Sheresheva, M. Y. and Johnston, W. J. (2013). Perspectives of SME innovation clusters development in Russia. *Journal of Business & Industrial Marketing*, 28(3), 240–59. doi:10.1108/08858621311302895.

Bettencourt, L.A., Ostrom, A.L., Brown, S.W. and Roundtree, R.I. (2002). Client co-production in knowledge-intensive business services, *California Management Review*, 44, pp. 100–28.

Bilderbeek, R., Hertog, P., Marklund, G. and Miles, I. (1998). Services in innovation: knowledge intensive business services (KIBS) as coproducers of innovation. *International Journal of Innovation Management*, 4(4), pp. 491–528.

Borodako, K., Berbeka, J., Niemczyk, A. and Seweryn, R. (2014a). *Influence of the meetings industry on the economy of Kraków*. Krakow: Foundation of the Cracow University of Economics (Foundation of CUE).

Borodako, K, Berbeka, J., and Rudnicki, M., (2014b). Cooperation as a source of innovation in the tourism sector of Krakow, in: Dias, F., Kosmaczewska, J. Dziedzic, E. and Magliulo, A. (eds) *Tourism research in a changing world*. Leiria: Tourism Research Group of Polytechnic Institute of Leiria, pp. 206–22.

Borodako, K., Berbeka, J. and Rudnicki, M. (2014c). *Professional services in business tourism* (Usługi specjalistyczne w turystyce biznesowej), Krakow: Foundation of the Cracow University of Economics (Foundation of CUE), (in Polish).

Borodako K., Berbeka J. and Rudnicki M. (2014d). Development of knowledge intensive business servicess in the context of business tourism changes in Krakow in a period of economic destabilisation, in: Baaken, T. and Teczke, J. (eds), *Managing disruption and destabilisation*, Kracow-Muenster: International Management Foundation, pp. 307–18.

Borodako, K., Berbeka, J. and Rudnicki, M. (2015). Management of innovations in meetings industry, C.H. Beck, Warszawa (in Polish).

Bowdin, G., Allen, J., O'Toole, W., Harris, R., McDonnell, I. (2006). *Events management* (2nd edn) Oxford: Elsevier.

Braun, B., and Rungeling, B. (1992). The relative economic impact of convention and tourist visitors on a regional economy: a case study. *International Journal of Hospitality Management*, 11(1), pp. 65–71.

Business Events Council of Australia. (2013), Wróblewski, S. (ed.) *Stowarzyszenia profesjonalne w tworzeniu przemysłu spotkań w Polsce*., SKKP, Warsaw.

Camacho, J.A., Rodriguez, M. (2005). How innovative are services? An empirical analysis for Spain, *Service Industries Journal*, 25(2), pp. 253–71.

Capik, P. and Drahokoupil, J. (2011). Foreign direct investments in business services: Transforming the Visegra´d four region into a knowledge-based economy. *European Planning Studies*, 19(9), pp. 1611–31.

Carlsen, J., Andersson, T.D., Ali-Knight, J., Jaeger, K. and Taylor, R. (2010). Festival management innovation and failure. *International Journal of Event and Festival Management*. 1(2), pp. 120–31.

Cheung, C. and Law, R. (2002). Virtual MICE promotion: a comparison of official web sites in Hong Kong and Singapore. *Journal of Convention & Exhibition Management*, 4, pp. 37–51.

Czarnitzki, D. and Spielkamp, A. (2003). Business services in Germany: bridges for innovation. *The Service Industries Journal*, 23(1), pp. 1–30.

Den Hertog, P. (2000). Knowledge-intensive business services as co-producers of innovation. *International Journal of Innovation Management*, 4(4), pp. 491–528.

Den Hertog, P. and Bilderbeek, R. (2000). The new knowledge infrastructure: the role of technology based knowledge-intensive business in national innovation systems, in: Boden, M. and Miles I. (eds). *Services and the knowledge-based economy*, London: Continuum, pp. 222–46.

Dunford, M., Louri, H., Rosenstock, M., (2001). Competition, competitiveness, and enterprise policies, *Munich Personal RePEc A(MPRA)* Paper No. 29971, pp. 109–46.

Fenich, G.G. (2014). *Meetings, expositions, events and conventions*, New Jersey: Prentice Hall.
Fenich, G.G. (2015). *Production and logistic in meetings, expositions, events and conventions*. New Jersey: Pearson.
Ferdinand, N. and Kitchin, P. J. (2012). *Events management: an international approach*. London: SAGE Publications Ltd.
Gadrey, J. and Gallouj, F. (1998). The provider-customer interface in business and professional services. *The Service Industries Journal*, 18(2), pp. 1–15.
Getz, D. (2005). *Event management and event tourism* (2nd edn) New York: Cognizant Communication Corp.
Goldblatt, J. (2005), *Special events: event leadership for a new world*. Hoboken, NJ: John Wiley & Sons.
Goldblatt, J. (2007). *Special events: the roots and wings of celebration* (5th edn) New York: Wiley & Sons.
Gorzelak, G. and Jałowiecki, B. (2000). Competitiveness of regions, *Studia regionalne i Lokalne*, 1(1), pp. 7–24.
Gracia, A., Magistris, T. D. and Albisu, L. M. (2011). Supply chain relationships and SME firms' competitiveness in the Spanish pig-to-cured ham chain. *Journal of International Food & Agribusiness Marketing*, 23, pp. 192–210, doi: 10.1080/08974438.2011.586908.
Grado, S., Strauss, C. and Lord B. (1998), Economic impacts of conferences and conventions. *Journal of Convention & Exhibition Management*, 1, pp. 19–33.
Hartshorn, J. and Wheeler, D. (2002). Facilities strategic business responses to sustainability. *Greener Management International*, 40(Winter), pp. 107–19.
He, Z. L., Wong, P. K. (2009). Knowledge interaction with manufacturing clients and innovation of knowledge-intensive business services firms. *Innovation: Management, Policy & Practice*, 11, pp. 264–78.
Hiller, R.S. (2015). The importance of quality: how music festivals achieved commercial success. *Journal of Cultural Economics*. doi: 10.1007/s10824-015-9249-2.
Hipp, C. and Grupp H. (2005). Innovation in the service sector: the demand for service-specific innovation measurement concepts and typologies. *Research Policy*, 34(4), pp. 517–35.
Hsueh, J. T., Lin, N. P. and Li, H. C. (2010). The effects of network embeddedness on service innovation performance. *The Service Industries Journal*, 30(10), pp. 1723–36.
Hu, C. and Hiemstra, S. (1996). Hybrid conjoint analysis as a research technique to measure meeting planners' preferences in hotel selection. *Journal of Travel Research*, 4, pp. 62–69.
Hu, T. S., Chang, S. L., Lin, C. Y. and Chien, H. T. (2006). Evolution of knowledge intensive services in a high-tech region – the case of Hsinchu, Taiwan. *European Planning Studies*, 14(10), pp. 1363–85.
Huggins, R. (2011). The growth of knowledge-intensive business services: innovation, *markets and networks, European Planning Studies*, 19(8), 1459–80.
Huggins, R. and Izushi, H. (2007). *Competing for knowledge: creating, connecting, and growing*. London: Routledge.
Huggins, R. and Izushi, H. (eds) (2011). *Competition, competitive advantage, and clusters: the ideas of Michael Porter*. Oxford: Oxford University Press.
Huggins, R. and Johnston, A. (2010). Knowledge flow and inter-firm networks: the influence of network resources, spatial proximity, and firm size. *Entrepreneurship and Regional Development*, 22(5), pp. 457–84.
Hung, Y., Huang, S., Lin, Q. and Tsai, M., (2005). Critical factors in adopting a knowledge management system for the pharmaceutical industry, *Industrial Management & Data Systems*, 105(2), pp. 164–83.
International Congress and Convention Association (2013), ICCA 50 statistics. A modern history of international associations meetings, pp. 1963–2012.

Jago, L., Dwyer, L., Lipman, G., van Lill, D. and Vorster, S. (2010). Optimising the potential of mega-events: an overview. *International Journal of Event and Festival Management*, 1(3), pp. 220–37.

Jansen, R. J. G., Curşeu, P.L., Vermeulen, P.A.M., Geurts, J.L.A. and Gibcus, P. (2011). Social capital as a decision aid in strategic decision-making in service organizations. *Management Decision*, 49(5), pp. 734–47.

Javalgi, R.G., Gross, A.C, Joseph, W.B., Granot, E. (2011). Assessing competitive advantage of emerging markets in knowledge intensive business services, *Journal of Business & Industrial Marketing*, 26(3), pp. 171–80.

Kam, W. P. and Singh, A. (2004). The pattern of innovation in the knowledge intensive business services sector of Singapore. *Singapore Management Review*, 26(1), pp. 21–44.

Karaev, A., Lenny Koh, S. C. and Szamosi, L. T. (2007). The cluster approach and SME competitiveness: a review. *Journal of Manufacturing Technology Management*, 18(7), 818–35. doi:10.1108/17410380710817273.

Kim, S., Chon, K., and Chung, K. Y. (2003). Convention industry in South Korea: an economic impact analysis. *Tourism Management*, 24(6), pp. 533–41.

Koschatzky, K. and Stahlecker, T. (2006). Structural couplings of young knowledge-intensive business service firms in a public-driven regional innovation system – the case of Bremen/ Germany, in: Fritsch, M. and Schmude, J. (eds), *Entrepreneurship in the Region*, Boston: Springer.

Kostopoulou, S., Vagionis, N. and Kourkouridis, D. (2013). Cultural festivals and regional economic development: perceptions of key interest groups, in: *Quantitative methods in tourism economics*, Berlin Heidelberg: Springer-Verlag, pp 175–94. doi: 10.1007/978-3-7908-2879-5_10.

Lamminmaki, D. (2008). Accounting and the management of outsourcing: an empirical study in the hotel industry. *Management Accounting Research*. 19, pp. 163–81.

Landry, R., Amara, N., Doloreux, D. (2012). Knowledge exchange strategies between KIBS firms and their clients. *The Service Industries Journal*. 32(2), pp. 291–320.

Leask, A., and Spiller, J. (2002). U.K. conference venues: past, present, and future. *Journal of Convention and Exhibition Management*, 4(1), pp. 29–54.

Lee, M. J. and Back, K. J. (2005). A review of convention and meeting management research 1990–2003: identification of statistical methods and subject areas. *Journal of Convention & Event Tourism*, 7, pp. 1–20.

Lee, T. H., and Park, J. (2002). Study on the degree of importance of convention service factors: focusing on the differences in perception between convention planners and participants. *Journal of Convention & Exhibition Management*, 3(4), pp. 69–85.

Leiponen, A. (2006). Organization of knowledge exchange: an empirical study of knowledge intensive business service relationships. *Economics of Innovation and New Technology*. 15(4–5), pp. 443–64.

Liua, P., Chenb, W. and Tsaia, C. (2004). An empirical study on the correlation between knowledge management capability and competitiveness in Taiwan's industries. *Technovation*, 24(12), pp. 971–77.

Lundberg, E. D., Stavenga, M. H., and Krishnamoorthy, M. (1995). *Tourism economics*. New York: John Wiley & Sons.

McCabe, V., Poole, B., Weeks, P. and Leiper N. (2000). *The business and management of conventions*. Milton, Qld: John Wiley and Sons.

Meetings Planners International MPI. (2014). Introduction: the real value of the meetings industry.

Miles, I. (2005). Knowledge intensive business services: prospects and policies. *Foresight: The Journal of Futures Studies, Strategic Thinking and Policy*, 7(6), pp. 39–63.

Miles, I. (2008). Miles, patterns of innovation in service industries. *IBM Systems Journal*, 47(1), pp. 115–28.

Miles, I., Kastrinos, N., Flanagan, K., Bilderbeek, R. and den Hertog, P.D. (1995). *Knowledge-intensive business services: users, carriers and sources of innovation*, European Innovation Monitoring Systems EIMS, Publication No. 15, Innovation Programme, DGXIII, Luxembourg. www.escholar.manchester.ac.uk/api/datastream?publicationPid=uk-ac-man scw:75252&datastreamId=FULL-TEXT.PDF.

Miozzo, M. and Grimshaw, D. (2005). Modularity and innovation in knowledge-intensive business services: IT outsourcing in Germany and the UK, *Research Policy*, 34, pp. 1419–39.

Montgomery, R. J. and Rutherford, D. G. (1994). A profile of convention-services professionals. *Cornell HRA Quaterly* (December), pp. 47–57.

Morrow, S. (1997). *The art of the show: an introduction to the study of exhibition management*. Dallas: International Association for Exhibition Management.

Muller, E. (2001). *Innovation interactions between knowledge-intensive business services and small and medium-sized enterprises. An analysis in terms of evolution, knowledge and territories*. Heidelberg: PhysicaVerlag.

Muller, E., Doloreux, D. (2007). The key dimensions of knowledge-intensive business services (KIBS) analysis: a decade of evolution, Working papers firms and region, No. U1/2007. Access from: Fraunhofer Institute for Systems and Innovation Research (Fraunhofer ISI), Department 'Innovation Services and Regional Development'. www.isi.fraunhofer.de/r/arbeitspapiere_u–r/ap_u1_2007.pdf.

Muller, E., Doloreux, D. (2009). What we should know about knowledge-intensive business services, *Technology in Society*, 31, pp. 64–72.

Muller, E., Zenker, A. (2001). Business services as actors of knowledge transformation: the role of KIBS in regional and national innovation systems, *Research Policy*, 30, pp. 1501–16.

National Science Board (1995). *Science and engineering indicators*. Washington, DC: National Science Foundation,.

O'Brien, E., Shaw, M. (2002). Independent meeting planners: a Canadian perspective. *Journal of Convention & Exhibition Management*, 3, pp. 37–68.

Opperman, M. (1996). Convention destination images: analysis of association meeting planners' perceptions. *Tourism Management*, 17(3), pp. 175–82.

Papadimitriou, D. (2013). Service quality components as antecedents of satisfaction and behavioral intentions: the case of a Greek carnival festival, *Journal of Convention and Event Tourism*, 4(1), pp. 42–64, doi: 10.1080/15470148.2012.755885.

Porter, M. (1990). *The competitive advantage of nations*. London: Macmillan.

Porter, M. E. (1985). *Competitive advantage: creating and sustaining superior performance*. New York: Free Press.

Ragsdell, G., Jepson, A. (2014). Knowledge sharing: insights from Campaign for Real Ale (CAMRA) festival volunteers. *International Journal of Event and Festival Management*, 5(3), pp. 279–96.

Rastogi, P.N. (1999). *Managing constant change*. New Delhi: Macmillan.

Rastogi, P.N. (2000). Sustaining enterprise competitiveness – is human capital the answer? *Human Systems Management*, 19, pp. 193–203.

Rogers, T. (1998). *Conferences: a twenty-first century industry*. Harlow, Essex: Adison Wesley Longman.

Rogers, T. (2003). *Conferences and conventions: a global industry*. Oxford: Butterworth-Heinemann.

Rogers T. (2013). *Conferences and conventions: a global industry*. London and New York: Routledge.

Romiti, A., Sarti, D. (2012). Managing knowledge networks for sporting events: evidence from an Italian case. Proceedings of the European Conference on Knowledge Management, ECKM, (ed) Cegarra, J. G., Cartagena: Universidad Politécnica de Cartagena, pp. 1016–26.

Sako, M. (2006). Outsourcing and offshoring: implications for productivity of business services, *Oxford Review of Economic Policy*, 22(4), pp. 499–512.

Sangpikul, A. and Kim, S. (2009). An overview and identification of barriers affecting the meeting and convention industry in Thailand. *Journal of Convention & Event Tourism*, 10, pp. 185–210.

Segal-Horn, S. (2006). Strategy in service organizations, in Faulkner, D.O. and Campbell, A. (eds) *Oxford handbook of strategy*, Oxford: Oxford University Press, pp. 472–506.

Setyawan, A., Isa, M., Wajdi, F. and Syamsudin, N., (2015). An assessment of SME competitiveness in Indonesia, *Journal of Competitiveness*, 7(2), pp. 60–74.

Shone, A. and Parry, B. (2001). *Successful event management*. London: Continuum.

Silvers, J. (2004). *Professional event coordination*. Hoboken, NJ: Wiley.

Simmie, J., Blake, N. Brownill, S., Glasson, J., Holt, R., Marshall, T. Martin, R. Westwood, A. and Wood, P. (2004). *Realizing the full economic potential of London and the core cities*. Report for Core Cities Group (ODPM) and Greater London Authority, London. www.corecities.com/coreDEV/Publications/LondonStudy.htm (accessed August 2006).

Smith, K., and Garnham, R. (2006). Distribution channels for convention tourism: association conventions in Wellington, New Zealand. *Journal of Convention and Event Tourism*, 8(1), pp. 1–30.

Starbuck, W. (1992). Learning by knowledge-intensive firms. *Journal of Management Studies*, 29(6), pp. 713–40.

Strambach, S. (2001). Innovation processes and the role of knowledge-intensive business services, Koschatzky, K., Zulicke, M. and Zenker, A. (eds), *Innovation networks: concepts and challenges in the European perspectives*, Heidelberg: Physica-Verlag.

Sundbo, J. and Gallouj F. (2000). Innovation as a loosely coupled system in services, in Metcalfe, J.S. and Miles, I. (eds) *Innovation systems in the services economy*. Boston: Kluwer Academic Publishers, pp. 43–68.

Supovitz, F., Goldblatt, J. (2004). *The sports event management and marketing handbook*. New York: Wiley.

Tassiopoulos, D. (2000). *Event management: a professional and developmental approach*. Lansdowne: Juta Education.

Tether, B.S. (2005). Do services innovate (differently)? Insights from the European Innobarometer Survey. *Industry and Innovation*, 12(2), pp. 153–84.

Tholons (2014). 2014 Tholons top 100 outsourcing destinations. www.tholons.com (accessed July 2015).

Toivonen, M. (2007). Innovation policy in services: the development of knowledge-intensive business services (kibs) in Finland. *Innovation: Management, Policy & Practice*, 9(3/4), pp. 249–61.

UFI The Global Association of the Exhibition Industry. (2010). Steinbeis Transfer Centre, University Ravensburg, in: Wróblewski, S. (ed.) Stowarzyszenia profesjonalne w tworzeniu przemysłu spotkań w Polsce, SKKP, Warsaw.

Van der Wagen, L. (2004). *Event management for tourism, cultural, business and sporting events* (2nd edn) Frenchs Forest, NSW: Pearson Education.

Windrum, P., Tomlinson, M. (1999). Knowledge-intensive services and international competitiveness: a four country comparison. *Technology Analysis & Strategic Management*, 11(3), pp. 391–408.

Wong, P.K., He, Z.L. (2005). A comparative study of innovation behavior in Singapore's KIBS and manufacturing firms. *The Service Industries Journal*, 25, pp. 23–42.

Wong, J., Wu, H. C. and Cheng, C. C. (2014). An empirical analysis of synthesizing the effects of festival quality, emotion, festival image and festival satisfaction on festival loyalty: a case study of macau food festival. *International Journal of Tourism Research*. doi: 10.1002/jtr.2011.

Wood, P. (2006). Urban development and knowledge-intensive business services: too many unanswered questions? *Growth and Change*, 37(3), pp. 335–61.
Yang, W., Yan, M. (2010). The policy to promote the innovative development of knowledge intensive business services. *International Journal of Business and Management*, 5(11), pp. 190–94.
Yoo, J., and Weber, K. (2005). Progress in convention tourism research. *Journal of Hospitality & Tourism Research*, 29(2), 194–222.

10 Interactive relationships, development effects and knowledge intermediaries among KIBS firms and their clients

A comparison of the Hsinchu and Tainan regions, Taiwan

Tai-Shan Hu, Chien-Yuan Lin and Hung-Nien Hsieh

Introduction

Over the past few decades, activities of Knowledge Intensive Business Services (KIBS) can be interpreted as an important trend in the economic evolution of industrialised nations. In fact, the growing importance of KIBS is a major feature of the so–called "knowledge economy" (Muller, 2001; Miles, 2005; Hu *et al.*, 2006; Toivonen, 2007). KIBS play important roles in the innovation system: as a type of economic activity, and also as professional knowledge providers for other businesses and organisations (Toivonen and Tuominen, 2009; Hu *et al.*, 2013). These two roles are important to specific geographic areas where learning and innovation are preferred. By strengthening the abilities of other system actors to use, generate, and manage local and global knowledge, KIBS can influence dynamic changes of regional innovation systems. This special role enables KIBS to mediate between different sized areas.

Although the literature has repeatedly emphasised the role of KIBS as knowledge and information source provider for other businesses and organisations, KIBS simultaneously acts as promoter, mediator, and enabler of client innovation (Wood, 2002; Simmie and Strambach, 2006; Miles, 2008). In empirical case studies, scholars have suggested that KIBS cluster in large metropolitan areas due to the agglomeration benefits they enjoy with their clients (Keeble and Nachum, 2002; Muller and Doloreux, 2009). However, few empirical studies have focused on interpreting the role of KIBS as knowledge intermediary to mediate and transmit knowledge among actors in different types of geographic innovation systems. The knowledge intermediary role of KIBS has compound characteristics and is also influenced by the circumstances of knowledge flow. Therefore, this intermediary function may vary when actors exchange different types of knowledge, or when a different knowledge type dominates a specific regional innovation system.

Furthermore, the increasing importance of learning and innovation also helps KIBS to serve as knowledge intermediaries. KIBS firms are private sector companies that provide professional knowledge to other businesses and organisations (Toivonen, 2004). KIBS are key sources for information, consultation and professional knowledge, and have expanded tremendously during the past decade. This expansion demonstrates increased utilisation of KIBS, and that their importance to innovations across economic domains will also increase over time (Miles, 2005).

Additionally, according to the previous literature (Scarborough & Lannon, 1989; Sundbo, 1997), KIBS firm innovation activities gradually became integrated into the projects of their customers. This characteristic was crucial to the provision of professional services by KIBS firms, and led to such firms acquiring characteristics of high-tech companies. While the expansion of KIBS is unquestioned, its economic impact on knowledge creation and accumulation requires further examination and research.

This investigation mainly attempts to link KIBS to the analytical structure of concepts, including regional innovation systems, knowledge exchanges and patterns of innovation. This study includes two parts. First, this study investigates the role and function of KIBS in an innovation system. This study explores the evolution capacities of KIBS firms and their key influence factors by examining the interactions between these service firms and their clients, and explores the interactive transformation associated with the different evolutionary stages of local industrial structure. Second, this study focuses on exploring the process of knowledge exchange in different geographic innovation systems. The analytical results are simultaneously used to analyse intermediary functions of KIBS in different metropolitan regions in southern and northern Taiwan. Restated, this study examines how KIBS firms act as intermediaries in different geographic and spatial conditions in Taiwan.

Literature review

The formation and knowledge characteristics of KIBS

The technological, economic and social importance of KIBS has gradually increased in developed countries at both the national and regional levels (Doloreux *et al.*, 2010). KIBS have rapidly grown since the 1970s, and were reviewed by interdisciplinary research in the early 1990s (Illeris, 1991; Miles *et al.*, 1995). These developments led to the recognition that KIBS are key components of an innovation system (Cooke and Leydesdorff, 2006), and are important to knowledge exchange within an economic system (Czarnitzki and Spielkamp, 2003; Muller and Zenker, 2001; Miles, 2008).

The transformation from the perspective of KIBS firms being focused on providing information-based services to that of them providing knowledge-driven services demonstrates that KIBS firms rely on professional expertise or skills related to specific technology or functions. These knowledge intensive businesses become the main sources of information and knowledge, or provide services that are inputs in products or production processes of other companies (Windrum and Tomlinson, 1999). This shows that KIBS are no longer considered conveyors of specific information, but rather knowledge intermediaries in interactions with their clients in relation to tacit knowledge and problem-solving (den Hertog, 2000; Miles, 2005).

Although KIBS firms mostly act as enablers, sometimes they are also innovators and develop new knowledge utilisation methods, production service methods, and service delivery methods (Camacho and Rodriguez, 2008). From the knowledge dimension, KIBS firms have gradually transformed from initially transferring professional information to their clients to become more than mere knowledge providers, also becoming influential partners to whom clients turn for assistance with problem solving.

Previous research on innovation has analysed numerous aspects of support for manufacturing technology innovation, particularly increased focus on KIBS innovation activities. Several studies have identified significant differences between the innovation model of KIBS firms and that of manufacturers (Tether, 2005; Wong and He, 2005; Freel, 2006; Doloreux and Shearmur, 2010). However, research on innovation-related interactions between KIBS and technology-based firms remains scarce, despite the differences in interaction during the dynamic aspects of local industrial structure.

The innovation and dynamics of KIBS

As for whether interactions between KIBS firms and their clients facilitate knowledge creation and diffusion in an innovation system, Muller and Zenker (2001) noted that the transfer of knowledge from KIBS firms to their clients is not simply a conveyance, but also involves further processing such knowledge via collaboration between KIBS and their clients. This conclusion is consistent with the perspective of Bettencourt *et al.*, (2002), who observed that value-added activities of KIBS mainly include the development of customised services to meet customer demands regarding knowledge accumulation, creation and dissemination. In terms of the knowledge dimension, KIBS firms gradually transform into influential partners of clients, not merely knowledge providers.

Based on the innovation perspective, KIBS may influence customer innovation, and increase customer added value and competitive advantage (Wood *et al.*, 1993; O'Farrell and Moffat, 1995). Therefore, the literature has gradually shifted from viewing KIBS firms as contributors to innovation to regarding them as agents of change for innovation, with a particular emphasis on the almost symbiotic relationship between KIBS firms and their clients (den Hertog, 2000). Researchers have gradually begun to regard KIBS firms as true innovators. Related empirical research also demonstrates that KIBS firms do not merely contribute to the innovation capacity of their customers, but also improve their own innovation capabilities through interacting with those customers (Muller and Zenker, 2001; Toivonen and Tuominen, 2009). Additionally, when considering the innovation dimension, KIBS tend to invest more heavily in innovation than similar companies that are less knowledge intensive. Examples of these investments include human capital and customised services related to research and development (R&D) expenditure, as well as information and communication technology

(Tether and Hipp, 2002; Wong and He, 2005). Based on this argument, KIBS are increasingly seen as the drivers of development. Such firms transform their roles in response to customer needs, contributing to KIBS growing faster than other sectors (Miles, 2005).

Finally, from the geographical dimension, relevant research typically pointed out that KIBS concentrated markedly in metropolitan areas (Wood *et al.*, 1993; Keeble and Nachum, 2002; Hu *et al.*, 2006). As for whether or not customer locations impacted service quality and performance, several empirical investigations failed to discover significant regional differences (O'Farell and Moffat, 1995). Some researchers believed that, through new information and communication technology, long-distance provision of KIBS is feasible, and could even incorporate proximity and diversity advantages (Antonelli, 1999), or the need to rely on proximity during different stages (Muller and Zenker, 2001; Rusten, *et al.*, 2005; Wood, 2006). However, some investigators expressed different viewpoints, namely that innovation activities of KIBS firms reflected their ability to interact with partners, and that the influence of innovation networks and proximity on KIBS differed regionally (Koschatzky, 1999).

Based on the above discussions, the roles of KIBS in the formation of globalisation and spatial polarisation are increasingly important. These roles are also related to the formation of the regional knowledge base. However, empirical studies designed to achieve a better understanding of interactions between KIBS firms and their clients are lacking and difficult to perform. This study thus attempts to explore and analyse these interactions between KIBS firms and their clients, as well as the interactive difference during the evolution in local industrial structure.

KIBS as knowledge intermediaries in a regional innovation system

As for regional innovation systems, Cooke *et al.*, (2000) pointed out two sub–systems. The first sub-system consists of companies in the primary industrial clusters in the region, their clients, suppliers, and the supporting industries of those suppliers. The second sub-system comprises supporting mechanisms, which include infrastructure to support innovation, such as universities and research centers (Hamdouch and Moulaert, 2006; Hu *et al.*, 2015).

A regional innovation system reveals its characteristics primarily through intensive interactions and continuous knowledge flows between and within its two sub-systems (Cooke, 2001). KIBS can stimulate knowledge flows between these two sub-systems in a regional innovation system. Thus, KIBS firms act as intermediaries to a certain degree. In other words, KIBS firms can 'transform' knowledge, which promotes collaborations of actors in a regional innovation system.

Furthermore, a narrowly defined regional innovation system includes R&D activities of universities, research institutions and companies. Such a concept involves innovation patterns that integrate science, technology, and innovation (Coenen and Asheim, 2006; Jensen *et al.*, 2007; Doloreux *et al.*, 2010). On the contrary, a more broadly defined regional innovation system includes all actors and activities in a region that influence learning, knowledge generation and innovation. This broader concept relates to innovation patterns that integrate doing, using, and interaction (Asheim and Gertler 2005; Jensen *et al.*, 2007). Thus, the regional innovation system approach stresses the development of an intensive interactive network of collaborations that stimulate innovations among innovators. The primary hypothesis of this approach is that the proximity between different actors and their environments gradually becomes a key determinant of knowledge generation and dissemination, and simultaneously, a region is essential to innovation (Wolfe and Gertler, 2004; Asheim and Gertler, 2005). Consequently, common cultural and social values allow actors in a region to attract, generate, and disseminate information. These values promote (or possibly hinder) social interactions of different actors involved in knowledge dissemination (Cooke *et al.*, 2000).

Therefore, the role of KIBS firms as knowledge intermediaries may be different based on how a regional innovation system is defined. KIBS firms usually play secondary roles in innovation patterns of a narrowly defined regional innovation system, and are considered replaceable sources of supplemental knowledge derived from firm non-core competitiveness. Thus, the bridging function of KIBS seems to be mainly horizontal, linking firms within and across industries (Tether and Hipp, 2002). Therefore, knowledge exchanges occur during the sales process, particularly when KIBS firms possess professional knowledge that their clients require during the innovation process, and the formation of both competitive relations and interactive networks is likely.

Characteristics of innovation systems and knowledge exchanges in different innovation patterns

The fundamental differences between different types of regional innovation systems are impacted mainly by leading regional innovation patterns (Asheim and Gertler, 2005; Aslesen and Isaksen, 2007a). Based on this concept, this study briefly explores two innovation patterns proposed by Jensen *et al.* (2007), STI (science, technology, and innovation) and DUI (doing, using, and interaction) patterns. First, the STI pattern emphasises types of company innovations that are dictated particularly by scientific knowledge (analytical knowledge) (Asheim and Coenen 2005; Coenen and Asheim, 2006). Most of these innovations occur in R&D departments of large companies, small R&D focused firms, universities, or research institutions. Knowledge generation mostly occurs through the development and testing of formal and scientific models. These activities depend primarily on analytical knowledge partially may also involve synthetic knowledge (Asheim and Coenen 2005; Lorenz and Valeyre, 2006).

Therefore, this argument implicitly demonstrates that the primary function of KIBS in STI industrial clusters is to support STI companies equipped with information and knowledge outside of core competition (Aslesen and Isaksen, 2007a). Relatively speaking, DUI patterns illustrate firm innovation processes. Whether a firm is guided by synthetic or codified knowledge, it innovates primarily to add value to its products and production processes. The knowledge the firm applies in innovation originates from the experiences and competitive advantages its internal staff acquire when they confront new questions. Such knowledge can be generated through collaborations with clients and/or suppliers (Jensen *et al.*, 2007), and includes important tacit factors and relies more on physical proximity. Synthetic knowledge flow mainly occurs at two levels: first, between consumers, producers, and suppliers, and second, internally within companies and local industrial social groups. Therefore, KIBS firms may be located in DUI clusters where they play major roles through converting experience-based and analytical knowledge into applicable knowledge for their clients.

Based on the above discussions, this study reviews dynamic and static knowledge exchanges as discussed by Aslesen and Isaksen (2007b), to explore how such exchanges occur between actors in two sub-systems of a regional innovation system. They noted that static knowledge exchanges involve the transfer of various pieces of complete information and knowledge between actors (Tödtling *et al.*, 2006). This interaction generates no new knowledge. Consequently, typical KIBS customise their expertise to provide tailored services and solutions to specific clients and solve their problems (Strambach, 2008). Generally, the collective learning of KIBS firms and their clients requires more frequent and intensive interactions than simple exchanges of information and standardised services (Toivonen, 2004). Although spatial proximity historically may have been vital in certain types of knowledge exchanges, recent research observed that companies and clusters also require exposure to knowledge across regions (Hsieh *et al.*, 2014); restated, knowledge outside cluster regions. This prevents stumbling blocks arising from obsolete technology and declining market niches (Bathelt *et al.*, 2004; Kautonen and Tukhunen, 2008). This implicitly reveals that KIBS firms are well suited to connect with global business intelligence networks. As a result, KIBS may act as intermediaries between the international and local levels.

Therefore, this study references two different patterns of innovation narrowly and broadly defined by the previous literature and applies them to two metropolitan regions in southern and northern Taiwan, to further analyse the knowledge exchange patterns of KIBS firms as intermediaries in regional innovation systems, especially in different geographic and spatial conditions.

Research design and data collection

Research design

This study analysed interactions among KIBS firms, their customers, and the innovation cycle. According to the literature, the interaction between KIBS firms and their customers possesses the following characteristics: first, innovation involves a complex learning process, knowledge interaction, and demonstration of the ability of firms to evolve; second, interactions with KIBS firms positively influence customer innovation capacity and development performance; third, interactions between KIBS firms and their customers also positively affect the innovation capacity and performance of those firms. Furthermore, interactive differences exist during different stages of the evolution in local industrial structure. Therefore, this study investigates the role and function of KIBS in an innovation system.

This study explores the evolution capacities of KIBS firms and their key influence factors by examining the interactions between these service firms and their clients, and explores the interactive transformation associated with the different evolutionary stages of local industrial structure. Additionally, This study references and applies knowledge dimensions and spatial scales theorised by Aslesen and Isaksen (2007b), Bathelt *et al.* (2004), Tödtling *et al.* (2006) and Wood (2006), to propose four knowledge exchange patterns. One spectrum runs from dynamic exchange to static exchange. Meanwhile, another spectrum runs from knowledge exchanges between KIBS firms and their clients in the same region, through to knowledge exchanges between actors in different regions. Local knowledge exchanges thus consist of local buzz and sparring relations. On the one hand, local buzz is a type of spillover of information and knowledge limited by geographic space. Companies can improve their learning by observing the activities and improvements of other companies. On the other hand, sparring relationships include service delivery and proactive parties at both ends of delivery. To an extent, these relationships result from collaborations between service providers and their clients.

Furthermore, trans-regional knowledge exchanges clarify the differences between sales relationships and pipelines. Sales relationships involve the delivery of standardised services from KIBS firms to their clients, which easily occur through long distance delivery. However, pipelines illustrate companies discretely establishing contacts, particularly with KIBS firms, both domestically and globally (Bathelt *et al.*, 2004; Kautonen and Tukhunen, 2008; Moodysson, 2008). Collective learning may occur when clients possess significant resources and simultaneously require intensive and continuous interactions with KIBS firms. In sum, this investigation references the previous literature and examines each aspect of KIBS firms acting as knowledge intermediaries in different environments.

Data collection

To comprehensively understand and analyse interaction roles and the mutual importance of KIBS firms and their customers in innovation activities, and analyse KIBS firms as intermediaries in regional innovation systems, this study establishes an analytical framework, and also attempts to investigate and analyse the characteristics of KIBS firms located in the local innovation systems of the Hsinchu and Tainan Science Parks and their vicinities. Hsinchu, located in northern Taiwan, is the core of the Taiwanese high-tech industry and home to the largest science park in Taiwan. Tainan, located in southern Taiwan, is a medium city with a smaller science park[1]. Hsinchu and Tainan in sector structure and many important location factors form a good case study with wider implications of our outcomes than only the Hsinchu case, although they are in the different stage of development. This study also explores how these KIBS firms act as knowledge intermediaries in these two regions within Taiwan. The study data primarily come from questionnaires. The survey was conducted in 2010. Although the survey subjects were primarily KIBS firms, they also included technology-based firms with

KIBS functions. The sample from the two study regions totaled approximately 1000 firms. The total number of valid returned questionnaires was 286, with a return ratio of 28.6 per cent. According to Cronbach α value is 0.82, the reliability is high; And according to content validity, the validity of questionnaire is reliable. Samples from the Hsinchu and Tainan regions accounted for 68 per cent and 32 per cent of the total sample, respectively. In the questionnaire, KIBS firms were assigned to the following sectors: legal and accounting services, engineering services, R&D services, market research services, computer system design services, technical services, specialised design services, management consulting services, and other (Industry, Commerce and Service Census in Taiwan, 2006). Technology-based firms with KIBS function included semiconductor industries and communication services. From the previous literature, the questionnaire content contains basic information on the surveyed firms as well as different aspects of various subjects, including firm establishment, knowledge base, innovation activities, networks and obstacles.

The emergence of specialisation interface in Hsinchu and Tainan regions

Hsinchu region: Taiwan's knowledge hub

Hsinchu Science-based Industry Park (HSIP) is Taiwan's knowledge hub. Besides a huge science and industry park, six Universities were established. In 2014, the park accommodated 489 enterprises, with more than 152,000 jobs and approximately annual sales of US$ 40 billion. Over 75 per cent of employees in the Hsinchu Science Park held at least the Junior college, and 2.3 per cent of employees had doctoral degrees. At present, the HSIP covers six locations – the Hsinchu, Zhunan, Tongluo, Longtan and Yilan parks as well as the Hsinchu Biomedical Science Park. Most enterprises are manufacturers involved in integrated circuits, optoelectronics, computer and peripherals, telecommunications, and precision machinery.

HSIP grew particularly rapid from 1991 to 2000. This period saw annual sales grow twelvefold, the number of companies double, and the number of employees quadruple. The presence of nearly 500 companies in HSIP created a strong magnet attracting technical labour, facilitating job mobility of professionals (Hu, 2008), and also enhancing talent quality via mobile and interactive learning. However, the generation and diffusion of innovation depends increasingly on new technologies, which are not merely generated through the learning of internal R&D laboratories, but increasingly are generated through high–frequency interactions, communications, networking and information exchange between companies and other academic and research institutions. KIBS play a key role in these interfaces for new knowledge generation.

The Industrial Technology Research Institute of Taiwan (ITRI), located near HSIP, provided HSIP with the most support during its embryonic stage[2]. ITRI is the Core Incubation force in the region. The factors of their success include stable operations body, complete industrial internal networks, careful screening of start–up businesses and excellent R&D professionals and spin–off (Hu *et al.*, 2015). Approximately 25 per cent of the 171 companies added during 1990 to 1999, the most rapid period of growth and expansion of HSIP, were spin–offs created through this knowledge infrastructure. This phenomenon was consistent with the theory of Brenner (2004), who described that the level of comprehensiveness of a supporting environment facilitated spin–off or start–up formation and survival, further strengthening regional self–growth. That is, KIBS act as a critical interface in creating a comprehensive supporting environment for self–growth. Additionally, out of the 21 companies successful incubated by the ITRI Incubator from 1997 to 2001, 60 percent chose to locate in the Hsinchu area (including HSIP). Of the 30 successfully incubated companies from 2002 to 2006, (nearly 80 per cent located in the Hsinchu area (including HSIP). By 2011, over 80 per cent of the 100 successfully incubated start–up companies chose to locate in HSIP or near ITRI and HSIP. In sum, start–up companies initially were mostly located in HSIP. Following 2005, start–up companies began to cluster near Hsinchu and then expanded northwards. This pattern demonstrates that spin–offs of start–up companies instigate the continuous local diffusion, accumulation and re–generation of new knowledge.

Thus, professional trained or start–up companies incubated by ITRI or universities all chose to locate in HSIP or nearby, causing the spatial boundaries of the Hsinchu cluster and related industries to continue to expand outwards. The analysis of various newly created local KIBS simultaneously with the growing number of companies in HSIP (Figure 10.1) also shows that the number of companies grew rapidly in HSIP from 1996 to 2011. Annual sales have increased significantly since 1994, heavily influenced by the proactive participation of ITRI and intensive spin–offs since 1986. The accumulation of such interactions stimulated demand for and dependence on KIBS, further contributing to demand for various types of specialised support.

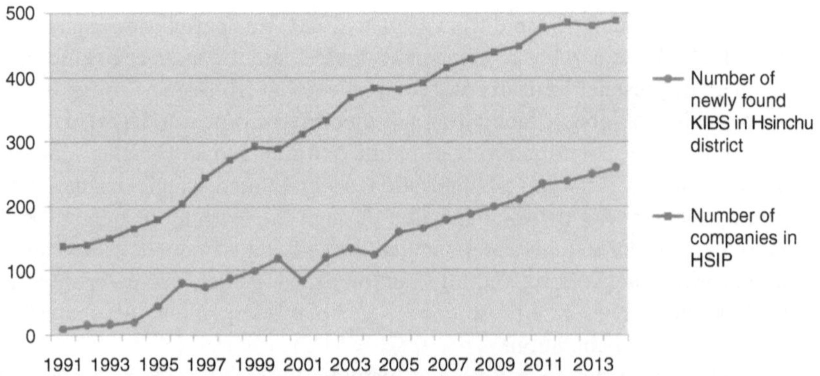

Figure 10.1 Growth of number of companies in HSIP and new establishment KIBS companies in Hsinchu

Tainan region

The Tainan Science-based Industrial Park (TSIP) was established in 1996. Over a decade (1998–2014) of development has seen TSIP attract 180 companies, with annual sales of US$ 20 billion in 2014. Most of these firms are involved in manufacturing, mainly in optoelectronics, integrated circuits and biotechnology. During its first decade of operations, TSIP has achieved faster annual sales growth than HSIP achieved during its first decade. This rapid growth has increased the number of local KIBS firms. During 2001–6, the number of new KIBS firms in Tainan increased by about 50 per cent.

In the embryonic stage, besides support from local resources, firms and talent from HSIP play key roles. Hu (2008) noted that in TSIP, R&D/design cooperation among firms comprised approximately 40 per cent. This figure was higher than for HSIP, where most firms engaged in R&D/design cooperation with other personnel in HSIP, since many were established as branches of TSIP firms. This phenomenon demonstrates the accumulation of such growing demand for and dependence on KIBS locally, further contributing to the formation of various types of specialised support.

Additionally, within the Tainan area, technology transfer from ITRI more than doubled in financial value from 1994 to 2003. By 2010, the TSIP Incubator and Cheng Kung University Incubator successfully incubated 23 start-up companies, 60 per cent of which chose to locate in TSIP. Furthermore, the analysis of numerous newly created local KIBS in correspondence with the growing number of companies in TSIP from 2000 to 2012 (Figure 10.2) suggests a relationship between them. However, whether this situation resembles HSIP requires further investigation.

Analysis of survey results

In this section, the authors present the survey results, and particularly emphasise KIBS in two different metropolitan regions, Hsinchu and Tainan. The authors first propose the main characteristics of KIBS sectors and the evolution

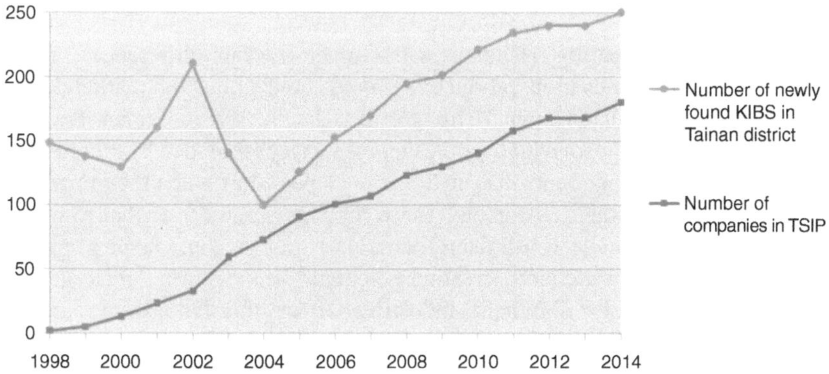

Figure 10.2 Growth of number of companies in TSIP and new establishment KIBS companies in Tainan

capacities based on the survey, and then illustrate how different or similar KIBS firms act as information and knowledge intermediaries between different actors and organisations and in different spatial dimensions.

Analysis of industrial innovation dimensions

Innovation activities of customers generated by innovation support

The survey shows that nearly 90 per cent of surveyed firms in both Hsinchu and Tainan interacted with customers and provided them with innovative support. The main customer innovation activity improved through these interactions and support in Hsinchu was product innovation (about 26 per cent), followed by process innovation, and market development for new products, while in Tainan it was product innovation (about 22 per cent), followed by market development for new products, and process innovation (Table 10.1).

Furthermore, industries requiring the most innovation support in Hsinchu included the semiconductor industry with 22 per cent of firms in this industry requiring innovation support, followed by computers and peripherals (15 per cent), communications (14 per cent), fiber optic electronics (12 per cent), electronic components (11 per cent), technical services (5 per cent), and management consultancy services (3 per cent); In Tainan, the fiber optic electronics industry had the greatest need of innovations support (17 per cent), followed by the precision machinery (13 per cent), electronic components (13 per cent), biotechnology (12 per cent), technical services (1.7 per cent), and management consultancy services (1.3 per cent). In the overall industrial structure of Hsinchu and Taiwan, companies with a high need for innovative support services were mostly manufacturing-oriented. Particularly, most such companies were export-oriented semiconductor, communications and IC design companies in Hsinchu with relatively high output value. Furthermore, the survey result also shows that semiconductor, communications and IC design companies with internal departments that fulfill KIBS functions also had strong demand for innovative support services.

Additionally, comparing Hsinchu with Tainan reveals differences. First, demand for product innovation, production process innovation, and market development for new products in much stronger in Hsinchu than in Tainan. Second, more KIBS firms also needed innovation support services provided by other KIBS firms in Hsinchu (27 per cent) than in Tainan (11 per cent) when they provided innovation support to their customers. These findings demonstrate that to maintain competitive advantage, KIBS firms, particularly in Hsinchu, pursue improvements together with their clients, creating potential for a symbiotic relationship. Furthermore, regionally dominant industries affect the innovation support services required by client firms.

Regarding customer services that facilitate innovation, the survey shows that the greatest customer need was cost reduction, followed by lower barriers to product functions, improved production flexibility, and provision of business knowledge (Table 10.2). For some manufacturers that have difficulty achieving a profit, reducing costs is the best method of increasing their net profit. However, information availability and transparency limits profit generated by reducing raw material costs, and thus companies do not rely entirely on direct reduction of costs to achieve competitive advantage. The survey results implied that the so-called "cost reduction" is actually a "cost innovation"; restated, an innovative business model is required, and must be able to reduce costs.

Customer location

As for the location characteristics of customers involved in innovation interactions or cooperation, the survey results revealed that customers were concentrated in the same areas as KIBS firms, facilitating implementation of innovation activities. That is, spatial proximity to facilitate customer interactions was emphasised (Table 10.3). This finding is consistent with previous research concluding that clustering benefits innovation (Baptista and Swann, 1998; Breschi, 2000). Additionally, spatial proximity benefited business interactions, and external benefits of clustering included more intense communications between companies. Particularly, face to face communications and interactions facilitated rapid exchange and adaptation of knowledge and thus strengthened competitive advantages and achieved the proximity learning effect, as referred to by Zucker *et al.* (1994). However, survey

Table 10.1 Clients acquiring innovation or support due to interactions (Unit: %)

	Product innovation	*Production process innovation*	*Market development for new products*	*Organizational innovation*	*Providing or assisting the development of conceptual innovation plans*
Hsinchu	26.0	23.2	21.3	18.6	4.2
Tainan	22.4	17.3	17.9	11.8	8.3
	Space and facilities of innovation environment	*Product exterior designs*	*Marketing and sales*	*Other*	
Hsinchu	2.2	1.8	1.5	1.2	
Tainan	4.8	8.1	3.3	6.1	

Table 10.2 The analysis of service provision to benefit clients' innovation production (Unit: %)

	Lower costs	*Reducing barriers in product functions*	*Improved flexibility*	*Business knowledge*	*Other*
Hsinchu	31	27	23	17	2
Tainan	35	29	20	16	0

Table 10.3 Innovation interaction customer location (Unit: %)

	Located in the same county/city as KIBSs	*Other East Asian countries*	*Outside East Asia*	*Located in nearby county/city where KIBSs were located*	*Other areas of Taiwan*
Hsinchu	30.1	21.2	19.4	17.2	12.4
Tainan	24.0	15.7	9.3	26.2	24.8

results from Hsinchu revealed that even though nearly 21 per cent of surveyed customers were located in other East Asian countries and nearly 19 per cent of surveyed customers were located in outside East Asia, as many as 30 per cent of them interacted frequently[3]. Further investigation revealed that possible reasons for this phenomena included: lack of suitable local KIBS; or business secrets made local services inappropriate. The latter implies that spatial proximity negatively impacts the consideration of certain factors, similar to the findings of Rusten *et al.*, (2005) and Broekel and Boschma (2012). This further implies that the evolution of industrial structure in Hsinchu cluster is in the transformation or regeneration stage, and globalisation and global competition are becoming significant triggers.

In contrast, the survey results for Tainan revealed that nearly 50 per cent of surveyed customers were located in the same or in a neighboring county/city, while nearly 25 per cent of surveyed customers were located elsewhere in Taiwan. Further investigation revealed that possible reasons for this included: some firms were start-up companies, or branch companies from Hsinchu or Northern Taiwan. Thus the major customers of these firms tended to be located in nearby areas with timely access to support services.

Factors impacting the success and effectiveness of customer interactions

In discussing influences on the success and effectiveness of customer interactions, Muller (2001) noted that interactions between KIBS and technology-based firms can improve mutual innovation capabilities. Although this study previously observed that spatial proximity of interactions could benefit innovation, it temporarily ignored the physical dimension, and interpreted influences on the success and effectiveness of cooperation using only the human dimension. In both Hsinchu and Tainan, 26 per cent of surveyed firms believe that the crucial factor is that customers have good knowledge in their respective professional fields, followed by mutual trust (19 per cent/24 per cent) and negotiator rank (17 per cent/19 per cent). Although the individual knowledge base is rated lower, it remains very important (Table 10.4). Based on these survey results, the author theorised that the flow of knowledge between KIBS firms and customers is two-way. Companies could obtain knowledge from clients, utilise such knowledge to recombine and strengthen their knowledge base, and further apply the increased knowledge obtained through interactions to provide solutions and synergy to other customers.

Table 10.4 Factors impacting success and effectiveness of customer interactions (Unit: %)

	Clients with good knowledge of the field	*Spatial proximity*	*Mutual trust*	*Ranking level of negotiators*	*Individual's knowledge base*	*Other*
Hsinchu	25.9	20.8	19.2	16.9	16.1	0.1
Tainan	26.4	10.0	24.6	18.8	19.2	1.0

As for the issue of the type of partners with whom cooperation most benefited innovation services, the survey indicated that approximately 49 per cent and 55 per cent of surveyed firms in Hsinchu and Tainan, respectively identified cooperation with partners providing equipment as the best form of cooperation for enhancing future innovation services, followed by cooperating with other service companies (34.8 per cent in Hsinchu and 26.2 per cent in Tainan). According to this study, interactions with these partners could strengthen firm knowledge base and enhance services to other clients through interactions generating increased knowledge. This proved again that the flow of knowledge in interactions between KIBS and their clients was two way, and benefited further accumulation of knowledge base.

Another important influence on the success and effectiveness of customer interactions is the connection and maintenance of professional networks (Dakhli and Clercq, 2004). This survey demonstrated that 60 per cent and 46 per cent of surveyed firms in Hsinchu and Tainan, respectively maintained interactions with partners that had influenced their work through interactions in the past. Regarding interactions involving firms in the same professional fields, 80 per cent of the interactions of surveyed firms from the Hsinchu sample were with non-competitors, while the figure for the Tainan sample was 68 per cent. The remaining 20 per cent and 32 per cent of interactions for the two samples were interactions with competitors. Thus interactions between non-competitors in the same professional fields were most frequent type of interactions. The author identified the following reasons for this phenomena: Non-competitors in the same professional fields usually included clients, suppliers or strategic partners who communicated and interacted with each other directly, whether face-to-face or via other communication platforms. These interactions and communications facilitated the development of solutions, allowed clients to instantly access services to reduce product manufacturing time and cost, and also helped clients improve their core competitiveness.

Analysis of internal innovation dimensions

Establishing service innovation types

The internal innovation dimension describes how companies provide service innovation from the perspectives of technical support or service department development. Table 10.5 illustrates that over 90 per cent of surveyed firms from the Hsinchu sample innovated internally. Forty three percent of these firms innovating internally did so through providing new or significantly improved services, while 51 per cent used new or uniquely advanced production processes to service clients. In contrast, 82 per cent of firms from the Tainan sample innovated internally. Furthermore, 55 per cent percent of these Tainan firms innovating internally used new or uniquely advanced production processes to service clients, while 43 per cent provided new or significantly improved services. Additionally, types of innovation that correspond with the market include enhancing service quality, cost innovation, expanding service scope and improving image (Table 10.6), particularly given that companies in Tainan are increasingly emphasizing improving service quality. This result shows that internal innovation is intended to enhance products/technologies or sustain product life cycle during different stages.

Obstacles and challenges

Lack of talent is the obstacle companies are most likely to encounter while implementing innovative programs, being encountered by 18.8 per cent and 20.1 per cent of companies Hsinchu and Tainan, respectively, followed by difficulties in estimating the cost of innovation programs, limited capital, organisational innovation procedures, lack of internal innovation factors, and unwillingness to cooperate with companies in the same professional field (Table 10.7). This

Table 10.5 Innovation types established in the past three years (Unit: %)

			Providing new or significantly improved services	*Use new or uniquely advanced production process to service clients*	*Other types of innovation*
Hsinchu	Innovation types established	90.6	43.0	41.1	6.5
	Innovation types not established	9.4	—	—	—
Tainan	Innovation types established	82	35	45	2
	Innovation types not established	18	—	—	—

Table 10.6 Service innovation types established previously and how they correlate with the market (Unit: %)

	Service quality enhancement	*Cost innovation*	*Expansion of scope of services*	*Image improvement*	*Other*
Hsinchu	28.4	25.3	24.2	18.0	4.1
Tainan	36.8	22.0	23.2	17.2	0.8

result is consistent with the finding of Nonaka *et al.* (2000). A company is an entity that produces knowledge to describe knowledge transformation, which may occur between companies and develop into functions that present firm knowledge generation capabilities. On the other hand, talents are knowledge actors. How companies retain and recruit talent to improve their competitiveness is extremely important, echoing the aforementioned importance of establishing and maintaining talent networks.

As mentioned above, difficulties in estimating the cost of innovative programs and limited capital are the second and third most commonly encountered problems. The survey shows that 33 per cent and 35 per cent of surveyed companies in Hsinchu and Tainan, respectively, have invested less than 5 per cent of their turnover in innovation over the past three years. However, 50 per cent and 39 per cent of surveyed companies in Hsinchu and Tainan, respectively, maintained their investment in innovation at 8 per cent or more of their turnover (Table 10.8). Additionally, despite obstacles existing in relation to talent and capital, the survey shows that some companies assumed challenges during the past three years. In Hsinchu, these companies frequently focused on developing strategies for increasing customer numbers, followed by in-house innovation strategies, research equipment upgrades, product marketing and exhibition strategies. In contrast, TSIP has only recently seen clustering, and the companies involved generally focused on research equipment upgrades and developing strategies for increasing customer numbers. This finding implies that the triggers for the interaction between KIBS firms and their clients differ during the different stages of the evolution of local industrial structure.

Table 10.7 Obstacles that companies mostly likely encounter when implementing innovation programs (Unit: %)

	Lack of talent	*Difficulties in estimating cost of innovation*	*Limited capital*	*Organizational innovation procedures*
Hsinchu	18.9	13.8	13.6	13.0
Tainan	20.1	18.2	18.0	11.0
	Lack of innovation internally	*Unwillingness to cooperate with companies in the same industry or same professional field*	*Unwillingness to cooperate with other companies*	*Other*
Hsinchu	11.8	11.7	11.6	5.7
Tainan	10.0	10.9	11.3	0.5

Table 10.8 Percentage of turnover invested in innovation over the past three years (Unit: %)

	More than 8%	*5~8%*	*3~5%*	*1~3%*	*Less than 1%*
Hsinchu	50	17	6	21	6
Tainan	39	26	20	10	5

Differences in competitive basis and innovation activities

Overall, nearly 60 per cent of KIBS employees had bachelors or higher degrees of education according to the survey result. However, there remain significant differences between metropolitan regions. Table 10.9 shows that over 70 per cent of KIBS employees in the Hsinchu area held higher education degrees and 4.8 per cent of employees had doctoral degrees. In contrast, only 45.6 per cent of KIBS employees in the medium-sized metropolitan region of Tainan had higher education degrees, and only one per cent had doctoral degrees. This reflects that the supply of talent with doctoral degrees is concentrated in northern Taiwan. Even though employees of KIBS firms in the northern metropolitan region of Taiwan had higher education levels, R&D activities were no more prominent in this region. This demonstrates that KIBS firms in northern Taiwan were no more reliant on R&D input. One possible contributing factor is that these KIBS were younger firms with highly educated employees that were knowledge–rich but capital-poor. The other is that some firms[4] which are towards STI benefit from their proximity to research institutions as this provides early pipeline to new research results and a pool of high-educated talent with high-mobility. This phenomenon also strengthens the above discussion. KIBS thus more closely resemble DUI industries activities[5], but with a few characteristics of STI.

Furthermore, this study examined the internal innovation dimension of the surveyed KIBS firms to examine how they innovate in the service process through technical services or the development of service departments. Table 10.5 shows that over 90 per cent of KIBS firms engaged in internal innovation. Types of innovation aimed at providing new or significantly improved services accounted for 43 per cent,followed by utilizing completely new or unique and superior production processes to serve their clients (41 per cent). Types of innovation corresponding to market demands included enhancing service quality, cost innovation, expanding service boundaries, and image promotion (Table 10.10). This result shows that

Table 10.9 Geographic distribution and educational levels of KIBS employees (Unit: %)

	Supply of talent with doctoral degrees[a]	*College/university and higher education[a]*	*With doctoral degrees[b]*
Northern regions of Taiwan	64.2	—	—
Hsinchu metropolitan region	16.3	70.3	4.8
Tainan metropolitan region	10.3	45.6	1.1

Notes:

a. Represents the nationwide percentage for comparison (Source: University and College Surveys in Counties and Cities of Taiwan, Bureau of Statistics, Department of Education, 2011).

b. Summarised from the survey of 286 samples in this study.

Data source: Summarised from survey results, Bureau of Statistics, Department of Education, 2011.

Table 10.10 Service innovation types established previously and how they correlate with the market (Unit: %)

	Service quality enhancement	*Cost innovation*	*Expansion of scope of services*	*Image improvement*	*Other*
Hsinchu	28.3	25.5	24.2	18.6	3.4
Tainan	36.8	22.0	23.2	17.2	0.8

Data source: Summarised from survey results

internally the intention of KIBS firms in innovating is to enhance products or technology to extend product life cycle. The Tainan metropolitan region differed from the Hsinchu metropolitan region in the innovation patterns involved in KIBS. In Tainan, KIBS primarily utilised new or unique and superior production process to serve their clients, and assigned less importance to providing new or improved services. This may indicate that KIBS firms located in medium-sized metropolitan region need to differentiate their products and services to serve diverse demands in the local market. Innovation patterns that corresponded to market demands in the Tainan region included, in descending order, enhancing service quality, expanding service boundaries, cost innovation, and image promotion.

According to the survey result of innovation achievements, KIBS firms do innovate. Specifically, the majority of KIBS firms believed that they innovated (Table 10.5). Innovation patterns of KIBS firms differed regionally, corresponding to the findings of the previous literature (Amara *et al.*, 2009). However, Table 10 shows differences between firms from different geographical areas in their demands for the incorporation and utilisation of atypical skills and knowledge. That is, KIBS firms located in the medium-sized metropolitan area outside of core regions are not as specialised as KIBS firms located in the core metropolitan region. Furthermore, KIBS firms in the core metropolitan region have bigger local markets, and thus can develop specialised products and services. On the one hand, the existence of large clients in core metropolitan region may reduce the incentive for firms to frequently alter products, services and production processes. On the other hand, KIBS firms in medium-sized metropolitan may need to differentiate their products and services to serve diverse demands in the local market.

Differences in knowledge exchange patterns

According to the survey comparing the Hsinchu and Tainan areas, besides KIBS firms participated in formal innovation collaboration, the former had a higher percentage of KIBS firms engaging in informal innovation collaboration (i.e. 32.8 per cent in Hsinchu and 15.5 per cent in Tainan). This finding is consistent with research by Hu (2008, p.171). That is, Hsinchu metropolitan regions have more informal innovation interactions, while formal exchanges of knowledge are utilised more frequently by KIBS firms in Tainan metropolitan regions. In this base, this finding is related to how KIBS firms in different types of regions actually sought and obtained external knowledge. Mutual observation of competitors, informal contacts with clients and suppliers, as well as recruitment of well suited talents, are more important to KIBS firms in Hsinchu metropolitan regions than to those in Tainan metropolitan regions. This demonstrates that KIBS firms in large or core metropolitan regions benefit from regional cluster effects.

Furthermore, the survey indicates that KIBS firms primarily collaborate with their suppliers, clients, universities and R&D organisations in innovation (Table 10.11). This demonstrates knowledge exchange relationships between KIBS firms, academic and research institutions with science-based knowledge, and actors with clients and suppliers that are more user-oriented and possess practical knowledge. In the era of the network economy, the expertise of KIBS firms in networking and internal sources is an important origin of innovation knowledge. Therefore, a higher percentage of KIBS firms in large or core metropolitan regions utilise knowledge of almost even origin type than do those in medium-sized metropolitan regions. However, discrete examinations of each type of knowledge origin and existing knowledge absorption capacities of companies are essential (Hu, 2008). The survey results show that KIBS firms in a core metropolitan region conducted more extensive research and knowledge applications, and simultaneously contacted diverse knowledge origins while acting as intermediaries. When KIBS firms involve into the innovation process with expert knowledge that often has to be adapted to the specific needs of the clients, sometimes KIBS firms (towards DUI) may need more standardised knowledge.

Additionally, Table 10.11 shows that differences also existed in the main innovation partners of KIBS firms in different types of metropolitan regions. A higher percentage of KIBS firms in the Hsinchu area collaborated with their clients and competitors in innovation, which revealed that these KIBS firms were focused on engaging innovation partners more extensively. This phenomenon also implicitly reflects the influence of proximity on the spatial distribution of KIBS in larger or core metropolitan regions; proximity prompts more open innovation strategies (Chesbrough, 2007; Parida *et al.*, 2012). Table 10.11 also demonstrates that a high percentage of KIBS firms in both the Hsinchu and Tainan metropolitan regions collaborated with universities and R&D institutions. This shows that KIBS acted as intermediaries between science-based knowledge, and user-oriented practical knowledge in both types of metropolitan regions.

Table 10.11 Major types of partners of KIBS firms engaged in innovation collaboration[a] (Unit: %)

	Clients	*Universities and research institutions*	*Suppliers*	*Other*	*Competitors*
Hsinchu metropolitan region	36.5	21.3	16.9	15.7	9.6
Tainan metropolitan region	22.8	29.4	25.6	18.0	4.2

Note:
a. Represents the number of samples engaged in innovation collaboration (N = 124)
Data source: Summarised from survey results

Differences of Spatial Intermediaries

Another consideration of KIBS firms as knowledge intermediaries is whether they are involved in knowledge exchanges with leading clients in the region, or involving clients outside the region. The survey result shows that, KIBS firms frequently sought innovation partners in the same region (Table 10.12). Additionally, the survey results show that the search of KIBS firms for innovation partners differed with type of metropolitan region. KIBS firms in Tainan metropolitan regions appear more focused on innovation collaboration with local and domestic companies than those in Hsinchu metropolitan regions, while KIBS firms in core or large metropolitan regions collaborated more with overseas partners. Table 10.12 demonstrates that KIBS firms in a Tainan metropolitan regions were more involved in local/regional innovation systems (approximately 70 per cent of this group of firms have such involvement). Accordingly, these KIBS firms functioned as regional and international knowledge intermediaries. In contrast, KIBS firms in Hsinchu metropolitan regions were more involved in international collaboration (nearly 50 per cent). This shows that KIBS firms in large regions act more as global knowledge intermediaries. This finding is consistent with previous literature that identified large or core metropolitan regions as nodes in global knowledge networks, and concluded that innovation was usually first reapplied and produced in large or core metropolis (Isaksen and Aslesen, 2001; Cooke and Memedovic, 2006). In these networks and processes, KIBS are essential to innovation. Policy-makers cannot ignore KIBS and their contribution to the competitiveness of their clients. Good physical local and regional accessibility, the offering of space (metropolitan suburban but nearby centers) and amenities are potentially useful investment strategies.

KIBS firms in different regions utilised various types of external actors and sources to access innovation knowledge (Table 10.13). KIBS firms in Tainan metropolitan region rely more on knowledge with local origins, while KIBS firms in a core or large metropolitan region rely less on locally developed knowledge and more on knowledge with overseas origins. The survey results also demonstrate that KIBS firms utilise formal or informal interactions to collect information and

Table 10.12 Locations of major partners of KIBS firms involved in innovation collaboration[a] (Unit: %)

	Local	*Other domestic areas*	*Neighboring overseas areas*	*Other overseas areas*
Hsinchu metropolitan region	38.9	14.2	20.1	26.8
Tainan metropolitan region	33.8	37.3	11.4	17.5

Note:
a. Represents the number of surveyed firms involved in innovation collaboration (N = 124)
Data source: Summarised from the survey results

Table 10.13 Primary sources of knowledge of KIBS firms (Unit: %)

	Local	*Other domestic areas*	*Neighboring overseas areas*	*Other overseas areas*
All surveyed firms (N = 286)				
Hsinchu metropolitan region	39.4	17.3	18.7	24.6
Tainan metropolitan region	34.5	33.1	12.2	20.2
Surveyed firms with innovation collaboration (N = 124)				
Hsinchu metropolitan region	40.6	12.8	16.4	30.2
Tainan metropolitan region	38.5	32.9	15.5	13.1

Data source: Summarised from the survey results

knowledge related to innovation. Differences also exist between pipelines utilised by KIBS firms in different metropolitan regions. The geographic locations of these different pipelines also vary among regions. They exhibit almost the same patterns as described above for the relationship between partnerships and knowledge origins and their locations. In core or large regions, few KIBS firms identified local pipelines as crucial to their knowledge exchange. This finding reveals once more that KIBS firms in a core or large metropolitan region function as global pipelines, while KIBS firms in an outlying region fulfill a domestic role.

Analysis of future development trend

According to the findings of the above survey on interactions between KIBS firms and their clients, interactions enabled firms to accumulate more know-how in diverse departments. Restated, specialised companies can collect business information during interactions. Companies then utilise this information to increase the quantity and scope of their knowledge, execute innovations by absorbing knowledge, and further provide innovation support services. Therefore, external service companies must be more innovative than customer internal activities. Consequently, companies that provide innovation support services must also constantly seek new market trends. Interactions in industrial clusters are one of the more advantageous methods of conducting this search. The above analysis also implies that large companies tend to handle corporate business internally because they encompass multiple departments and functions, particularly in relation to technology law. Not only do large companies have their own patent and legal departments, legal services related to their technology are also closely tied to their core businesses. These services are not easy to outsource like other functions, and solutions are developed with the aim of benefiting company innovation activities.

Therefore, in Hsinchu, the survey shows that companies believe the main trends in their professional fields during the next five to ten years include the increasing importance of business know-how of specific customers, closer links to customer strategies, increasing demand for external professional services, with a stronger emphasis on their suitability, expanded scope of customer services, and diversification of international industrial activities (Table 10.14). In Tainan, companies believe the main trends include closer ties to customer strategies and increasing demand for external professional services with stronger emphasis on their suitability. Furthermore, comparing Hsinchu with Tainan reveals that the main trends are closer ties to customer strategies and providing professional services. Although diversification of international industrial activities currently is not significant, it is emerging and becoming an important trigger factor while mature industrial clusters such as Hsinchu require transformation and regeneration.

As for customers involved in knowledge intensive business industries, the surveyed firms wish to strengthen their future collaborations. Table 10.15 lists that approximately 35 per cent of the surveyed firms in Hsinchu and 30 per cent of those in Tainan hope to collaborate in future with technical/R&D services, computer system design services, and management consultant services. This finding implies that KIBS is increasingly important in industrial innovation. Furthermore, regarding the enhancements the surveyed firms expected to achieve through strengthened collaborations with these future partners, Table 10.16 shows that approximately 28 per cent of the surveyed firms in Hsinchu and 30 per cent of those in Tainan anticipated enhanced professional service content and quality. This result implies that KIBS such as technical/R&D services play an intermediary role in enhancing customer content and professional service quality. In the process these services also improve themselves and generate a substantial innovation cycle.

Table 10.14 Future development trends (Unit: %)

	Include increasing importance of business know-how of specific customers	*Closer ties to customers strategies*	*Increasing demand for external professional services with stronger emphasis on their suitability*	*Expanded scope of services for customers*	*Diversification of international of industrial activities*
Hsinchu	24.7	23.8	21.3	15.1	15.1
Tainan	20.8	24.7	23.8	18.8	11.9

Table 10.15 Industries of customers that survey companies intend to strengthen future collaborations (Unit: %)

	Technical/ R&D services	*Fiber optic electronics*	*Semiconductor*	*Communication*	*Computer system designs*
Hsinchu	18.6	17.2	13.9	12.9	8.8
Tainan	23.1	17.9	10.6	4.5	3.3
	Computer and peripherals	*Management consultant services*	*Biotechnology*	*Precision machinery*	*Other*
Hsinchu	7.5	7.3	7.0	6.8	—
Tainan	4.1	4.0	13.8	17.1	1.6

Table 10.16 Improvement brought by customers that companies intend to strengthen in future collaborations with (Unit: %)

	Enhance content and quality of professional services	*Expand strategic ties with clients*	*Acquire business know-how of specific clients*	*Enhance companies' diversity of industrial activities*	*Other*
Hsinchu	28.4	26.9	22.9	20.8	1.0
Tainan	30.2	27.0	17.3	23.5	2.0

Conclusions

Since the 1990s, service industries have become increasingly important in local or global economic development. This study explores the roles and functions of KIBS in the evolution of an area innovation system. Second, this investigation analysed actual interactions between KIBS firms and their clients, and how KIBS firms act as information and knowledge intermediaries in different regional innovation systems guided by different knowledge bases. The manufacturing-orientation in the Hsinchu and Tainan Areas shifted towards R&D and incubation-orientation, and gradually focused on innovation of production processes, products and business management, which stimulated the emergence and formation of KIBS. To accommodate the evolution of the industrial structure in HSIP and TSIP, innovation support services provided by KIBS also transformed from finance, law and architecture to technical R&D, consulting, and design services. Additionally, this study finds that such cumulative interactions also increased demand for and reliance on KIBS, and created different demand for functions and types of KIBS during the evolution of local industrial structure. Particularly in HSIP and the surrounding area, which has three decades of development, the evolution of industrial structure led to the spatial boundaries of the Hsinchu cluster and related theme industries continuing to expand outwards. Furthermore, this situation reintegrates various specialised and diversified knowledge intensive industries into this boundary.

This positively impacts knowledge development of KIBS firms and their clients, and also enhances their abilities to evolve and sustain their innovation and development. However, TSIP has just 15 years of development, and so TSIP and its vicinity remain in the initial stage of investment in knowledge infrastructure, and have difficulty reaching such stage and achieving synergies.

This investigation analysed how KIBS firms act as information and knowledge intermediaries in different regional innovation systems guided by different knowledge bases. Briefly, KIBS firms in a core or large metropolitan region exhibit more potential to act as multi-functional and multi-spatial knowledge intermediaries, which means that KIBS firms in such regions utilise knowledge with more diverse origins, have more innovation partners, apply more multi-dimensional pipelines, and engage more in more international knowledge exchanges compared to KIBS firms in other regions. Although economic policy in many cases focuses on stimulating manufacturing-related R&D within industrial clusters (Bristow, 2005), policy-makers cannot ignore KIBS and their contribution to the competitiveness of their clients. Future research should also focus on the different stage of regional development between their various KIBS functions. In contrast, although KIBS firms in medium metropolitan regions do not have significantly different methods of knowledge exchange than KIBS firms in large metropolitan regions, their collaborators and knowledge origins are less diverse, and they are also less involved in international knowledge exchanges. Restated, this type of non-core region supports simpler roles for knowledge intermediaries, in terms of both function and spatial dimensions. Consequently, KIBS firms in such regions rely less on informal knowledge origins, and focus mainly on local intermediary functions. This heterogeneity requires to be developed on various levels of the city and region simultaneously, according to their characteristics and development stage, which can be promoted by local governments as well.

Acknowledgement

The authors would like to thank the Ministry of Science and Technology of Taiwan for Partially financial supporting this research under Contract Numbers NSC 98-2221-E-216-041-MY2 and NSC 100-2410-H-216- 009-MY2.

Notes

1 The Northern Region includes Taipei, New Taipei, Keelung and Hsinchu Cities, and Taoyuan, Hsinchu and Yilan counties; the population is 10.3 million; there are 27,685 KIBS firms in Northern Region. The Southern Region includes Kaohsiung, Tainan and Chiayi Cities, and Chiayi, Pingtung and Penghu Counties; the population is 6.4 million; there are 9,698 KIBS firms in Southern Region.

2 The Industrial Technology Research Institute was created in 1973. Since its inception, it has trained more than 70 chief executive officers, incubated 165 start-ups and accumulated over 10,000 patents. Currently, it has 13 research units and centers and over 5,000 researchers and professionals.

3 The study defines high frequency interaction as weekly interaction between KIBS firms and clients during a sustained period, usually 1–3 months based on the initial or uncertain stage of the cooperation.

4 The kinds of firms are either younger spin-offs from surrounding research institutions/universities or larger firms. As Klepper and Sleeper (2005) pointed out that spin-offs will have a better chance of survival and growth than start-ups because they will likely follow the routines from the parent company/institution.

5 KIBS often plays a minor role in an STI innovation pattern, primarily because knowledge exchanges often occur in direct communications between researchers. Relatively speaking, KIBS are used as sources of supplemental knowledge, which is established on experience-based knowledge. In the innovation process of a DUI innovation pattern, KIBS must constantly adjust to meet specific needs of their clients. As a result, KIBS act as various types of knowledge intermediaries in the process (Aslesen and Isaksen, 2007a).

References

Amara, N., R. Landry and D. Doloreux (2009) "Patterns of innovation in knowledge-intensive business services", *The Service Industries Journal*, 29(4): 407–30.

Antonelli, C. (1999) "The evolution of the industrial organization of the production of knowledge", *Cambridge Journal of Economics*, 23: 243–60.

Asheim, B.T. and L. Coenen (2005) "Knowledge bases and regional innovation systems: comparing Nordic clusters", *Research Policy*, 34(8): 1173–90.

Asheim, B.T. and M. Gertler (2005) "The geography of innovation: regional innovation systems", in Fagerberg, J., D.C. Mowery and R.R. Nelson (eds) *The Oxford handbook of innovation*. Oxford: Oxford University Press, pp. 291–317.

Aslesen, H.W. and A. Isaksen (2007a) "Knowledge intensive business services and urban industrial development", *The Service Industries Journal*, 27(3): 321–38.

Aslesen, H.W. and A. Isaksen (2007b) "New perspectives on knowledge-intensive services and innovation", *Geogr. Ann.*, 89(s1): 45–58.

Baptista, R. and P. Swann (1998) "Do firms in clusters innovate more?", *Research Policy*, 27: 525–40.

Bathelt, H., A. Malmberg and P. Maskell (2004) "Clusters and knowledge: local buzz, global pipelines and the process of knowledge creation", *Progress in Human Geography*, 28: 31–56.

Bettencourt, L.A., A.L. Ostrom, S.W. Brown, and R.I. Roundtree (2002) "Client co-production in knowledge-intensive business services", *California Management Review*, 44: 100–28.

Brenner, T. (2004) *Local industrial clusters: existence, emergence and evolution,* London: Routledge.

Breschi, S. (2000) "The geography of innovation: a cross-sector analysis", *Regional Studies*, 34(3): 213–29.

Bristow, G. (2005) "Everyone's a 'winner': problematizing the discourse of regional competitiveness", *Journal of Economic Geography*, 5: 285–304.

Broekel, T. and R. Boschma (2012) "Knowledge networks in the Dutch aviation industry: the proximity paradox", *Journal of Economic Geography,* 12(2): 409–33.

Camacho, J. A. and M. Rodriguez (2008) "Patterns of innovation in the service sector: some insights from the Spanish innovation survey", *Economics of Innovation and New Technology*, 17(5): 459–71.

Chesbrough, H. (2007) "The market for innovation: Implications for corporate strategy", *California Management Review*, 49(3): 45–66.

Coenen, L. and B.T. Asheim (2006) "Constructing regional advantage at the northern edge", in Cooke, P. and A. Piccoluga (eds) *Regional development in the knowledge economy*. London: Routledge, pp. 84–111.

Cooke, P. (2001) "Regional innovation system, clusters, and the knowledge economy", *Industrial and Corporate Change*, 10(4): 945–74.

Cooke, P., P. Boekholt and F. Tödtling (2000) *The governance of innovation in Europe*. London: Pinter.

Cooke, P. and L. Leydesdorff (2006) "Regional development in the knowledge-based economy: the construction of advantage", *Journal of Technology Transfer*, 31: 5–15.

Cooke, P. and O. Memedovic (2006) *Regional innovation systems as public goods*, Unites Nations Industrial Development Organization, Working Paper.

Czarnitzki, D. and A. Spielkamp (2003) "Business services in Germany: bridges for innovation", *The Service Industries Journal*, 23: 1–30.

Dakhli, M. and D. de Clercq (2004) "Human capital, social capital, and innovation: a multi-country study", *Entrepreneurship & Regional Development*, 16(2): 107–28.

den Hertog, P. (2000) "Knowledge-intensive business services as co-producers of innovation", *International Journal of Innovation Management*, 4: 491–528.

Doloreux, D., M. Freel and R. Shearmur (eds) (2010) *Knowledge intensive business services: geography and innovation*. Farnham, UK: Ashgate.

Doloreux, D. and R. Shearmur (2010) "Exploring and comparing innovation patterns across different knowledge business services", *Economics of Innovation and New Technology*, 19: 605–25.

Freel, M. S. (2006) "Patterns of technological innovation in knowledge-intensive business services", *Industry and Innovation*, 13: 335–58.

Hamdouch, A. and F. Moulaert (2006) "Knowledge infrastructures, innovation dynamics and knowledge creation/diffusion/accumulation processes: a comparative institutional perspective", *Innovation – the European Journal of Social Science Research*, 19(1): 25–50.

Hsieh, H. N., T. S. Hu, P. C. Chia and C. C. Liu (2014) "Knowledge patterns and spatial dynamics of industrial districts in knowledge cities: a case study of Hsinchu, Taiwan", *Expert Systems with Applications*, 41(12): 5587–96.

Hu, T. S. (2008) "Interaction among high-tech talent and its impact on innovation performance: a comparison of Taiwanese science parks at different stages of development", *European Planning Studies*, 16(2): 163–87.

Hu, T. S., S. L. Chang and K. C. Chen (2015) "Incubators, networks, and their performance: an in-depth case study in Taiwan", *Int. J. Business Environment*, 7(3): 281–301.

Hu, T. S., S. L. Chang, C. Y. Lin and H. T. Chien (2006) "Evolution of knowledge intensive services in a high-tech region – the case of Hsinchu, Taiwan", *European Planning Studies*, 14(10): 1363–85.

Hu, T. S., C. Y. Lin and S. L. Chang (2013) "Knowledge intensive business services and of client innovation", *The Service Industries Journal*, 33(15–16): 1435–55.

Illeris, S. (1991) "Location of services in a service society", in Daniels F, and P. W. Moulaert (eds) *The changing geography of advanced producer services*, London: Wiley, 91–109.

Industry, Commerce and Service Census in Taiwan. (2006) The Report on 2006 Industry, Commerce and Service Census in Taiwan, The Republic of China. Retrieved June 2–10, 2010 from www.dgbas.gov.tw/ct.asp?xItem=13759&ctNode=3374.

Isaksen, A. and H. W. Aslesen (2001) "Oslo: in what way an innovative city?" *European Planning Studies*, 9(7): 871–87.

Jensen, M.B., B. Johnson, Edward Lorenz, B. Å. Lundvall (2007) "Forms of knowledge and modes of innovation", *Research Policy*, 36(5): 680–93.

Kautonen, M. and A. Tuhkunen (2008) "Intermediating between the international and local levels: business consultancy and advertising firms and their clients in Finland", *International Journal of Services Technology and Management*, 10(2/3/4): 235–53.

Keeble, D. and L. Nachum (2002) "Why do business service firms cluster? Small consultancies, clustering and decentralization in London and southern England", *Transactions of the Institute of British Geographers*, 27: 67–90.

Klepper, S. and S. Sleeper (2005) "Entry by spinoffs", *Management Science*, 51(8): 1291–1306.

Koschatzky, K. (1999) "Innovation networks of industry and business-related services: relations between innovation intensity of firms and regional inter-firm cooperation", *European Planning Studies*, 7(6): 737–57.

Lorenz, E. and A. Valeyre (2006) "Organisational forms and innovative performance: a comparison of the EU-15", in Lorenz, E. and B. Å. Lundvall (eds) *How Europe's economies learn: coordinating competing models*. Oxford: Oxford University Press.

Miles, I. (2005) "Knowledge intensive business services: prospects and policies", *Foresight: the Journal of Futures Studies, Strategic Thinking and Policy*, 7(6): 39–63.

Miles, I. (2008) "Miles, patterns of innovation in service industries", *IBM Systems Journal*, 47: 115–28.

Miles, I., N. Kastrinos, K. Flanagan, R. Bilderbeek and P. den Hertog (1995) *Knowledge intensive business services – users, carriers and sources of innovation*. Manchester: PREST.

Moodysson, J. (2008) "Principles and practices of knowledge creation: on the organization of 'buzz' and 'pipelines' in life science communities", *Economic Geography*, 84(4): 449–69.

Muller, E. (2001) *Innovation interactions between knowledge-intensive business services and small and medium-sized enterprises: an analysis in terms of evolution, knowledge and territories*. Heidelberg: Physica-Verlag.

Muller, E. and A. Zenker (2001) "Business services as actors of knowledge transformation: the role of KIBS in regional and national innovation systems", *Research Policy*, 30(9), 1501–16.

Muller, E. and D. Doloreux (2009) "What we should know about knowledge-intensive business services", *Technology in Society*, 31: 64–72.

Nonaka, I., R. Toyama and A. Nagata (2000) "A firm as a knowledge-creating entity: a new perspective on the theory of the firm", *Industrial and Corporate Change*, 9(1): 1–20.

O'Farrell, P.N. and L.A.R. Moffat (1995) "Business services and their impact upon client performance: an exploratory interregional analysis", *Regional Studies*, 29(2): 111–24.

Parida, V., M. Westerberg, and J. Frishammar (2012) "Inbound open innovation activities in high-tech SMEs: the impact on innovation performance", *Journal of Small Business Management,* 50(2): 283–309.

Rusten, G., J. R. Bryson and H Gammelsater (2005) "Dislocated versus local business service expertise and knowledge: the acquisition of external management consultancy expertise by small and medium-sized enterprises in Norway", *Geoforum*, 36: 525–39.

Scarborough, H. and R. Lannon (1989) "The management of innovation in the financial services sector: a case study", *Journal of Marketing Management*, 5(1): 51–62.

Simmie, J. and S. Strambach (2006) "The contribution of KIBS to innovation in cities: an evolutionary and institutional perspective", *Journal of Knowledge Management*, 10: 26–40.

Strambach, S. (2008) "Knowledge-intensive business services (KIBS) as drivers of multilevel knowledge dynamics", *International Journal of Services Technology and Management*, 10(2/3/4): 152–74.

Sundbo, J. (1997) "Management of innovation in services", *The Service Industries Journal*, 17(3): 432–55.

Tether, B. S. (2005) "Do services innovate (differently)? Insights from the European Innobarometer survey," *Industry and Innovation*, 12(2): 153–84.

Tether, B. S. and C. Hipp (2002) "Knowledge intensive, technical and other services: patterns of competitiveness and innovation compared", *Technology Analysis & Strategic Management*, 14(2), 163–82.

Tödtling, F., P. Lehner and M. Trippl (2006) "Innovation in knowledge intensive industries: the nature and geography of knowledge links", *European Planning Studies*, 14(8): 1035–58.

Toivonen, M. (2004) *Expertise as business: long-term development and future prospects of knowledge – intensive business services (KIBS)*, unpublished Doctoral Dissertation, the Department of Industrial Engineering and Management, Helsinki University of Technology.

Toivonen, M. (2007) "Innovation policy in services: the development of knowledge-intensive business services (KIBS) in Finland", *Innovation: Management, Policy & Practice*, 9: 249–61.

Toivonen, M. and T. Tuominen (2009) "Emergence of innovations in services", *The Service Industries Journal*, 29(7): 887–902.

Windrum, P. and M. Tomlinson (1999) "Knowledge-intensive services and international competitiveness: a four country comparison", *Technology Analysis & Strategic Management*, 11: 391–408.

Wolfe, D A. and M. S. Gertler (2004) "Clusters from the inside and out: local dynamics and global linkages", *Urban Studies*, 41(5/6): 1071–93.

Wong, P. K. and Z. L. He (2005) "A comparative study of innovation behaviour in Singapore's KIBS and manufacturing firms", *The Service Industries Journal*, 25(1): 23–42.

Wood, P. (2002) "How may consultancies be innovative", in Wood, P. (ed) *Consultancy and innovation: the business service revolution in Europe.* London: Routledge, pp. 72–89.

Wood, P. (2006) "The regional significance of knowledge-intensive services in Europe", *Innovation*, 19(1): 51–66.

Wood, P. A., J. Bryson and D. Keeble (1993) "Regional patterns of small firm development in the business services: evidence from the United Kingdom", *Environment and Planning A*, 25: 677–700.

Zucker, L. G., M. R. Darby and J. Armstrong (1994) "Intellectual capital and the firm: the technology of geographically localized knowledge spillovers", *NBER Working Paper* No. 4946.

11 Regional competitiveness and localised Knowledge Intensive Business Services

The case of the Gold Coast, Australia

Vanessa Ratten

Introduction

The competitiveness of a region depends on the entrepreneurship and innovation embedded within a society. Increasingly more regions are focusing on entrepreneurial innovation as a source of their competitiveness but also to connect with Knowledge Intensive Business Services. This is due to regions requiring entrepreneurial thinking but also innovative sources in order to differentiation their geographic area from competing regions (Ferreira, Fernandes and Raposo, 2015). The purpose of this chapter is to analyse from a qualitative approach, the relationship between entrepreneurial innovation and regional competitiveness in the region of the Gold Coast (Australia). The case study analysis is premised on the entrepreneurial innovation literature to discuss how institutions, social and industry context affect regional economic growth. The research questions that are address in this chapter are: (1) what institutions affect entrepreneurial innovation in the Gold Coast? (2) how does the social context drive entrepreneurial innovation and regional development in the Gold Coast? and (3) how do industry and technology within the Gold Coast affect regional competitiveness? To answer these questions, a case study method is used to study the Gold Coast region of Australia.

The main findings of the study suggest that business and government policy is greatly influenced by institutions that exist in a region and investment in infrastructure development is important for regional competitiveness. The research also highlights the influence of entrepreneurial innovation on industry, specifically through the generation of new business, jobs and tourism. This research has important practical, theoretical and policy implications for understanding entrepreneurial innovation and regional competitiveness. Theoretically, the chapter advances the application of entrepreneurial innovation to the field of regional economics and development. Empirically, the chapter is useful for the design of government policies towards infrastructure development and for international best practice behaviour.

After the introduction, this chapter is structured as follows. First, the literature about entrepreneurial innovation is discussed in terms of how it bridges the gap between entrepreneurship and innovation management study. Second, the role of regional development is stated for its impact on entrepreneurial innovation. Third, the conceptual framework focusing on institutional, social and industry factors is stated, which includes a set of research propositions. The methodology of the chapter is defined followed by a discussion of the findings for the research propositions. Finally, the theoretical and managerial implications of entrepreneurial innovation for regional competitiveness are highlighted. The chapter concludes by stating the important role that entrepreneurial innovation has for regional competitiveness.

Knowledge-based economy and innovation

A knowledge-based economy is complex due to the dynamic nature of innovation particularly technology that is continuously improving in terms of service application and market need (Lindberg, Lindgren and Packendorff, 2014). The knowledge-based economy centres around innovation due to the importance placed on creativity and development of business ideas. In knowledge-based economies, innovative ideas are fostered by the distribution, production and use of knowledge (Antonelli, Patrucco and Quatraro, 2011).The systems of innovation approach has grown in popularity and largely replaced the traditional way of looking at firms innovating in isolation (Wang, Sutherland, Ning and Pan, 2015). Prior to the systems view of innovation, the Schumpeterian view of firms proposed that firms innovated by themselves in a linear manner without any interaction with the external environment (Todtling and Trippl, 2005). The key premise of the systems view of innovation is that innovation requires communication between different people in an environment including other firms, suppliers, customers, governments and universities. This communication enables collaboration to occur so that feedback and information is disseminated in a more holistic manner. Information is shared in the environment that acts as an ecosystem for knowledge generation and diffusion (Freeman, 1995). This knowledge can be applied to innovation at the regional, national and international level (Carlsson, Jacobsson, Hohmen and Rickne, 2002). Much of the focus on innovation has focused on the technological and sectoral level because of the regional disparities in competitiveness structures (Carlsson, 1997). The regional nature of innovation systems is important as it affects the type of industry located in the region (Wang *et al.*, 2015). This means that regions differ depending on the level of government, university and firm collaboration that affects knowledge spillovers (Jaffe, Trajtenberg and Fogarty, 2004) As knowledge plays a key role in the innovation process the way a regional ecosystem develops is important to innovation performance.

All regions in the global learning economy need to place emphasis on knowledge and the spillovers they have to other industry contexts (Wang *et al.*, 2015). The role of regional industrial clusters has been affected by the spatial dimensions of innovation processes (Binz, Truffer and Coenen, 2014). The spatial dimensions examine the interrelationships between industries, universities and government (Wang *et al.*, 2015). Innovation-incentivising policies are helpful in generation activity within firms and their environment (Rieu, 2014). Part of innovation requires interaction between agents in the environment that leads to the generation of knowledge. Innovation system paths vary depending on the level of government support and composition of a region (Wang *et al.*, 2015). This is impacted by the capabilities of firms in a region and the industry environment facilitating innovation (Gao, Guo, Sylvan and Guan, 2010).

Knowledge can be integrated into regions using a generation or application approach to the ecosystem (Wang and Li-Ying, 2014). Generating innovation is part of the change in the traditional system of innovation as it incorporates both internal and external environmental factors (Rieu, 2014). This means that generating innovation relates to the knowledge that comes from internal and external sources within society (Chesbrough, 2003). Ideas that lead to innovation often come from research and development outside a firm's internal environment (Samara, Georgiadis and Bakouros, 2012). The application of innovation focuses on how interaction between consumers, governments and business can influence ideation. This process of ideation then impacts on the development of innovation and the fostering of a society based on exchanging ideas (Huizingh, 2011). Ideas can come from a variety of sources depending on their origins and how they are generated into innovative activity.

Conceptual framework

Entrepreneurial innovation and regional competitiveness

Most literature narrowly focuses on innovation as being associated with technological processes linked with invention. This has meant less attention has been paid to other forms of innovation that focus on organisational and business model application (Autio, Kenney, Mustar, Siegel and Wright, 2014). Entrepreneurship incorporates a broad view of innovation to include technology, organisation and business types that involve service and processes (Autio *et al.*, 2014). The act of entrepreneurship incorporates a range of behaviour and actions that are linked to innovation. Most entrepreneurship studies are centred on individuals and organisations with less concern about the environmental context (Autio and Acs, 2010). However, the entrepreneurial context is important for firms that focus on Knowledge Intensive Business Services. Huggins and Thompson (2015, p. 104) define entrepreneurial firms as "firms with significant entrepreneurial traits such as being opportunity seeking, growth orientated and alert to new ideas". The individual behaviour and choices of entrepreneurial firms are based on the contextual process (Leydesdorff and Park, 2014). This means that human action occurs in context that influences choices and outcomes of behaviour (Zahra, Wright and Abdelgawad, 2014). The contextual influencers on entrepreneurial behaviour affect the understanding of the way entrepreneurial activities function in the environment. This is helpful with identifying how entrepreneurial action is determined by industry, market, social, organisational and ownership structure (Autio *et al.*, 2014).

Autio *et al.* (2014, p. 1094) defines entrepreneurial innovation as "involving the disruption of existing industries and creation of new ones through multi-level processes and stakeholders, multiple actors and multiple contexts that constitute different entrepreneurial ecosystems". An important component of entrepreneurial innovation is the existence of ecosystems as they shape business development (Isenberg, 2010). This contextual influence on entrepreneurial innovation is the existence of organisations in terms of their experience and expertise in the market (Autio *et al.*, 2014). Some organisations have more practice and skill at developing innovation that helps them build a better ecosystem. Organisations can have different types of knowledge that is build on practising certain behaviour (Nanda and Sorensen, 2010).

Entrepreneurial firms have developed from their association with innovative activity. This is due to entrepreneurship and innovation being strongly related due to their close association with creativity and new products, processes and services in the marketplace (Autio *et al.*, 2014). Entrepreneurial innovation is the source of national competitive advantage because entrepreneurs can introduce novel ventures. Entrepreneurs produce radical ideas that challenge the status quo by establishing new ways of doing things (Autio *et al.*, 2014). Firms that are entrepreneurial have different development paths that encourage the growth of new competences. Local, regional and national policy has been aimed at stimulating innovations by entrepreneurial firms (Grimaldi, Kenney, Siegel and Wright, 2011). Despite the importance of innovation by entrepreneurs not all entrepreneurial ventures include innovation. This is due to some new ventures not being innovative but rather catering to a different market based on geographical location. Location is important for entrepreneurial innovation as it can differ depending on the setting and industry structure. Entrepreneurial ecosystems that develop in certain regions influence the direction and quality of innovation (Autio *et al.*, 2014). The next section will further discuss entrepreneurial innovation from a regional development perspective.

Regional development from an innovation and entrepreneurship perspective: proposed propositions

Regional development needs public intervention to help with network formation (Asheim and Isaksen, 2003). There are two core types of regional innovation policies: systems oriented and firm oriented. System oriented policies focus on the networks, clusters and cooperation needed for regional innovation (Huggins and Thompson, 2015). This means the development of innovation systems from the brokering of networks and building of clusters is important. Firm-oriented systems focus on the resources needed for innovation and they include human, financial and physical capital (Huggins and Thompson, 2015). Entrepreneurial firms will use business advice to see how they can access loans and technology centres to support regional innovation. Incubators are a source of advice for entrepreneurial firms as they can access knowledge and resources in a specific geographic setting.

The regions are important to understanding innovation as there is a specific geographic area that can be developed (Storper, 1997). Regional environments are helpful in fostering innovation due to the attitudes and success of businesses in a certain geographic area. Entrepreneurial attitudes influence the regional dimension of innovation by encouraging business growth. This enables regions to be incubators of new ideas by providing opportunities for entrepreneurship (Huggins and Thompson, 2015). Regions enable the discovery of new knowledge, which can be disseminated to different stakeholders within a certain location. The investment from regional innovation nurtures further entrepreneurship and brings people together for a common cause. Business can be attracted to a region due to the activities taking place that enables access to knowledge (Audretsch and Lehmann, 2005). The capacity of entrepreneurs to influence regional growth is related to their ability to exploit knowledge that leads to innovation. Regional economic growth has innovation and entrepreneurship as its base because it can rearrange resources in a valuable manner. This ability to take resources stimulates competition but also cooperation in regional settings.

Regional economic growth derives from the combination of innovation and knowledge in an environment (Audretsch, 2000). The conversion of knowledge into economic outcomes this multiples the effects of the knowledge in a region. The selection process for accessing knowledge means that entrepreneurs influence the market processes by further diversifying the knowledge (Huggins and Thompson, 2015). Firms that produce more entrepreneurial products and services facilitate better commercialisation of the knowledge (Acs and Plummer, 2005).

Innovation helps regions foster growth and competitiveness (Huggins and Thompson, 2015). This is important as increasingly regions are sources of economic activity as they spawn ventures and increase global competitiveness. This occurs when the creation and dissemination of knowledge is turned into regional innovation (Huggins and Thompson, 2015). Networks are a source of innovation as they facilitate knowledge flows in a region. This comes from the regional variety of innovation systems being a factor in managing effectively innovation as it encourages knowledge to be disseminated. An important part of information dissemination is the capability of entrepreneurial firms in regions to utilise networks (Huggins and Thompson, 2015). Network capital enables entrepreneurial firms to have the capacity to innovate and incorporates strategic relations with other firms. In order to access knowledge, network capital can help firms make better economic returns.

Entrepreneurial connections to knowledge encourage regional innovation depending on the dynamics of the environment. Some more innovative regions have more knowledge-based firms that enable better networks for facilitating information flow (Hayter, 2013). By collaborating and sourcing appropriate knowledge, regions can further develop their innovative capabilities. Local connections are important in monitoring the entrepreneurial network formation in a region (Linder and Strulik, 2014). This is due to knowledge acting as a catalyst for intellectual exchange and transfer (Glaeser, Kallal, Scheinkman and Shleifer, 1992). The dynamics of knowledge sharing means that the networks can also be global as information flows into different sources.

Acs, Braunerhjelm, Audretsch and Carlsson (2009) propose that knowledge spills over to other individuals, firms and entities in a region. This makes knowledge spillovers a public good as it enables others to access opportunities that might have been previously unavailable. The knowledge spillover effect depends on entrepreneurs as it is not automatic but needs visionary people to identify the opportunities (Audretsch, 2000). The knowledge spillovers process is driven by economic agents in the form of entrepreneurs that convert the knowledge to business opportunity (Audrestch and Keilbach, 2004). Part of this process comes from the network capital, which helps to explain how relationships affect knowledge spillovers. Networks interact with different capital structures to produce value. This value can be horizontal or vertical depending on the customers, firms and universities involved (Fitjar and Huber, 2014). The knowledge spillover theory of entrepreneurship incorporates the role of knowledge access and entrepreneurship in facilitating regional innovation (Huggins and Thompson, 2015).

The key presence of the knowledge spillover theory of entrepreneurship is that knowledge flows into other sources of economic activity (Audretsch and Lehmann, 2005). This has meant commercialised knowledge from an organisation filters to generate entrepreneurial opportunities. The knowledge spillovers can then lead to regional innovation particularly when networks are involved. Inter-organisational networks and knowledge are sources for regional innovation as they sustain competitiveness (Knoben and Oerleman, 2006). Knowledge spillovers are helpful in gaining access to innovation. Firms that access new knowledge whether it is economic or commercialisable are able to innovate in their region. This enables unintentional and intentional efforts at stimulating innovation in the regional environment. The localisation of knowledge can be spatially related in close proximity to the knowledge source. Sometimes access to the knowledge is a complex process when the source is geographically close to the localisation of the knowledge spillovers (Huggins and Thompson, 2015).

Research propositions

Based on the previous discussion of entrepreneurial innovation and knowledge spillovers, a number of research propositions are stated, which focus on the institutional, social and industry context. These research propositions are now discussed in terms of how they facilitate entrepreneurial innovation and regional competitiveness.

Institutional context

Institutional environments are important for supporting entrepreneurial innovation as they provide the governance structures to help business succeed (Autio *et al.*, 2014). There are formal and informal institutions that affect innovation depending on the environmental context and level of economic development in the country (Kenney and Patton, 2005). Formal institutions include the regulation of business entry into the market and determination of taxes (Levie and Autio, 2011).

For most companies the formal structures governing competition, law and property protection are important for the stability of the economy (Autio and Acs, 2010). Some regions have special formal institutions based on the industry existing in the area to help with the growth of business (Ferreira *et al.*, 2015). Informal institutions are based on social norms existing in a region about what is expected in terms of business behaviour. These social norms are established on informal institutions to increase legitimacy of business transactions (Autio *et al.*, 2014).

Social desirability is enhanced when informal institutions act as governance mechanisms regulating behaviour. Informal institutions do this by assisting businesses with cultural requirements and appropriate standards of behaviour. Some cultural attitudes are dictated by peer groups who designate what is appropriate behaviour given the context. Part of the institutional context is looking at innovation as a systemic process in which university, industry and government policy interacts with civil society to promote learning. This emphasises the role society has in the way knowledge is integrated into learning outcomes. The institutional context within civil society brings a new perspective about how knowledge develops in a community setting. The institutional perspective takes into account how society and economy combine to encourage better interaction with knowledge production and innovation systems.

The institutional perspective helps explain the interactions between university, business and public sector partnerships in innovation processes (Rosenlund, Hoglund, Johansson and Seddo, 2015). The key ingredient in this approach is knowledge as it drives regional development and innovation ecosystems. An institutional approach to regional development has become a way to evaluate innovative projects that involve collaboration. Regional innovation is a more action-orientated than descriptive innovation models as several organisations work together in the approach. Institutions in the form health, education and tourism providers are important components of international cooperation in knowledge (Villareal and Calvo, 2015). This is due to the increased importance placed on knowledge and innovation as part of international relationships existing between industry, government and education providers. This leads to the first research proposition:

Proposition 1: Institutional and policy context influences entrepreneurial innovation in regional economies.

Social context

The social context is important in allowing interactions to occur between different entities enabling knowledge dissemination (Autio *et al.*, 2014). The exchange of knowledge means that more entitles can acquire knowledge that helps them in business practices. Entrepreneurs who create new ideas can disperse knowledge that enables better performance. Organisations that are entrepreneurial develop resources around their social networks in order to discover better management processes. This is important for organisational development as more networks are enabled through social connections (Autio *et al.*, 2014).

The social context of a region is understood by the dynamic interactions between government, industry and university in terms of facilitating innovation and entrepreneurship (Leydesdorff and Etzkowitz, 1998). The social elements within a region are focused on three main actors (industry, government, university) in an innovations ecosystem rather than the numerous other actors involved in economic policy. The emphasis of this approach was to focus on the interaction between industry, government and university, which emphasises the importance of education to innovation. This means that education and interaction with institutions in a region delivered the best conditions for facilitating innovation and social interaction. Social interactions within universities instigate the generation and transfer of knowledge. As part of this emphasis on social interaction, universities develop knowledge that then can be transferred to society by interacting with industry (Villarreal and Calvo, 2015). Universities that have continuing relationships with industry are then able to develop ideas, processes and technology that incorporate innovation. Chesbrough (2003) proposed innovation transfer as it includes innovation generated from experimentation amongst consumers, firms, governments and universities. Entrepreneurial innovation has increased in popularity due to its incorporation of final users as the basis for how interaction between a group of stakeholders leads to better innovation performance (Villarreal and Calvo, 2015). Therefore, it is proposed that the social context of a region will have an affect on entrepreneurial innovation and the following proposition is proposed:

Proposition 2: Social context influences entrepreneurial innovation in regional economies

Industry and technological context

The industry and technological context of a region affects its entrepreneurial orientation (Autio *et al.*, 2014). The industry existing in a region can take a variety of forms from manufacturing to high–tech business. The various types of industry affecting knowledge transfer is based on the industry structural conditions that exist in the business environment. Structural conditions shape resource acquisition and the sharing of knowledge. In some industries resources influence entrepreneurship as more innovative businesses relocate to be closer to other firms. This creates an imitation effect as other firms follow the same innovation path. Firms focus on how their industry integrates innovation by looking at entry modes in the market. Industry life cycles are important in analysing these entry modes as entrepreneurial activity changes (Autio *et al.*, 2014). In the early stages of an industry life cycle there can be more entrepreneurial activity as innovation is encouraged. This emphasis on innovation then affects the ability of an industry to introduce new product features and additional services (Autio *et al.*, 2014).

Technology is an important part of industry context depending on the level of innovation. Some technological innovation is shaped by the entrepreneurial activities of other stakeholders in the environment (Autio *et al.*, 2014). Industries that focus on networked environments made possible by location advantages can lend to further technological innovation developing and fuelling economic growth rates. Firms that share technological capabilities and resources in their industry can influence innovative activity. The location of industry also has an effect on the development of entrepreneurial innovation as an ecosystem develops from the linkages between stakeholders in the environment. Therefore, the following proposition is proposed:

Proposition 3: Industry and technology context influences entrepreneurial innovation in regional economies

Methodology

This chapter is based on a case study analysis that enables a detailed view of the entrepreneurial innovation existing in a region. Case study analysis enables the understanding of research propositions that are derived from theory to help understand a phenomenon. Case studies enable the use of a multiple analysis design with rich data drawn from the methodology. In this chapter, an embedded design for the case study is utilised seeking to conform or counter the research propositions. The data was analysed using the conceptual framework from the literature review. The data focused on the research propositions that were derived from the theory and important to regional innovation. The research data analysis was designed to see how entrepreneurial innovation occurred in a regional context. The interviews, case studies and secondary data were analysed to obtain findings for the research propositions. The next section will focus on how entrepreneurial innovation affects regional competitiveness of the Gold Coast region of Australia.

Research setting: characteristics of the Gold Coast (Australia)

The Gold Coast has a population of 591,472 (2010), which makes it the sixth biggest city in Australia. It covers 414.3 km^2 and is located 69 km from Brisbane. It is a coastal city on the east coast of Australia and is the most populous non-capital city in Australia. The Gold Coast metropolitan area converges with Brisbane to form a population of over 3 million people. Local councils in the Gold Coast include Albert, Broadwater, Burleigh, Coomera, Currumbin, Gaven, Mermaid Beach, Mudgeeraba, Southport and Surfers Paradise. The Gold Coast is a tourism destination known for its warm climate, surf beaches and night life. It has a canal and waterway system dominated by high-rise buildings closer to the beach. The Gold Coast has theme parks including SeaWorld and MovieWorld with a rainforest hinterland area on the edges of the city. The Gold Coast is Australia's fastest growing city and in the South-East growth corridor. Most industry in the Gold Coast has focused on tourism, retail and construction but it has diversified to include information communication technology, education and health.

Case studies

Case 1

Firm 1 is an information technology service provider in the Southport region of the Gold Coast that focuses on enterprise information services for business clients. The firm was established in early 2000 from a small government grant and had its initial headquarters in Brisbane but relocated to the Gold Coast to be close to Griffith University and Bond University. The focus of the firm is on business enterprise solutions for the education market but it also has a growing interest in sports technology and data analytics.

Case 2

Firm 2 is a health services company that specialises in mobile home aids and healthcare products. The firm is located in the industrial region of Broadbeach in the Gold Coast with similar types of firms. The firm has recently focused more on elderly customers as the Gold Coast is a region where there is a large number of retirees. The core geographic area of the firm is the Gold Coast region but it also services the Brisbane region of Queensland and has offices in Sydney and Melbourne.

Case 3

Firm 3 is an education consulting company that was founded in 2006 due to the founders networks with local universities in the Gold Coast region. The firm specialises in providing business and language programs for students studying in the region with a focus on university students.

Findings and discussion

In this section, the case studies are described in the context of the Gold Coast region. The main findings are related to the literature review focusing on innovation systems to answer the research propositions. Each of the propositions are discussed with quotes and analysis from the three main firms that comprised the case studies.

The Managing Director of firm 2 associated the institutions in the form of the local Gold Coast council with innovation, and some of the observations show that the council positively affects entrepreneurial innovation. This is seen in their comment "the Gold Coast has a good institutional framework for investing in healthcare. They just built a new University Hospital, which is bringing healthcare practitioners to the area and encouraging more medical businesses to set up in the area." There was an emphasis on institutions for regional development both from the local council to the state government. During interviews it was observed that businesses from the region were attracted to the area because of the institutions supporting entrepreneurship such as the universities and technology centres. The owner of firm 3 stated "I think it's very important what the Queensland government is doing for the Gold Coast. They just built the new light rail link. Public transport is crucial for our business. Students in particular need public transport. The rail link has opened the area up." Strong institutional structures have been associated with entrepreneurial innovation due to the infrastructure that is built based on government support. The Managing Director of firm 1 stated "unlike other regions, the Gold Coast has grown quickly because of the massive investments in infrastructure. We have the airport rail link to Brisbane. We have light rail. We have the freeway. All these make the Gold Coast accessible and open for business." Thus, the cases show that institutions do have a positive relationship with entrepreneurial innovation in the Gold Coast. As observed, there was some degree of difference about the idea of institutions affecting infrastructure development. Overall, the cases support proposition 1 (institutional context affects entrepreneurial innovation and regional competitiveness) and is supported by the findings.

In the research it was observed that the social context was important for the cases but it might not have an effect on entrepreneurial innovation. For example, the owner of firm 1 is from a family that has a number of businesses, and this influenced his ability to start his Information Technology firm on the Gold Coast as he had family connections in the area. However, as he stated "social relationships are important to me as business is about money. I'm here at the Gold Coast because there are government grants for tech companies to relocate here. The bonus is my family is close by." This finding is aligned to the suggestion that social networks influence business venture creation but may not be as influential towards creating an entrepreneurial business culture. In the case of firm 2, the social context was a driver of innovation in terms of coming up with new business ideas and creations for the website of their firm. As entrepreneurial innovation involves both change but business creation firm 3 highlighted in this quote how social factors influence creativity but might not lead to regional competitiveness "We are educators. That is what we do. We constantly learn but so does everyone else. The Gold Coast is able to grow because of the international students coming to the area. But are we entrepreneurial and innovative? I think a bit but it is because of our industry". Based on these results and the literature review, there is a weak support for the second proposition (social factors influence entrepreneurial innovation and regional competitiveness).

In order to answer the third research proposition, we observed how firms in the region connected with industry and how this linked to entrepreneurial innovation. For example, most of the staff of firm 2 came from overseas and were drawn to the Gold Coast because of the climate and lifestyle. This helps to attract people in the healthcare industry to the area in addition to the interstate migration of retirees to the region. This is important to highlight because of the similarities to Huggins and Thompson's (2015) suggestion that global talent affects regional innovation. From this connection it is observed that the industry develops based on the people located in a region and the skill set they bring. The owner of firm 1 stated "technology industry has developed here as the government is trying to make the region like Silicon Valley. Near Bond University and Varsity Lakes there are a number of technology firms and supporting industries located there. There is start up week there and a business incubator. We say Varsity Lakes is Silicon Lakes as it is so full of technology firms". The findings of the interviews and analysis of the case studies suggest there is strong support for the connection between industry factors, entrepreneurial innovation and regional competitiveness. Therefore, proposition 3 (industry factors affect entrepreneurial innovation and regional competitiveness) is supported by the findings of this research.

Conclusions

This chapter has focused on the role of entrepreneurial innovation to regional competitiveness using a case study approach of the Gold Coast region of Australia. As was described in the introduction section innovation systems are important in fostering Knowledge Intensive Business Services. The research was based on the application of entrepreneurial innovation for the study of regional competitiveness (Fitjar and Huber, 2014). The research methodology was qualitative, and led to an in-depth discussion of the Gold Coast region, which is Australia's fastest growing area. The regional competitiveness of the Gold Coast was analysed using a contextual approach focusing on institutions, social and industry factors. Research propositions were stated based on the previously discussed literature review that were analysed in terms of the contextual approach. The influence of entrepreneurial innovation on regional competitiveness using the specific case of the Gold Coast was stated.

The most important evidence from this chapter demonstrates that industry and technology policy have the most relevant influence on regional competitiveness in the Gold Coast. The analysis suggested that whilst institutions and social context can influence entrepreneurial innovation, the industry context is the most important driver of regional competitiveness. During the case study analysis a weak link was found between social connections and entrepreneurial innovation but business was more likely to rely on industry connections in the Gold Coast region. The importance of institutional development was observed and this was seen in the importance placed by the Gold Coast government on building hospitals, universities and rail links to support business. The different types of institutional development from the construction of university hospitals and regional light rail is important in the Gold Coast. The results support the proposition that institutional policy affects entrepreneurial innovation in regional economic growth. This is an important finding as this supports the high level of government funding both the Queensland state government and the Australian government have placed on building infrastructures. The weak finding for social context influencing entrepreneurial innovation means the view of social networks stimulating regional competitiveness is not strong. Thus, it can be considered that social context has a greater influence on individual and family well-being rather than on regional competitiveness. The strong support for industry and technological development mirrors the view of the knowledge economy placing importance on high-tech areas. These findings are in line with the literature, which states that the value of the knowledge economy is seen in the long run. Another result suggests that a focus on special Knowledge Intensive Business Services such as health and education may have an impact on the economic development of a region. This shows that the research question of how entrepreneurial innovation influences regional competitiveness has been answered by our findings of the contextual factors of the Gold Coast. It is important to observe that the Gold Coast has a particular geographic advantage being on the eastern seaboard side of Australia and close to Asia, yet it is geographically far from larger markets in Europe and North America.

Policy intervention is needed in regions that have low levels of innovation and the results of the analysis reported in this chapter show how a region such as the Gold Coast can change to be more orientated towards Knowledge Intensive Business Services. Support is needed to help firms develop their networks to access knowledge that can lead to more innovation. Policymakers should be aware of the needs of entrepreneurial firms to establish better sources of knowledge. This can be accomplished by using relevant knowledge that influences innovation (Freel, 2000). Regional policy should ensure there is enough human capital in the form of expertise to help build entrepreneurial firms (Giuliani, 2005).

Future research could increase the number of case studies to capture more date from different sectors of the Gold Coast. More comparative research is needed about the Gold Coast to see how it compares to other Australian regions including the Sydney-Newcastle hub and greater Geelong area in Melbourne. Comparing the regional competitiveness of the Gold Coast with other Australian regions would help see whether the institutions, social and industry context shape entrepreneurial innovation policy in the same way. A quantitative survey of small to medium sized enterprises could be conducted in each of these regions to see how the Gold Coast compares. This future research work would help improve our understanding of entrepreneurial innovation and regional competitiveness but also promote better Knowledge Intensive Business Services. Finally, more specific government detail and research is needed about the formal and informal institutional influences of regional competitiveness in the Gold Coast region of Australia.

References

Acs, Z. J., Braunerhjelm, P., Audretsch, D. B., and Carlsson, B. (2009). The knowledge spillover theory of entrepreneurship. *Small Business Economics*, 32(1), 15–30.

Acs, Z. J., and Plummer, L. A. (2005). Penetrating the knowledge filter in regional economies. *The Annals of Regional Science*, 39(3), 439–56.

Antonelli, C., Patrucco, P., and Quatraro, A. (2011). Productivity growth and pecuniary knowledge externalities: an empirical analysis of agglomeration economies in European regions. *Economic Geography*, 87(1), 23–50.

Asheim, B., and Isaksen, A. (2003). SMEs and the regional dimension of innovation. In B. Asheim, A. Isaksen, C. Nauwelaers, and F. Tödtling (eds), *Regional innovation policy for small-medium enterprises* (pp. 21–48). London: Edward Elgar.

Audretsch, D. B. (2000). Knowledge, globalization, and regions: an economist's perspective. In J. H. Dunning (ed.), *Regions, globalization, and the knowledge-based economy* (pp. 63–81). Oxford: Oxford University Press.

Audretsch, D. B., and Keilbach, M. (2004). Entrepreneurship and regional growth: an evolutionary interpretation. *Journal of Evolutionary Economics*, 14(5), 605–16.

Audretsch, D. B., and Lehmann, E. E. (2005). Does the knowledge spillover theory of entrepreneurship hold for regions?, *Research Policy*, 34(8), 1191–1202.

Autio, E. and Acs, Z. (2010). Intellectual property protection and the formation of entrepreneurial growth aspirations, *Strategic Entrepreneurship Journal,* 4(3): 234–251.

Autio, E., Kenney, M., Mustar, P., Siegel, D. and Wright, M. (2014). Entrepreneurial innovation: the importance of context. *Research Policy*, 43(7): 1097–1108.

Binz, C., Truffer, B. and Coenen, L., (2014). Why space matters in technological innovation systemsmapping global knowledge dynamics of membrane bioreactor technology. *Research Policy*, 43(1), 138–55.

Carlsson, B., (1997). On and off the beaten path: The evolution of four technological systems in Sweden. *International Journal of Industrial Organization*, 15(6), 775–99.

Carlsson, B., Jacobsson, S., Holmén, M. and Rickne, A. (2002). Innovation systems: analytical and methodological issues. *Research Policy*, 31(2), 233–45.

Chesbrough, H. (2003). *Open innovation: the new imperative for creating and profiting from technology*. Boston, MA: Harvard Business School Press.

Ferreira, J.M., Fernandes, C.I. and Raposo, M.L. (2015). The effects of location on firm innovation capacity. *Journal of Knowledge Economy*, Online 28 July 2015.

Fitjar, R.D. and Rodriguez-Pose, A. (2014). The geographical dimension of innovation collaboration: networking and innovation in Norway. *Urban Studies*, 51(12): 2572–95.

Freeman, C., (1995). The national innovation systems in historical perspective. *Cambridge Journal of Economics*, 19(1).

Gao, X., Guo, X., Sylvan, K.J. and Guan, J., (2010). The Chinese innovation system during economic transition: a scale-independent view. *Journal of Informetrics*, 4, 618–28.

Glaeser, E. L., Kallal, H. D., Scheinkman, J. A., and Shleifer, A. (1992). Growth in cities. *Journal of Political Economy*, 100(6), 1126–52.

Grimaldi, R., Kenney, M., Siegel, D. and Wright, M., (2011). 30 years after Bayh–Dole: reassessing academic entrepreneurship. *Research Policy* 40(8), 1045–57.

Hayter, C. (2013). Conceptualizing knowledge-based entrepreneurship networks: perspectives from the literature. *Small Business Economics*, 41(4), 899–911.

Huggins, R. and Thompson, P. (2015). Entrepreneurship, innovation and regional growth: a network theory. *Small Business Economics*, 45: 103–28.

Huizingh, E. (2011). Open innovation: state of the art and future perspectives. *Technovation* 31, 2–9.

Isenberg, D., (2010). The big idea: how to start an entrepreneurial revolution. *Harvard Business Review*, 88(6), 40–50.

Jaffe, A., Trajtenberg, M. and Fogarty, M., (2004). Knowledge spillovers and patent citations: evidence from a survey of inventors. *American Economic Review*, 90(2), 215–18.

Leydesdorff, L. and Park, H.W. (2014). Can synergy in triple helix relations be quantified? A review of the development of the triple helix indicator. *Triple Helix*, 1(4): 1–18.

Lindberg, M., Lindgren, M. and Packendorff, J. (2014). Quadruple helix as a way to bridge the gender gap in entrepreneurship: the case of an innovation system project in the Baltic sea region. *Journal of Knowledge Economy*, 5: 94–113.

Linder, I., and Strulik, H. (2014). From tradition to modernity economic growth in a small world. *Journal of Development Economics*, 109, 17–29.

Nanda, R. and Sorensen, J., (2010). Workplace peers and entrepreneurship. *Management Science* 56(7), 1116–26.

Rieu, A. (2014). Innovation today: The triple helix and research diversity. *Triple Helix*, 1(8): 1–22.

Rosenlund, J., Hoglund, W., Johansson, A.W. and Seddo, J. (2015). A cross-national environmental cluster collaboration: shifting between an analytical and management level of the triple helix. *Science and Public Policy*, 42: 583–93.

Samara, E., Georgiadis, P. and Bakouros, I., (2012). The impact of innovation policies on the performance of national innovation systems: a system dynamics analysis. *Technovation* 32, 624–38.

Storper, M. (1997). *The regional world: territorial development in a global economy*. New York, NY: Guilford Press.

Tödtling, F. and Trippl, M., (2005). One size fits all? Towards a differentiated regional innovation policy approach. *Research Policy* 34(8), 1203–19.

Villareal, O. and Calvo, N. (2015). From the triple helix model to the global open innovation model: A case study based on international cooperation for innovation in the Dominican republic. *Journal of Engineering Technology Management*, 35: 71–92.

Wang, Y., Sutherland, D., Ning, L. and Pan, X. (2015). The evolving nature of China's regional innovation systems: insights from an exploration-exploitation approach. *Technological Forecasting & Social Change*, in press.

Zahra, S., Wright, M. and Abdelgawad, S., (2014). Contextualization and the advancement of entrepreneurship research. *International Small Business Journal*, forthcoming.

Part IV

KIBS and public policy

12 Skills, competitiveness and regional policy

Knowledge Intensive Business Services in the West Midlands, UK

John R. Bryson and Peter W. Daniels

Introduction

From 1966 the economy of the UK entered another round of economic restructuring (Bryson and Ronayne, 2014). The driver for this was the beginning of a new wave of globalisation that undermined the competitiveness of British manufacturing (Bryson *et al.*, 2013). The deindustrialisation of the UK was a complex process. On the one hand, output of manufactured products continued to grow but, on the other hand, employment in manufacturing declined. These two trends are directly related as new forms of competition combined with new production technologies led to productivity improvements within British manufacturing firms. Fewer manufacturing workers were directly involved in the manufacture of goods (Bryson and Ronayne, 2013; Mulhall and Bryson 2013).

Nevertheless, there is an added complication in that the restructuring of British manufacturing was also part of an alteration in the spatial division of labour as well as the division of labour (Bryson *et al.*, 2004; Bryson and Rusten, 2011). Developments in the spatial division of labour that supported manufacturing involved the creation of global value chains combined with the relocation of some manufacturing activity to low-cost production locations (Gereffi *et al.*, 2005). Labour costs were a critical driver behind the relocation of British manufacturing to lower labour cost locations. British manufacturing firms closed plants in the UK or relocated production processes, but also began to focus on more high-value processes and products (Bryson *et al.*, 2008a; Bryson and Rusten, 2011). Alterations in the division of labour led to the emergence of a new definition of manufacturing (Bryson *et al.*, 2013; Bryson *et al.*, 2015). Academics and policy-makers have begun to adopt a more complex definition of manufacturing informed by relatively sophisticated accounts of commodity or value chains (Bryson and Rusten, 2011; Gereffi, 2001; Gereffi *et al.*, 2005). In this context, the concept of value chain fragmentation has become an important conceptual tool for understanding the evolving structure and new geographies of production systems. This concept of 'fragmentation' highlights the fact that manufacturing involves more than just fabrication, but also includes service functions that are integral to production processes that are designed to produce goods (Damesick, 1986). Academics and policy-makers have begun to shift their attention away from a narrow fabrication view of manufacturing to one in which manufacturing includes research and development, design functions, marketing and advertising, Knowledge Intensive Business Services (KIBS) that support production processes and a set of services that have been created to support customers' experiences of a product. Manufactured goods should now be conceptualised as products that contain different quantities of service inputs; some of these service inputs are wrapped into a good during the production process and some are wrapped around a completed product (Daniels and Bryson, 2002; Livesey, 2006; Pryce, 2008). The implication is that the competitiveness of many manufacturing firms is partly dependent upon expertise that lies within KIBS or business and professional service firms. KIBS provide intermediate inputs (information, expertise, knowledge) to private and public sector clients (Bryson and Daniels, 2015b).

The shift in employment towards services in all developed market economies reflects alterations in consumer behaviour (both business-to-business and business-to-consumer) and in the structure of production systems (Bryson *et al.* 2004; Bryson and Daniels 2015a; High Level Group on Business Services, 2014). The growth of service functions also reflects alterations in the skill sets required to support economic activity. In some accounts this employment shift is considered to indicate the rise of a service economy in which service functions and outputs become increasingly dominant (Illeris, 1996; Bryson *et al.*, 2004; Bryson and Daniels, 2007). There is no question that services as a category, let alone the subcategory of KIBS, have not occupied a prominent place in UK regional policy. It is also the case that, until quite recently, any government initiatives targeted at service industries have been about rectifying spatial imbalances in unemployment or the structure of economic activity rather than enhancing education and training in ways that improve skills in the labour force available to service employers in the cities and regions of the UK (Marshall, 1985).

The decline of manufacturing employment in the UK led to a period of policy neglect that ended with the onset of the 2008 recession (Begg, 1993; Bryson *et al.*, 2015). Manufacturing was considered as one solution to rebalancing the British economy away from the perceived domination of financial and consumer services. The emphasis placed on 'rebalancing' the UK economy also included a discussion regarding a spatial rebalancing of the economy away from the South East. This included a discussion regarding the centralisation of political control in London and the devolution of policy development and implementation to city-regions (Heseltine, 2012). The devolution agenda is based on the assumption that policy development and implementation should be informed by local understanding rather than directed centrally. This has led to a political debate regarding the development of policies to create a Northern Power House in the UK centred on Manchester and a Midlands Engine of Growth centred on Birmingham.

By 2003 there was a notable shift towards the view that modern regional policy 'must focus on improving the economic performance of every nation and region, by tackling the diverse market and social failures that are hindering their performance, and promoting opportunities for all' (Department of Trade and Industry, 2003; Fothergill, 2005). Not only is regional policy seen as locally led but involving policy delivery providers also operating regionally and nationally, but it is now driven by the need to rebalance regional disparities shaped by productivity differentials rather than differences in unemployment levels or the uneven distribution of dynamic economic activities such as KIBS. Improvements in the key areas influencing productivity included maximising the contribution of human capital to growth by improving the skills base, enabling the potential of new ways of working and of new technologies by supporting innovation, facilitating new business start-ups by promoting an enterprise culture, and as a result improving competition that encouraged an improvement in the quality of services (Department of Trade and Industry, 2003, 12). KIBS or business and professional services are not specific targets for this more nuanced approach to UK policy for the regions but the successful implementation of policies that address the underlying causes of regional disparity are in fact more appropriate. Intercalated with these new emphases has been the drive towards the UK as an economy where the 'use of knowledge has come to play the predominant role in the creation of national wealth, achieved by effective use and application of all types of knowledge and technology, in all manner of economic activity' (Rogers, 2006, 5). Defining and measuring the knowledge economy continues to be a serious obstacle to identifying effective policies but there is certainly a consensus that improving the skills, abilities and motivation of the UK labour force is a key component (Brinkley, 2006; Rogers, 2006).

One such approach focuses on the innovation capacity of services which is increasingly recognised as important but for which any policies have their roots in advanced manufacturing rather than services (Abreu *et al.*, 2008). The scope of research on innovation in services has escalated enormously (Department of Trade and Industry, 2007; Department for Innovation, Universities and Skills, 2008) but because innovation may be 'softer' in services, such as the way in which networks of firms collaborate for the production of a services or how the expertise embodied in people is used in new and innovative ways rather than based on better use of technology, the evidence base for devising policy is weaker (Abreu *et al.*, 2008; Roper *et al.*, 2009; Love *et al.*, 2011). While by no means the largest share of expenditure by services, training of staff and the acquisition of external knowledge (which is largely exchanged and managed by people rather than the physical resources within firms) are an important source for services innovation with some 'hidden' aspects because it is not easy to capture or measure simply because it is embodied in the people that are so crucial to success in activities such as KIBS. Indeed, 'the skills and knowledge of employees with degrees in both 'hard' (science based) and 'soft' (non-science based) disciplines ….*and*…the quality of in-house expertise is of critical importance to innovation in every firm' (Abreu *et al.*, 2008: 32). This recognition is reflected in a raft or government publications since 2002 devoted to identifying skills needs, specifying policies and strategies, and establishing organisations for delivering education and training that is demand and employer-led in circumstances where 'Our nation's skills are not world class and we run the risk that this will undermine the UK's long-term prosperity' (Leitch, 2006, 1).

Further, 'Our skills base compares poorly and, critically, all of our comparators are improving' (Leitch, 2006, 2). While the skills problem exists across the economy as a whole, in reality it is a particular challenge with respect to servicessince they employ upwards of 70 per cent of the labour force. It is not just about employees' skills either; the leadership and management skills required to improve the performance of service firms also requires urgent attention (Abreu *et al.*, 2008).

The on-going debate regarding devolution in the UK increasingly recognises the interdependencies that exist within local economies. Thus, the changing nature of manufacturing implies that a manufacturing policy has to be simultaneously a service policy. This is to recognise the critical role that KIBS functions play in the competitiveness of manufacturing firms and also in regional competitiveness (Bryson and Daniels, 2015c). Independent KIBS firms are important providers of advice and expertise to clients. Thus, the competitiveness of a local economy is partly related to the quality of local KIBS expertise or the ability of potential client firms to access such expertise from elsewhere. Saxenian (2006) for example has assembled detailed time-series evidence from Silicon Valley, California, that suggests that regional advantage requires the recombination of both local and distant know-how via local knowledge-producing and circulating mechanisms. There is a challenge in that it is difficult to develop policies intended to enhance local provision or availability of KIBS expertise. Early studies of KIBS and regional policy have noted that for service sector regional policy to be successful requires access to significant financial resources, flexibility and must emphasise existing services in less-prosperous area (Marshall, 1985). This early account of services and regional policy places too little emphasis on the importance of embodied expertise in the creation and delivery of service expertise.

KIBS firms sell advice, expertise and knowledge (Quader, 2007). This implies that their core asset rests on the quality of their people. Skills, capabilities and competencies are critical for the competitiveness of KIBS firms. This chapter explores KIBS firms and skills in the West Midlands region of the UK. At the centre of the West Midlands region is the Birmingham conurbation; the second city in the UK. The West Midlands includes many cathedral cities or market towns and two conurbations (Birmingham and Stoke-on-Trent). The focus of this chapter is on exploring skills, hard-to-fill vacancies, technical skills versus soft skills, commercial skills and the relationship between KIBS and the types of policy interventions that might be appropriate. Part of the analysis is to identify skills and capabilities as core elements of the competitiveness of KIBS firms. Our analysis draws upon a detailed telephone survey of 1198 KIBS firms located in the West Midlands. Quota sampling based on SIC codes and firm size distribution was used to construct a representative sampling frame of key KIBS sectors (Table 12.1). Additionally, 208 in-depth face-to-face interviews were undertaken and six focus groups. The research was undertaken in 2006.The depth interview transcripts provide a wealth of insights on the attitudes, role, challenges, and needs of KIBS firms with respect to training and enhancing the skills of professional and

Table 12.1 Characteristics of KIBS firms in Birmingham and Solihull, UK

	Number	*%*
Business activity (SIC divisions and selected groups)		
65 Financial intermediation	51	*9.4*
66 Insurance and pension funding	23	*4.1*
67 Activities auxiliary to financial mediation	31	*5.6*
70 Real estate, renting and business activities	66	*12.1*
71–73 Computing and Research	10	*1.8*
74 Other business activities	366	*66.9*
of which		
74.1 Legal, accounting, business consultancy	142	*25.9*
74.2 Architectural and engineering activities	108	*19.7*
74.5 Labour recruitment	85	*15.5*
Organisational type		
Only site	278	*50.9*
HQ, regional or divisional HQ	75	*13.7*
HQ elsewhere in UK	178	*32.6*
HQ outside UK	15	*2.8*
Number of on-site workers		
1 – 10	421	*77.0*
11 – 24	59	*10.9*
25 – 99	44	*8.1*
100+	22	*4.0*

support staff. The chapter is an overview of some of the key findings of this large project. The objective is to explore key issues that were identified from the quantitative and qualitative research and to bring these issues together into a narrative that explores the relationship between skills, KIBS and policy in the context of a British city-region.

Knowledge intensive business and services in the West Midlands, UK

Birmingham and Solihull has the largest concentration of KIBS firms and employment in the West Midlands with some 5460 establishments forming the most diversified professional services cluster in the region (AWM, 1999, 2001, 2014, 2007). The firms in this cluster have developed expertise appropriate to the needs of firms in the local as well as regional economy, but also in response to demand from other firms located in other regions and countries. The diversity is reflected in a population structure that consists of a large number of small and medium-sized firms (4,512 [83 per cent] employ less than 10) as well as a smaller number of much larger firms that have either grown to support the needs of larger local clients or are branches of national and international service suppliers. Many of the latter have been opened in the region during the last 10 years and have experienced high employment growth rates. The number of very small firms combined with significant proportions of branch offices raises a series of questions regarding the ability of these firms to innovate. Branch offices may only provide services to a regional client base with innovation occurring elsewhere in the firm's branch network, while many smaller firms do not have the time or resources to innovate.

KIBS firms are established by individuals who are capitalising on their expertise, personality, reputation and network of contacts (Bryson, 1997; Bryson *et al.*, 2004). Much KIBS work is performed at the client's premises or even remotely via information communications technology (ICT) (Bryson, 2007). Clients may not have to visit a KIBS provider's offices. This means that KIBS firms may be located in relatively remote locations as well as in large cities. ICT combined with work that is often undertaken in clients' premises enables services to be provided from rural locations. This means KIBS firms can be established in relatively remote locations and still provide services to local, national and even sometimes to international clients (Bryson, 2007).

An expert labour force? kibs and skill problems

The qualitative interviews highlighted that the competitive success of a KIBS firm is attributable to three interrelated factors:

1 Individuals and firms must acquire and continue to develop technical expertise.
2 There needs to be an ability to transform technical expertise into something that has commercial value by providing distinctive inputs into the activities of client firms.
3 People-focussed skills involving presentation and communication techniques are key skills; these can be subsumed under the term impression management or client relationship building.

The first factor is relatively easily acquired whilst the second might come with experience, but the third factor is difficult to develop in someone that does not have the right personality. All three factors are combined together to form a reputation (Bryson *et al.*, 2004; Bryson and Daniels, 2015a). It is difficult to make an objective appraisal of the effectiveness of the contribution a KIBS firm makes to client companies. The services that are provided are often intangible and it can be difficult for a potential client to compare KIBS firms. Third-party referral is often used to select a KIBS firm, as is previous experience. This reputational affect is usually related to a named individual with an established reputation in a particular network and/or place. Firm-level reputations are also established based on long-term investment in a brand. Ultimately, the reputations are based on the quality of an individual or group of individuals. Fundamental to KIBS competitiveness is the ability of firms and professionals to lock client companies into their business on the basis of an established personal relationship that is founded on friendship and trust that has developed over a period of time (Bryson 2015).

The KIBS sector is heterogeneous and extremely diverse. Small firms or sole practitioners have different characteristics to large firms. The KIBS term also includes many different types of services including accounting, law, design, market research and many different types of consultancy. Each type of KIBS has different characteristics (Von Nordenflycht, 2010). The diversity of KIBS activities within the West Midlands does not make it easy to arrive at generalisations about the complete population of firms and the policies for skills and training that reflect a collective need.

The assumption in much of the academic literature is that KIBS firms are knowledge intensive organisations that employ expert staff (Bryson and Daniels, 2007, 2015a). This is an assumption that needs to be challenged and further research

Table 12.2 Staff proficiency levels by main occupation groups (SOC2000)

Occupation	*Number of firms with staff not fully proficient*	*Firms lacking full proficiency as a % of all offering the occupation*
Managers	58	5.4
Professional	56	11.2
Associate Professional and Technical	62	14.2
Administrative and Secretarial	89	13.2
Sales and Customer Service	33	14.3
Personal Service	3	14.3
Elementary	8	8.0

is required to explore the competencies and capabilities of KIBS firms and their employees (Bryson and Daniels, 2015b). The telephone interviews indentified skill deficiencies in 11.2 per cent of firms with professional occupations (Table 12.2). Failures in training are regarded as most serious for managers and administrative and secretarial occupations. A lack of experience or recent recruitment is viewed as significant in virtually all cases where proficiency is a problem for sales and customer occupations. The main causes of under–skilling were explored further (Table 12.3). Lack of experience and/or recent recruitment is easily the single most important cause of skills shortages. This suggests that proficiency problems are largely transitional but this conclusion needs to be tempered with the recognition that inadequacies in training are an important secondary factor. Culpability of employees seems to be another strand contributing to proficiency levels; the third and fourth often-cited causes are the inability of the workforce to keep up with change and its lack of motivation. Shortcomings in the educational system represent another important factor although there is no clear agreement about which level(s) within the educational system are most responsible. The relative importance of these causes does not vary greatly from one type of firm to another. Where differences do exist, the sample sizes are generally too small to establish statistical significance. The exceptions are that large employers (100+) are much less likely to cite recruitment problems as a cause of skill problems while medium-sized firms (especially those with 25–99 employees) are most likely to find their workforce unable to keep up with change. What both sets of results indicate is that there is plenty of scope for enhanced training. Approximately a quarter of the firms recognise that training inadequacies are directly responsible for the skill limitations of their staff. It is difficult to believe that carefully designed

Table 12.3 Main causes for why some KIBS staff in the West Midlands were not fully skilled (firms with skill gaps)

Main causes for why some staff were not fully skilled	*Number of firms with skill gaps*	*% of all firms with skill gaps*
Failure to train and develop staff	55	24.2
Recruitment problems	40	17.8
High staff turnover	20	8.9
Inability of workforce to keep up with change	43	18.9
Lack of experience/recently recruited	155	68.4
Staff lack motivation	49	21.6
Other	2	0.7
No particular causes	38	16.6

programmes would not help overcome poor staff motivation, assist staff to keep up with change, and speed up the induction process for new staff.

Respondents were asked to indicate which of a number of skills they thought needed improvement to bring occupations previously identified as not fully proficient to the required level of expertise. A great variety of skills elicited substantial support and twelve out of the eighteen identified in the survey instrument were selected by at a least a quarter of the firms (Table 12.4). Communication

Table 12.4 Skills needed to bring KIBS staff in the West Midlands to full proficiency

Skills that need improving	*Number of firms*	*% of all occupations with skill gaps*
Communication skills	120	42.8
Technical and practical skills	120	42.6
Customer handling skills	115	41.0
Practice/general management skills	114	40.5
Personal skills	112	40.0
Problem solving skills	112	39.7
Office admin. skills	106	37.7
Team working skills	96	34.2
General IT skills	88	31.5
Leadership skills	80	28.6
Data analysis	73	26.1
IT professional skills	71	25.3
Accounts/finance	60	21.3
Technical skills related to the occupation	58	20.5
Human resource management skills	47	16.7
Literacy	39	14.0
Numeracy skills	31	11.0
Foreign language skills	22	8.0
No particular skills	20	7.1
Other skills	12	4.4

skills topped the list (mentioned by 42.8 per cent of firms with skill gaps), along with other skills relating to social interaction which included customer handling (41.0 per cent), personal skills (40.0 per cent) and team working (34.2 per cent) also scored highly. Another common theme could be described as managerial, as reflected in the high percentages given to general management (40.5 per cent), problem solving (39.7 per cent), office administration (37.7 per cent) and leadership (28.6 per cent). Perhaps unsurprisingly, technical and occupation specific skills generally were not cited frequently as exemplified by accounts/finance (21.3 per cent), technical skills related to occupation (20.5 per cent) and foreign language (8.0 per cent). General office-related skills received more mentions, but perhaps less than expected. For example, general IT skills were only cited by 31.5 per cent of firms, while literacy and numeracy were selected by just 14 per cent and 11 per cent of firms respectively. This is at odds with the emphasis often given by business leaders to such factors but it may be that these problems are only acute among younger employees.

Skills – large KIBS firms versus small firms

Each KIBS establishment, firm, or organisation has a unique mix of human resources at its disposal. There is some scope however, for understanding skill needs by adopting a *segmentation approach* that highlights, for example, different organisational forms such as micro single person firms with no employees; small single office local firms; small firms with more than one office; branch offices of medium-sized firms; and branch offices of large firms.

Small and medium-sized firms far outnumber much larger firms that have either grown to support the needs of larger local client firms or are the branch offices of national and international service suppliers. Many of the latter appear to have entered the West Midlands market place over the last 10 years and have experienced high growth rates. Larger KIBS firms in the West Midlands are also much more complex. They can be budget or cost centres in their own right, have near independence from the parent headquarters or be completely controlled from outside the region. There are many different combinations of legal structure. Most, however, are responsible for the recruitment of support staff while, in some cases, the recruitment of professional staff is dealt with by a national HR division that is likely to be located outside the West Midlands.

Legal and organisational diversity, and therefore skills and training needs, is not confined to large KIBS firms. For example, there is a marketing company that operates as a collective in which professional 'employees' are self-employed video producers, web designers, animators and illustrators who are all accessible through one website which makes the firm look larger than is actually the case. There is also a virtual human resources company providing outsourced HR functions for SMEs that has a small dedicated central base with the professional staff 'in the field' providing dedicated advice at the site of the client company. A number of home-based KIBS firms were interviewed; micro-firms or sole practitioners are quite prominent, alongside the virtual firms and collectives that rely on established relationships with self-employed associates. These examples demonstrate the importance of organisational flexibility for KIBS firm. This, in turn, has important implications for training and skill development. Such firms have limited requirements for support staff and, in many cases, professional staff will be self-employed. Micro-firms that operate using associates have very different skill needs to larger firms. Many do not employ support staff. This highlights the combinational skill problem experienced by many small companies. Such firms may be technically competent in their specialist areas, but are underperforming because of poorly developed support skills (time management, organisation, sophisticated understanding of computer packages, etc.). Flexibility can also mean that self employed associates are responsible for fulfilling their own training and development needs. It can also involve stretching the working day to meet the needs of clients and it can also involve manipulating salary levels.

KIBS firms also operate in a number of different market segments. These range from the very local to the national and sometimes international. Firms delivering services to each of these market segments will experience different skills and training requirements and will possess different dynamics in relation to the organisation of their business activities. Segmentation is also of course relevant to exploring skills and training for broad categories of occupations within KIBS. The key segments are professional staff, paraprofessionals, and support staff. To highlight the different types of training and skills needed by each segment one of the surveyed firms graded employees into: "finders, minders and grinders". Grinders never meet clients since not all employees have the ability or personality to deal with clients and these employees become grinders. It is worth noting that it would appear that the number of grinders in KIBS firms is shrinking as more employees are becoming fee earners. 'Finders' are involved in obtaining new clients through networking and the identification of business opportunities. The 'minder's' are engaged with working with 'finders' to co-create services with clients.

When exploring recruitment by KIBS firms it must be recognised that for many it is an activity that is engaged in only rarely, if at all. The recruitment experiences of KIBS firms can be divided into three types: firms that do not recruit, those that recruit infrequently and, those that recruit on an annual basis. If this is mapped on to the types of organisational forms typical of KIBS i.e. the relatively small number of medium–sized and large firms, it will be appreciated that formal recruitment processes are the exception rather than the norm. Extant relationships and referrals are at the centre of the KIBS recruitment process. They are the dominant modes of recruitment amongst small and medium-sized enterprises and are by no means absent in the recruitment activities of larger companies. In relation to the latter it is perhaps important to distinguish between branch offices of firms with headquarters outside the region and local firms. In the majority of cases support staff are recruited locally via advertisements in the local press or word of mouth.

The challenge of attracting and retaining professional staff is shared by most firms across the region, but especially in those more rural counties such as Shropshire or Staffordshire that do not have a critical mass of KIBS activities. Professionals are more geographically mobile than support staff and this has revealed an interesting relationship between the London KIBS market and that of a regional economy such as the West Midlands. Professionals may be trained in London, work there for some years and then relocate to work in other cities. It may also be that the best locally trained KIBS professionals migrate from the West Midlands to London to advance their careers. However, the evidence to support this is limited and further research into the career biographies of KIBS professionals, for example, would be useful. Many of the larger firms highlighted the fact that professionals moved between firms in the area and that the market could be very competitive; sometimes larger firms engaged in a sort of 'Dutch auction' to attract professionals.

Effective training changes employees' attitudes. There is a relationship between training and motivation; good training means that employees have enhanced confidence. This can come from on-the-job training, in-house or external courses, but employers should be able to notice a difference in the employee – in attitude, confidence and technical ability. Training in KIBS is therefore not just about achieving the appropriate qualifications as required by professional bodies or for employees to be able to perform particular roles or to gain promotion. KIBS managers also need to be trained in how to identify the training needs of their employees. The survey identified that many KIBS owners and managers lack professional management skills. The training needs of a KIBS firm are identified and assessed by individuals who have limited understanding of the required management skills in this area. Many have a very limited, poorly developed and perhaps outdated understanding of staff appraisal and development processes.

The distinction between support staff training and that of professionals was stressed by all the survey respondents. For support staff there were numerous references to a requirement for good communication skills, technical expertise related to their area, for example legal secretaries or knowledge of relevant ICT. Professionals need sophisticated client relationship management expertise, technical competence and all this is supported by continuous professional development (CPD). The requirement for CPD, which is forced upon professionals, in some cases, is considered to be a mere box ticking exercise. However, for most firms CPD is an essential part of the business, as without it the professional staff will be out-of-date and not aware of current developments (legislation, etc) that have an impact on their clients' activities.

For most of the businesses in this study it does not matter if training is provided by private or public trainers. The key concerns are quality and cost. It was noted that, even though they are used more extensively than public sector trainers, there are some very bad private trainers. Firms look at the quality first and for some the location of the training provision is unimportant. For others, however, there is an important relationship between point specific provision of key training courses and access to the providers. This is a difficult problem to solve as it is not commercially viable to develop a heavily dispersed network of training providers. However, it would be possible to combine face-to-face training provision with student-centred or distance learning with the balance between formats depending on the trainees' location.

It is perhaps worth making reference to the succession issue which is a common feature of small KIBS firms. It is especially important amongst KIBS firms some of which are, for example, established around the expertise of a key professional; an individual with an established reputation and network for a particular type of expertise as well as set of established client relationships. In many cases, the established relationships might be more important than the expertise. The implication is that some small KIBS firms wither away following the retirement or death of their key professional. There is thus constant churn within the KIBS SME community. It also means that firms change as mergers occur that are part of succession planning strategies. Some of these mergers are visible, resulting in an alteration to the name of a firm, whilst others are invisible with a retiring professional being subsumed into the receiving firm as a temporary short-term measure that will assist their retirement while also ensuring the transfer of clients to the receiving firms.

Unsurprisingly, the evidence from this study confirms that the competitiveness of KIBS firms is derived from a combination of technical competence, the personality of individual professional and support staff, and a set of soft skills. This implies that academic excellence and world class technical competence does not provide the platform for commercial success; it is also crucially founded upon relationship building, soft skills and personality. At one level, it is difficult and perhaps impossible to develop advanced soft skills in some personality types but a country's national educational system should be designed to develop minimum levels of competence in a set of essential KIBS skills.

The interviews identified four essential KIBS skills:

1 Verbal dexterity.
2 An ability to relate to people from a wide variety of backgrounds and cultures (gender, ethnicity, age, nationality).
3 An ability to listen (to try to understand what the client wants and then to be able to frame an appropriate response).
4 Appreciation of body language (its interpretation and projection); impression management.

In addition, much of activity engaged in by KIBS is about creative writing. Therefore, advanced literacy skills are required, as well as a level of numeracy.

KIBS firms and training: a triple penalty

Although not exclusively an issue for support staff in KIBS firms, there is evidence pointing to poorly developed soft skills, including managing relationships – internal and external to firms. This deficiency does not just arise from any inadequacies in the monitoring and training regimes used by employers. It is complicated by a failure on the part of new employees to appreciate some of the survival techniques that are required in the world of employment. This includes an appreciation of the nature of the work environment in KIBS, understanding how as individuals they relate to other employees (as well as clients), and the requirements associated with working as part of a team. Within KIBS these are critical skills as the success of a firm is founded upon the projection of a consistent professional image that is constructed around, and from, the activities of all staff members.

There are three cost penalties arising from a commitment by KIBS firms to training:

1 Management time spent identifying skills and training needs (staff appraisal).
2 Time and effort invested in identifying suitable courses.
3 Fee earning or support staff time lost as a result of training.

There is a fourth less visible cost – the opportunity costs associated with a firm employing less than proficient staff. On the job training produces a double cost penalty arising from the pre-occupation of the internal trainer and the trainee with the training task in hand and the tendency, in many instances, for the training not to be concluded because everyday business tasks distract the participants.

Just over a quarter of firms believed that they did not have time to train staff. One remedy is to engage external training providers but firms had serious concerns about local availability. Just under a third of respondents were worried about the relevance of local provision (32.5 per cent), 23.3 per cent about the quality of training programmes, and 22 per cent by the absence of appropriate training in the local area. It also seems reasonable to assume that the 35 per cent who found it difficult to access training at suitable times were also dissatisfied with local provision. It may be that the perceived inadequacies of training provision at the local level are a function of incomplete information or imperfect knowledge; a conclusion supported by the comparatively high ranking attached to "lack of information on training" as a barrier to staff development and training. There was also some evidence of scepticism about the value of training.

KIBS professionals require a stimulating work and residential environment. This means that it is difficult for places to attract and retain highly trained professional staff. During one of the focus groups held in Coventry, Warwickshire, as part of this project it was noted by a participant that:

> Birmingham is now seen to be the place to be for business and professional service firms. We are competing with Birmingham rather than London, and Birmingham firms consider that they can cover this area from Birmingham. Leamington has a quality of life and perception advantage over Coventry. People want to come and work there, they have a better quality of life and are also able to obtain Birmingham business'.

This highlights the importance of place-based assets or resources in attracting and retaining KIBS professional expertise. Firms located in places with hard-to-fill vacancies claimed that salary levels were comparable (and sometimes enhanced) with those in Birmingham, but they still found it difficult to attract suitable highly skilled staff. Family ties and previous place-based associations are some of the most effective mechanisms for attracting and retaining people in this area.

Support staff attraction and retention is a lesser problem but poor soft skills as well as uneven Further Education provision of key courses, for example the absence of suitable secretarial training in Coventry and Warwickshire, was identified as a source of difficulty. The concentration of some support staff training suppliers in Birmingham was considered by some firms to be a major problem, especially when account is taken of transportation infrastructure problems. Poorly paid support staff were often unwilling or unable to travel far to undertake formal training. This was seen as an invisible barrier preventing many smaller KIBS firms from persuading members of their support staff to undertake training. In one case a partner of a small firm noted that junior support staff were unable to afford reliable cars and were reluctant to travel far on wet winter evenings. The impact of such 'invisible' barriers to sourcing training by KIBS firms should not be underestimated (Quader, 2007).

There are significant structural differences across the West Midlands region, especially when the qualitative rather than the quantitative survey evidence is examined closely. A key structural difference is the density of KIBS firms within a local or regional economy. This is a key issue that should not be under emphasised as it makes it difficult to attract staff as well as reducing the local provision of training. Places such as Coventry, Stoke-on-Trent, Hereford, and Worcester have low representations of the larger practices that tend to be the primary training grounds for some KIBS professionals (Bryson, 1997). This means that KIBS firms outside Birmingham are relatively disadvantaged as a consequence of the absence of large firms that sometimes operate as a lobbying group for the KIBS sector and often as a training ground.

Larger firms have the resources to develop, implement and manage employees' appraisal and training schemes The employees in small KIBS firms encounter constant demands for multi-tasking and do not have the spare capacity, time and people that are needed to develop effective training regimes. This means that in large-firm economies, such as Birmingham, Further Education colleges must fit in with the training schemes of the larger companies while in the regional or smaller town economies they must be more proactive in engaging with companies and ensuring that they identify and fill local skills deficiencies and gaps.

Commercial skills and KIBS firms

KIBS firms found it was difficult to recruit commercially aware professionals as well as customer focussed and, in some cases, work focussed individuals. Commitment to client needs was seen as an important factor especially in terms of working after 5pm. Many of the firms identified a skills gap between 'self skills', management skills and technical skills. The technical skills are not an issue, but the development of management skills and self or softer skills lags behind the acquisition and refinement of technical expertise. One law firm had experienced difficulties with a secretary who was not performing at the required level. She was still a probationer and her line managers wanted to get rid of her. However, a performance review was undertaken, benchmarks set and the skills gaps identified and a training programme, including mentoring, developed. The skills gap was IT related and in general administrative skills. The probationary period was extended and eventually she became a permanent employee. This illustrates that the recruitment process can fail and people can be recruited who do not have the required skill set, and that the skill deficiencies can be in relatively simple areas. It also shows that appraisal and training programmes can be successful mechanisms for creating useful employees.

The key skill shortage is in the area of more intangible skills many of which are related to personality types. One property consultancy noted that hard-to-fill vacancies:

> are the biggest problem that we have. To the point whereby it does have an influence on how much work we take on. So you could say that it's a limitation to growth. And it's as fundamental as that. If you have to put it in a strategic context, it is preventing growth and it's not just my office, not just this office it's right through the business and when I meet colleagues in other businesses they've got exactly the same problem. It's not a question of there not being enough people out there who would like to work for us, but there's not enough people of the right quality out there and available. There's a shortage of supply *(BS54/Acc/25–99)*.

The key issue concerns people who have wider expertise which enable them to appreciate the wider commercial implications of a range of decisions. A key skill for this company was problem solving and the ability to sort data and manage complex projects. Much more research needs to be undertaken on understanding the nature of the commercial skills that exist within KIBS firms. On the one hand, some of these skills involve client identification and relationship management. On the other hand, a key skill is the translation of an idea into a commercially valuable product.

Another company identified a similar skills issue by highlighting the difference between technical and commercial skills:

> technical skills, that is first awareness of the market – so it is personality, being able to communicate, ideas, lateral thinking, business acumen. So the ability to see opportunity, go and find it, go and acquire it and at the end of the day we're a business, we're about making money for our shareholders as well as providing a cracking service to our clients and we need those skills in our surveyors. They're not there just to do a technical job, you can only do a technical job once you've got the clients and the jobs to do it on and it is our job to get in high quality work, process it, give the best advice, look for opportunities and maximise our position in the market and make money. That requires people with a skill set which is above just being able to do the technical work, that is where we run into problems *(BS33/Real/25–99)*.

A representative from a branch office of a worldwide property consultancy made a similar point when they argued that:

> I would say generally that not enough people have a sufficient degree of self confidence to enable them to what I call cold call face-to-face. Just go and introduce yourself to somebody that you don't know. We have a client cocktail party every year and I would say to people please don't let me see you standing in a group of more than two of you without anybody else talking to you because if we have invited a load of people here and we are buying a few drinks and we are having a chat with them, all of which could be clients or good contacts of ours, why are you talking amongst yourselves. You can do that all day, so tonight go and talk to people and if you see somebody on their own whether you know them or not go and say hello. And a lot of it is we are quite a confident business but that's mainly the people that are market facing all the time. Some other people aren't quite like that
>
> *(BS82/Real/25–99)*.

This highlights the mix of skills needed within the KIBS part of an economy. Technical expertise is required, but effective communication skills and an ability to develop long–term relationships are essential capabilities. This mix of technical with soft and commercial skills is very different to that required in manufacturing. This difference is, in part, explained by the difficulty clients have in differentiating between competing services provided by different KIBS firms. The intangible nature of the KIBS product means that relationships and trust play a critical role at the centre of the client-KIBS relationship.

Discussion and conclusions: KIBS, skills and regional policy

KIBS workers are equated with knowledge intensive positions or occupations that are found in the full range of industry sectors but, in particular, in advanced or knowledge intensive services such as business and professional services. The knowledge intensive class of accountants, lawyers, consultants, designers and advertising executives amongst others are prominent in business and professional services (BPS) and a high degree of skill acquisition by formal education and other means is a pre-requisite. Developing policies to support KIBS is difficult given the heterogeneous nature of the sector combined with the intangible nature of the products provided to clients.

Regional policy targeted at KIBS should focus on three related areas. First, the relationship between the supply and demand for skilled workers as well as the local availability of skills training, in this case for KIBS firms, should be central to both regional as well as national public policy. This policy needs to focus both on higher education institutions and colleges of further education (Taylor, 2008). It also needs to recognise the training needed by KIBS support staff and the professional or fee-earning employees. For support staff the key issues are cost, availability and accessibility. The training of KIBS professionals is much harder to address using regional policy. There are a number of issues that may be difficult to incorporate into policy. These include softer skills that are linked to personality and also the development of a reputation; technical skills appear to be relatively easily acquired via formal education and the professional bodies, but the softer and commercial skills are much more challenging.

Second, KIBS firms are established by individuals with a reputation for technical expertise rather than management expertise. This means that there are problems with the management skills of KIBS firms across the West Midlands. This deficiency is illustrated by the fact that 28 per cent of firms did not have a business plan while just over 40 per cent of firms did not have training plans in which the level and type of training given to employees is specified in advance. The importance of appraisal was acknowledged with 70 per cent of respondents holding annual performance reviews and, although over half the firms included all their employees, in some cases only 10 per cent of the workforce was incorporated into some appraisal exercises. Developing policy-led interventions to enhance the management capabilities of KIBS firm owners and managers should form an essential element of any KIBS policy.

Third, KIBS firms have a double significance: they create wealth in their own right, but they can also enhance wealth creation in their client companies (Greenfield, 1966). This means that the activities of KIBS firms contribute to two types of gross value added (GVA). First, GVA produced directly by their own activities and, second, indirect GVA that is produced by client firms that can be attributed to the activities of KIBS firms. It is very difficult to measure the impact KIBS firms have on the competitiveness and profitability of client companies, but in some cases enhancement to GVA occurs (Bryson *et al.*, 1999 a and b). One difficulty is identifying a direct simple linear relationship between the activities of a KIBS firm in a client firm and impact. Time complicates the assessment and measurement of such impacts as a KIBS project might produce an impact over a long time period (Bryson *et al.*, 1999a and b). The double impact created by KIBS suggests that it is important that clients are able to effectively manage the relationship with KIBS firms. This is an area for policy intervention. It emphasises the importance of ensuring that all industrial policies contain interventions intended to enhance the value of the interactions and processes of co-production that occurs between KIBS professionals and client companies. The competitiveness of regional and national economies is, in part, founded upon skills, capabilities and competencies. At the centre of these relationships is the co-production of innovation and expertise that occurs between KIBS and their clients.

Acknowledgements

The research reported in this paper was funded by the European Social Fund and was managed by the Learning and Skills Council, Birmingham

References

Abreu, M., Grinevich, V., Kitson, M., Savona, M. (2008). 'Taking services seriously: How policy can stimulate the "hidden innovation" in the UK's services economy', NESTA Research Report. London: National Endowment for Science Technology and the Arts.

AWM (1999). *Creating advantage*. Birmingham: AWM.

AWM (2001). *Agenda for action*. Birmingham: AWM.

AWM (2004). *Delivering advantage*. Birmingham: AWM.

AWM (2007). *Connecting to success*. Birmingham: AWM.

Begg, I. (1993). 'The service sector in regional-development', *Regional Studies*, 27, 8: 817–25.

Pryce, V. (2008). *Manufacturing: new challenges, new opportunities*. London: BERR.

Brinkley, I. (2006). *Defining the knowledge economy*. London: The Work Foundation.

Bryson, J.R. (1997). 'Business service firms, service space and the management of change', *Entrepreneurship and Regional Development,* 9: 93–111.

Bryson, J.R. (2007). 'Lone eagles and high flyers: rural-based business and professional service firms and information communication technology', in Rusten, G. and Skerratt, S. (eds) *Information and communication technologies in rural society: being rural in a digital age*. London: Routledge, 36–60.

Bryson, J.R. (2015). 'Business and professional service firms and the management and control of talent and reputations: retaining expert employees and client relationship management', in Bryson J.R. and Daniels P.W. (2015), *Handbook of service business: management, marketing, innovation and internationalisation*. Cheltenham: Edward Elgar, 316–29.

Bryson, J.R., Clark J. and Mulhall, R. (2013). *The competitiveness and evolving geography of British manufacturing: where is manufacturing tied locally and how might this change?*, London: Department of Business Innovation and Skills.

Bryson, J.R., Clark, J. and Vanchan, V. (Both US), (2015). *Handbook of manufacturing in the world economy*, Cheltenham: Edward Elgar.

Bryson, J.R. and Daniels, P.W, eds (2007). *The handbook of service industries in the global economy*. Cheltenham: Edward Elgar.

Bryson J.R. and Daniels P.W. (2015a). *Handbook of service business: management, marketing, innovation and internationalisation*. Cheltenham: Edward Elgar.

Bryson J.R. and Daniels, P.W. (2015b). 'Developing the agenda for research on knowledge-intensive services: problems and opportunities', in Bryson J.R. and Daniels P.W. (2015), *Handbook of service business: management, marketing, innovation and internationalisation*. Cheltenham: Edward Elgar, 417–37.

Bryson J.R. and Daniels, P.W. (2015c). 'Service business: growth innovation, competitiveness', in Bryson J.R. and Daniels P.W. (2015), *Handbook of service business: management, marketing, innovation and internationalisation*. Cheltenham: Edward Elgar, 1–20.

Bryson, J.R., Daniels, P.W. and Ingram, D.R. (1999a). 'Evaluating the impact of business link on the performance and profitability of SMEs in the United Kingdom', *Policy Studies*, 20, 2: 95–105.

Bryson, J.R., Daniels, P.W. and Ingram, D.R. (1999b). 'Methodological problems and economic geography: the case of business services', *The Service Industries Journal*, 19, 4: 1–17.

Bryson, J.R., Daniels, P.W. and Warf, B. (2004). *Service worlds: people, organizations, technologies*. London: Routledge.

Bryson. J.R. and Ronayne, M. (2014). 'Manufacturing carpets and technical textiles: routines, resources, capabilities, adaptation, innovation and the evolution of the British textile industry', *Cambridge Journal of Regions, Society and Economy*, 7: 471–88.

Bryson, J.R. and Rusten, G. (2011). *Design economies and the changing world economy: innovation, production and competitiveness*. London: Routledge.

Bryson, J.R., Taylor, M. and Daniels, P.W. (2008a). 'Commercializing "creative" expertise: business and professional services and regional economic development in the West Midlands, UK', *Politics and Policy*, 36, 2: 306–28.

Damesick, P.J. (1986). 'Service industries, employment and regional development in Britain: a review of recent trends and issues', *Transactions of the Institute of British Geographers, New Series*, 11, 2: 212–26.

Daniels, P.W. and Bryson, J.R. (2002). 'Manufacturing services and servicing manufacturing: changing forms of production in advanced capitalist economies', *Urban Studies*, 39, 5–6: 977–91.

Department for Innovation, Universities and Skills (2008). *Innovation nation*. London: The Stationery Office.

Department of Trade and Industry (2003). *A modern regional policy for the United Kingdom*. Norwich: HMSO.

Department of Trade and Industry (2007). *Innovation in services, occasional paper no. 9*. London: DTI.

Fothergill, S (2005). 'A new regional policy for the UK' *Regional Studies*, 39, 5: 659–67.

Gereffi G. (2001). 'Shifting governance structures in global commodity chains, with special reference to the internet', *American Behavioural Scientist,* 44:1616–37.

Gereffi G, Humphrey J., Sturgeon T. (2005). 'The governance of global value chains', *Review of International Political Economy*, 12:78–104.

Greenfield, H. I. (1966). *Manpower and the growth and producer services*. New York: Columbia University Press.

Heseltine, M. (2012). *No stone unturned: in pursuit of growth*. London: Department of Innovation, Business and Skills.

High Level Group on Business Services (2014). *High level group on business services – final report*. Brussels: European Union.

Illeris, S. (1996). *The service economy: a geographical approach*. Chichester: John Wiley.

Livesey, F. (2006). *Defining high value manufacturing*. Cambridge: Institute for Manufacturing.

Leitch, S. (2006). *Prosperity for all in the global economy: world class skills*. Norwich: The Stationery Office.

Love, J., Roper, S. and Bryson, J. R. (2011). 'Openness, knowledge, innovation and growth in UK business services', *Research Policy*: 40, 10: 1438–52.

Marshall, J.N.I (1985). 'Business services, the regions and regional policy', *Regional Studies*, 19, 4: 353–63.

Mulhall R. and Bryson, J. R. (2013). 'The energy hot potato and governance of value chains: power, risk and organizational adjustment in intermediate manufacturing firms', *Economic Geography* 89, 4: 395–419.

Quader, M.S. (2007). 'Human resource management issues as growth barriers in professional service firm SMEs', *Journal of Service Research* 7, 2: 115–61.

Rogers, B. (2006). *Navigating the new economy: defining and measuring progress towards the UK knowledge economy*. London: Intellect.

Roper, S., Hales, C., Bryson, J.R. and Love, J. (2009). *Measuring sectoral innovation capability in nine area of the UK economy*. London: NESTA.

Saxenian, A. (2006). *The new Argonauts: regional advantage in a global economy*. Boston, Mass.: Harvard University Press.

Taylor, M., Plummer P., Bryson J. R. and Garlick S. (2008). 'The role of universities in building local economic capacities', *Politics and Policy*, 36, 2, 216–31.

Von Nordenflycht, A. (2010). 'What is a professional service firm? towards a theory and taxonomy of knowledge-intensive firms', *Academy of Management Review* 35, 1: 155–74.

13 Prospects and policies in the development of Knowledge Intensive Business Services in Europe

Marja Toivonen and Antonella Caru

Introduction

Knowledge Intensive Business Services (KIBS) have been one of the most intensively studied service sectors since the mid-1990s. The reason lies in the tight linkage of this sector to the contemporary knowledge economy. KIBS are expert companies whose offering consists of knowledge inputs to other companies and organisations, and in order to be successful and competitive, KIBS have to continuously accumulate and update their own knowledge base (den Hertog, 2000). The rapid adoption and application of new knowledge are the lifeblood of these companies. Thus, KIBS have been suggested to be a core actor in the promotion of innovations in two ways: *innovations emerge in KIBS and through the use of KIBS* (Gallouj, 2002). The former refers to KIBS as active innovators and the latter to their roles as facilitators of innovation activities in their client organisations and as disseminators of innovative outcomes in the economy (Hipp and Grupp, 2005; Miles *et al.*, 1995).

KIBS provide typically design and consultancy-type services: IT services, R&D services, technical consultancy, legal, financial and management consultancy, and marketing communications services (Miles *et al.*, 1995; Toivonen *et al.*, 2008). To depict the high educational level of their personnel, these companies are also called professional service firms (Suddaby and Greenwood, 2001; Wagner *et al.*, 2014). Knowledge repositories in KIBS contain elements from a range of diverse knowledge domains (Wolpert, 2002). However, the question is not only about the amount of knowledge or expertise, but co-learning with clients plays a central role in the fostering of innovation (Miles *et al.*, 1995). In addition, KIBS have an important role as nodes in innovation networks, which today reach from the local level to national and global levels (Miozzo and Miles, 2003; Tomlinson, 2001; Wood, 2002).

The role of KIBS has been highlighted at the regional level, in particular. In addition to the creation of value added and employment, KIBS have been argued to *enhance the competitiveness of the local economy*. Advanced manufacturing industries depend on the local presence of qualified services: a high quality service environment attracts new investments. In addition, other service firms increasingly demand external expertise (Shearmur and Doloreux, 2014; Toivonen, 2007).

From this viewpoint, it is problematic that the geographical distribution of KIBS is typically uneven: these services are concentrated in big cities in which the economic development is dynamic. The growing economy needs more services and the services further promote the growth. In peripheral regions, the absence of external services depresses the demand and is especially disastrous for small firms that have no access to advanced and specialised services. As a consequence, a sharp contrast develops between central and peripheral regions (Muller and Zenker, 2001; Strambach, 2008).

These observations have led to concerns among policy makers about the sufficient supply of Knowledge Intensive Business Services in different countries and regions, and about the skills level of service providers. The functioning of KIBS as knowledge intermediaries enables the facilitation of innovation and favours the emergence of innovations, but does not automatically lead to innovativeness in individual companies. The realisation of innovative practices also depends on the skillful use of knowledge intensive services, implying that the development of both supply and demand is needed (OECD, 2006). Studies on KIBS and their clients have brought to the fore several problems. The KIBS market is divided into a few big conglomerates and the majority of small companies, with too scarce mutual relationships (Toivonen, 2007). As companies prefer service providers whose size is near to their own size (Contractor *et al.*, 2010), polarisation has slowed down the development of the middle market. Demanding clients, which would push forward growth and development, purchase services from big providers. Small KIBS lack resources for the acquisition of newest knowledge and for the systematic improvement of the knowledge base. Among the small clients, locally restricted search for knowledge and deficiencies in absorptive capacity cause problems (Escribano *et al.*, 2009). A general problem is price-driven and transaction-based procurement, which does not foster the utilisation of the benefits of expert services (Wagner *et al.*, 2014).

Currently, new issues are emerging together with the aforementioned challenges. Digitalisation affects all economic activities and its essential part is the tightening of networked business. While KIBS have been among the main users and supporters of the information and communication technologies (ICT) since the 1990s (Antonelli, 1999), they will meet new kinds of requirements in the era of digitalisation. Besides high-level expert services, KIBS have to pay an increasing attention to their business models and positioning in the networks, which involve – not only clients – but also the clients' partners and end-users. This means that the practices of collaboration and co-learning, that earlier have focused on the provider-customer dyad, have to be carried out in much more demanding context: in a multi-actor environment (Leminen *et al.*, 2014). Simultaneously, the nature of innovations is changing. In addition to individual product, process and service innovations, there is a growing need for complex systemic renewals (Geels, 2005). Discussion on the ways in which the creation of these kinds of innovations can be facilitated is beginning.

In this chapter, we analyse in more detail the current opportunities and challenges in the KIBS sector and anticipate their further development. The chapter is based on the scenarios and policy options built and recognised in the working group for innovation in the High Level Expert Group on Business Services – an initiative of European Commission 2013–14 (HLG, 2014). The working group for innovation was chaired by the first author of this chapter, and the second author was the rapporteur of the group. The members of the group represented both academia and business life. In order to elaborate the policy-focused work of HLG for scientific purposes, the produced material has been re-analysed and its linkages to relevant literature have been substantially strengthened. However, the basic structure and contents of the scenarios as well as the drivers behind them have been concluded in the working group, using the method of expert-based foresight (Kuusi, 1999). The focus of the policy analysis is on the European KIBS sector, but the recognised prospects and challenges are also more widely generalisable.

The chapter is conceptual, supplemented with a few illustrative examples, but without an empirical study. The structure of the chapter is as follows. The second and third sections deepen the views briefly summarised in this introduction: the central role of KIBS in innovation and challenges in the realisation of the promise included in this role. The fourth section describes the method in which the scenarios on the futures of KIBS were built in the working group of HLG. The fifth section presents the results: four scenarios, two of which are driven by topical developments in the operational environment (digitalisation and systemic changes), and the other two by developments in the business of KIBS and their clients respectively. Based on these views about the future, the sixth section identifies policy options that the stakeholders at the regional, national and international level could apply in order to tackle the foreseen challenges and to strengthen the innovative impact of KIBS in the economy. The final section evaluates the work and raises ideas for further studies.

The central role of KIBS in innovation – background of the policy interest

Innovation is today considered the main way to achieve economic growth and competitiveness and to foster employment and welfare in a sustainable manner (Porter, 2000). Openness of firms to external knowledge sources is an important element when evaluating their innovative potential (Caloghirou *et al.*, 2004; Chesbrough and Crowther, 2006). With the growing complexity of products and services, the companies no longer have all the diverse components of knowledge within their own organisation to be competitive in research, production and marketing (Bilbao-Osorio and Rodríguez-Pose, 2004; Contractor *et al.*, 2010).

The development of ICT has increased the need for external assistance. While ICT have drastically facilitated handling, storing and moving of information, they have also raised the question how to find essentials and analyse and interpret them. The importance of expertise linked with locating and selecting the relevant information has notably grown (Preissl, 2000). In addition, organisations are only rarely able to adopt external information into practice as such and by themselves. They need professional help in converting this input into such types of knowledge that enable the development of new products, processes and services (Caloghirou *et al.*, 2004).

The idea that KIBS, in particular, may support other companies and organisations to fulfill the need for outside sources of cognition is based on the fact that knowledge constitutes both the main input and output in them (den Hertog, 2000, Gallouj, 2002). The specialisation of KIBS in certain expert areas results in the level of know-how for a particular task that the clients cannot achieve. Aggregating similar work over many customers means a continuous accumulation of experience and the respective improvement of solutions regarding the problems of clients (Wolpert, 2002; Zhang and Li, 2010). It also provides KIBS with an 'objective' perspective that can be used to facilitate necessary organisational changes of clients (Bessant and Rush, 1995).

On the other hand, KIBS professionals have to collaborate with their clients in order to find answers to the context-specific issues (Bettencourt *et al.*, 2002). This collaboration creates a basis for a joint learning process – learning takes place both in KIBS and in the client organisation (Miles *et al.*, 1995). For the emergence of innovations, learning is essential. In the KIBS transactions, it becomes concrete in the interplay between the generic knowledge accumulated by experts and the tacit knowledge buried in the daily practices of clients (Greenwood *et al.*, 2005; Boone *et al.*, 2008). KIBS act as mediators in the conversions between different forms of knowledge: their diagnostic and training activities support clients in the transfer of tacit knowledge into explicit, and vice versa (den Hertog, 2002). Diagnosis and problem-clarification as such may be crucial: it helps clients define and articulate their needs in such a way that external resources and opportunities can be effectively used (Bessant and Rush, 1995).

Thus, the information flows between KIBS and their clients are bidirectional – a linear transfer of expert knowledge from the provider to the client supports the whole but is not sufficient in the solving the complex problems (Bettencourt *et al.*, 2002). On the other hand, KIBS also have an important role in the markets of codified knowledge. Codification has increased the divisibility of information, which together with the enhanced accessibility has made the commodity nature of information more marked. This has increased the commercial potential of information and been one factor that has promoted the outsourcing of expert services (Contractor *et al.*, 2010).

In addition to the knowledge functions which facilitate the creation of innovations within client organisations, KIBS act as 'bridging intermediaries' between various clients (Corrocher and Cusmano, 2014). They benchmark business practices and help establishing relationships between different stakeholders (Bessant and Rush, 1995). These activities of KIBS have led researchers to conclude that while serving numerous clients, KIBS carry ideas from one context into another and disseminate innovations. KIBS could be particularly important vehicles in the spread of innovations from larger firms to small and medium-sized enterprises (OECD, 1999).

Some researchers have suggested an even more central role to KIBS in innovation, based on the versatile contacts that the bridging and brokering functions offer. These suggestions include the possibility that KIBS could develop into a 'second', private knowledge infrastructure in the society, complementing the public infrastructure, i.e. educational and research institutions (Corrocher and Cusmano, 2014; den Hertog and Bilderbeek, 2000). KIBS could form nodes in the collaborative systems of clients, partners, public institutions and R&D establishments (den Hertog, 2000). They could orchestrate innovations and innovation networks, and integrate knowledge flows between regional, national and global levels (Howells and Roberts, 2000; Miles, 2000). If realised, these kinds of roles would essentially promote the connectivity of innovation systems in different sectors and at different levels. However, more research is needed to gain empirical evidence of KIBS' roles and the ways in which they vary in different operational environments (Corrocher and Cusmano, 2014).

In order to perform the roles of facilitators and carriers of innovation, KIBS have to continuously upkeep their own knowledge base, i.e. innovation is a necessary prerequisite for their business success. The repeated interactions with clients enable generalisation on the basis of problem-specific insights: the cognitive inputs behind unique, tailor-made solutions are often widely applicable (Preissl, 2000). The nature of KIBS as active innovators has been confirmed in several studies. European Community Innovation Surveys (CIS) that include data on technology-based KIBS (technical consultancy and ICT companies) have shown that these companies are actually very similar to high-tech industrial companies in their innovation activities (Eurostat). Case studies have provided similar evidence from managerial KIBS, e.g. advertising agencies and management consultancies (Leiponen, 2001; Santos-Vijande, 2013). However, the degree of innovativeness varies in KIBS (Corrocher and Cusmano, 2014; Freel, 2006), and its potential may be underutilised due to the problems of clientele (OECD, 2006). Next, we open up these issues in more detail.

Seizing the promise of KIBS – challenges in supply and demand

Since the mid-1980s, KIBS have been the most rapidly growing economic sector in the advanced economies (Simmie and Strambach, 2006). However, the development has been geographically uneven: KIBS tend to concentrate very strongly to the biggest cities worldwide and to the metropolitan areas in each country (Muller and Zenker, 2001; Shearmur and Doloreux, 2014; Strambach, 2001). In urban contexts, KIBS have multifaceted opportunities for using, not only localised knowledge, but also national and international knowledge inputs. Agglomeration also includes advantages such as high quality communication infrastructure, flexible labour markets, and connections to domestic as well as foreign clients. Correspondingly, the advanced regions benefit from the availability of KIBS as the competitiveness of companies increasingly depends on the knowledge provided by specialised suppliers (Strambach, 2008). In peripheral regions, a 'vicious circle' is a threat: the insufficient supply of external services slows down the development, which in turn depresses the demand (Macpherson, 1997). The situation may be particularly problematic in small firms which have not enough internal resources to compensate the service shortage.

On the other hand, the relationship between the share of KIBS and the innovation performance of a region is not causal or linear. Recent studies have questioned the idea that the use of KIBS requires regional proximity. KIBS typically maintain an extensive network of geographically diverse clients (Zhang and Li, 2010). From the viewpoint of users, it is today easy to identify providers and interact with them via electronic means of communication irrespective of the location (Shearmur and Doloreux, 2014). There is also another reason that questions the linear connection. Cross-country studies have indicated that the inter-linkages between KIBS and other sectors play a central role in the positive impacts that the abundant supply of KIBS enables (Windrum and Tomlinson, 1999). This highlights the institutional structures and arrangements surrounding KIBS. The governance of the networked systems, which KIBS are part of, should foster interactive learning. The resulting ability to transfer both codified and tacit knowledge allows the constituent institutions to adapt to changing external circumstances (Simmie and Strambach, 2006).

In addition to the quantitatively sufficient and 'institutionally embedded' supply, the content and quality of services are essential. The service experience of the client is increasingly dependable on the ability of KIBS to show deep understanding as regards the client's value creation process and strategy (Fosstenløkken, 2003). The demand is growing for solution-oriented services, and is challenging in small KIBS which, due to quite a specific focus, cannot provide broad service packages. The challenge is maintained by the fact that collaboration between SMEs is often too loosely organised and there is insufficient interaction between big and small companies in know-how sharing (Toivonen, 2007). A further problem is the lack of companies that would be capable (or willing) to take the role of a business integrator: to synchronise several resource domains. The service content expected from KIBS also includes totally new areas of know-how; a topical example is services needed for the promotion of environmental sustainability (Miles, 2005).

Like generally in service sectors, the issues of productivity and efficiency are coming to the fore in KIBS and are linked to the company size. In big companies, customer applications are usually based on systematised – sometimes modularised – service concepts that enable scaling up and international tradability (Strambach, 2008). Small companies often rely on fully tailor-made services as a way to offer a positive experience in the customer encounter. This approach hinders the achievement of an efficient business strategy because it is highly dependent on the skills of individual professionals. It also weakens the elaboration of ideas into full-blown innovations: the broader insights behind individual solutions are not exploited. Thus, the development of replicable offerings should be emphasised, not only due to profitability reasons, but also due to its importance from the viewpoint of innovation (Gallouj, 2002). Visible service concepts and shared understanding among professionals are a meaningful alternative to standardisation and mechanical input–output considerations.

Knowledgeable clients are the other side of the coin in the realisation of the potential of KIBS (Fosstenløkken, 2003). Here, too, the question is both of quantity and quality: whether the clients are aware of the benefits of KIBS in the solution of expert issues, whether they know where to buy services, and whether they are able to skillfully organise the interaction with the service provider. The first point concerns the 'make or buy' solution that includes specific characteristics when the target of acquisition is expert service. Several researchers have found out that the main reason for outsourcing and collaboration with KIBS is the upgrading of services (Kox, 2002; Strambach, 2001). The second point is linked to the context and area within which a suitable provider is searched. Some recent studies indicate that involving more distant knowledge can increase the innovation performance of organisations (Miller *et al.*, 2007). Conducting the search in a wide area is purposeful in the use of knowledge repositories, in particular, as they do not require face-to-face contact (Wagner *et al.*, 2013). The third point highlights the importance of dedicating qualified human capital to the actual interaction with KIBS in order to secure the integration of their knowledge into the firm (Simmie and Strambach, 2006).

The concept of absorptive capacity describes the extent to which a firm is able to utilise available knowledge flows. The concept includes two dimensions: the ability to identify and evaluate knowledge, and the ability to exploit and realise the potential of knowledge. Thus, both the amount of external knowledge that the firm perceives and the degree to which it derives benefits from a given quantity of external knowledge are functions of its absorptive capacity (Escribano *et al.*, 2009). Absorptive capacity is weakened, among others, by the lack of sophistication in the procurement function of the client organisation. It is also important to note that the KIBS interventions can improve absorptive capacity: the role of KIBS as 'bridging intermediaries' helps in 'know where' issues – in the location of the sources of knowledge (Gallouj, 2002)

Summarizing the literature, we can state that during the last twenty years more and more evidence has been accumulated on the important role of KIBS in innovation. Simultaneously, several challenges have been identified in the supply and demand of knowledge intensive services that may slow down the realisation of the benefits of KIBS. Not all KIBS are driven by a clearly defined orientation towards innovation, and within innovative KIBS, innovation takes place in various forms as a result of different competitive strategies and produces different impacts based on the interaction with the business environment (Corrocher and Cusmano, 2014; Freel, 2006). In the following sections, we aim to contribute to a deeper understanding of these issues by building four different scenarios on the development of KIBS in the coming years.

Methodology in scenario building

A futures perspective is essential for the policy activities that support promising new possibilities in various sectors. As society and the economy are nowadays developing in increasingly faster cycles, comprising a number of uncertainty factors, the conventional forecasting methods have proved ineffective in the acquisition of futures intelligence. The approach, whose primary aim is not to identify the most probable state of affairs in the future but to understand new emerging processes, has gained ground. This approach is called 'foresight' and its starting point is preparedness for many different futures (Godet, 2000; Miles, 2005).

Methods based on expert knowledge are commonly used in foresight (Kuusi, 1999). Scenario building, applied in our case, belongs to the 'core repertoire' of these methods – other well-known methods include road mapping, trends analysis, and weak signals detection, for instance (Popper, 2008). In line with the basic idea of foresight, scenarios are not predictions of the future, but purposeful 'stories' about how the phenomena under study could develop over time. They consist of three elements: a description of the possible end states in a horizon year; an interpretation of current events and their propagation into the future; and an internally consistent account of how a future may unfold. The analysis of interconnections and mutual impacts of different factors reveals consequences of the dynamic interplay of uncertainties and their emergent resolutions (Burt *et al.*, 2006).

Both quantitative and qualitative approaches can be applied in the building of scenarios. Our approach was qualitative as the creation of meaning to events and perceptions in the context of KIBS was our main aim (cf. Popper, 2008). A division into three alternatives – the best, the worst and an intermediate alternative – is usually unfavourable in scenario building because it leads to unidimensional analysis. Four alternatives keep the number of scenarios modest while helping to break simplistic thought patterns. Including a different driving force in each scenario favours creative thinking and profound sensemaking. On the other hand, a policy-oriented foresight exercise is targeted at the production of implementable options and therefore requires a view of the desirable and avoidable futures (Wright and Goodwin, 2009). In our exercise, we reconciled these aims by constructing four scenarios with different drivers, and *within* each scenario explored the desirable and avoidable alternatives. The time perspective was until 2020.

Expert groups in specific areas and issues are a typical way to produce futures intelligence from the thematic field in question; the networks of experts supplement the creation of futures views (Kuusi, 1999). The expert group in our case consisted of eleven members of the HLG working group on innovation. A few members were researchers while the majority represented business: individual companies, professional associations and development agencies in the KIBS sector. The practical activity was carried out in two workshops for which each individual member collected relevant documentary material and case examples from his/her own background organisation and networks. This material was combined with relevant literature in the construction of scenarios. The results were presented twice to the core team of the High Level Group. The feedback given helped to develop the scenarios further and to link the policy conclusions to the topics discussed in the other HLG working groups. (There were altogether five working groups: besides the innovation group, the other groups examined the futures of KIBS from the viewpoints of internationalisation, skills, internal markets and instruments.)

Four scenarios on the development of KIBS

In this section, we present the results of the work of the HLG innovation group: four scenarios with four different focuses. These scenarios are not excluding each other, but show complementary developments based on different main drivers. The first scenario is technology-driven and includes the deepening and broadening of digitalisation as its core phenomenon. The second scenario is linked to the former but concentrates on the increasing importance of systemic phenomena and networking. The third scenario focuses on the business models in KIBS companies: the prospect that these companies increasingly highlight the customer value and extend the analysis of this value until the end-user experience. The fourth scenario also anticipates a change in business models but focuses on the clients of KIBS: it anticipates the development of new types of procurement practices and examines their impact on the interaction between the provider and the client. Table 13.1 summarises the desirable and avoidable alternatives within each scenario. Based on the main driver, the scenarios are named as technology-driven, network- and systems-driven, service-driven and procurement-driven.

Table 13.1 Four scenarios on the development of the European KIBS sector until 2020

Scenario and its main driver	*Desirable alternative*	*Avoidable alternative*
Technology-driven scenario	Awareness grows of the importance of service infrastructure in the context of digitalisation. KIBS are actively used for the analysis of big data. New KIBS are established based on open data.	Digitalisation is understood narrowly: as a development of the ICT sector. Risks linked to data security and privacy will be realised and stop the advancement of new practices.
Network- and systems-driven scenario	Service providers and client organisations essentially increase collaboration. Policy actors support this development (incl. the construction of systemic innovation indicators).	Demand for KIBS in the context of systemic innovations develops slowly, because systemic problems are more difficult to tackle than developments in individual products and services.
Service-driven scenario	Use value as the focus in KIBS' business models. Value considerations reach the whole business chain until the end user. Acceleration in innovation processes.	Value-based practices are seen as opposite to profit striving. Clients are not appreciated as value co-creators, which leads to underdevelopment of collaborative competences.
Procurement-driven scenario	The spread of a new procurement model: transparent process, performance-related specifications, multiple selection criteria and long-term relationships.	Current procurement practices persist. In the regulatory framework, the focus continues to be on the form instead of the content.

Technology-driven scenario

During the coming years, the economic development will be increasingly dependent on technological knowledge. From the viewpoint of KIBS, a particularly important phenomenon is digitalisation, and our technology-driven scenario concentrates on the analysis of its impacts. It is interesting to note that during the 'discovery of KIBS' in the mid-1990s, the active use of ICT in these companies was pointed out as a main expression of their innovativeness. The KIBS sector was even divided into two sub-sectors: technology-based KIBS and traditional KIBS (Miles *et al.*, 1995). Later on, the one-sided technological emphasis has given way to views that highlight the interaction between technology and services (Gallouj, 2002). These views are essential when an attention to the relationship between ICT and KIBS is again topical – from a new perspective: ubiquitous presence of digital technology.

In the digital era, the accumulation of data is no longer limited to human creation, but computers, phones and other devices and sensors collect, store and transfer data automatically, reflecting the phenomenon of the 'internet of things' – Haller, 2009). The concept of 'big data' refers to the huge, unstructured mass of data emerging in this way (Boyd and Crawford, 2012). It is characterised by three dimensions: increasing volume (amount of data), velocity (speed of data in and out), and variety (range of data types and sources). A lifelong log of human individuals and product life-cycle management will be possible when everything can be connected easily.

However, a big portion of the data is still in silos: enterprise data, financial data, health data etc. The analysis of the content of these silos would reveal new potential for both sector based and issue based activities. It would make the growing data reserves as the core of innovation resources of companies. The realisation of this potential requires services for integrating, analysing and interpreting the data. Connecting and combining big data is a business opportunity for KIBS firms. KIBS could also encourage their customer enterprises to use up-to-date and forward-looking information instead of merely analyzing actual trends. Consultancy based on big data highlights the creative use of ICT. For instance, social data mining from Facebook, Twitter, etc. is becoming part of this data (Boyd and Crawford, 2012).

Another important trend is open data. It includes the idea that certain data is freely available to everyone to use and republish, without restrictions from copyright, patents or other mechanisms of control (Janssen, 2012). Open public data in particular is an important resource: many governmental organisations and cities collect a broad range of different types of data in order to perform their tasks. Recently, open data has gained popularity with the launch of public initiatives in several countries. Key to this approach is that public service providers develop new collaborative ways of working with data users, including commercial users – and where necessary actively engage in the market to stimulate demand for data. The premise for growth is that public service providers do not charge users for their data, but enable access to it, so that the data can be used as a platform for innovation or enterprise. Examples of application areas are legislation data, health data, energy data, meteorological data and transport data. The following case illustrates how new services can be developed in the last mentioned area:

> ITO World Ltd is a UK based KIBS that specializes in mapping and visualising transport data. Founded in 2006, the company has collaborated with several public organisations and private companies. For instance, it supports

> the provision of public transport journey planning for London using Google Maps. This service was based on official Transport for London data, released as open data. In 2012, ITO World integrated real-time information about disruptions on the London Underground into their service. In the event of any service interruption, travellers are presented with alternative route options and estimated travel times based on real-time data.
>
> (HLG, 2014)

Besides big data and open data, the technology (digitalisation) -driven scenario includes a third interesting phenomenon: crowdsourcing. It is a collaboration model enabled by people-centric web technologies to solve individual, organisational, and societal problems using a dynamically formed crowd of interested people who respond to an open call for participation (Pedersen *et al.*, 2013). Two of the most common crowdsourcing models include one-time challenges and on-going communities. Crowdsourcing provides a grassroots perspective on how technological facilitation alters the realm of collective innovation. Collaborative service innovation has traditionally taken place in dyadic relationships between the service provider and the user. The emergence of the internet and social media has significantly lowered the cost of involving masses of users via virtual platforms. Crowdsourcing has been argued to have an important positive impact on the emergence of novelties: social interactions trigger new interpretations and new discoveries that individual actors' thinking alone could not have generated (Hargadon and Bechky, 2006). The following example illustrates how a service provider can facilitate crowdsourcing:

> InnoCentive provides a third-party online platform that firms can use to 'outsource' their innovation problems, leveraging on a network of talented solvers all over the world. An expert from InnoCentive facilitates the process, helping the client to formulate the problem and the specific criteria against which solutions will be evaluated. This is an effective solution for SMEs, in particular, as they typically cannot afford large, fixed investments in R&D activities.
>
> (HLG, 2014)

Summarizing the technology-driven scenario, we can state that digitalisation is 'propelling' our world towards a reality in which intelligent services will constitute one of the most important infrastructures of society. These services must be available anywhere and anytime. In the future development of this scenario, the most desirable alternative is awareness of the significance of KIBS for the exploitation of advancements like big data, open data and crowdsourcing. Correspondingly, skillful creation of new services is needed within KIBS. Making sense out of huge data reserves and flows requires new, improved methods of communication between humans and machines, and between machines.

Whereas the pressure from markets may be sufficient to promote the general development, the promotion of open data in particular requires encouraging activities of policy actors. Policies are also needed for necessary regulation to manage the risks of openness. The development of data management and data governance methods – standards, practices, and applications – is essential because along with the opportunities, also the risks of misuse of information will grow. The avoidable alternative is that some risks linked to data security and privacy will be realised and stop the advancement of new practices. Also narrow views that see digitalisation as the next step in the development of the ICT sector slows down the progress.

Network- and systems-driven scenario

In recent years, it has become apparent that many economic, environmental and social challenges are too big to be solved via individual product or service innovations created in individual organisations. In these cases, innovation has to be studied as a systemic phenomenon and this broader perspective is also needed in the related policy measures. A system innovation is based on the simultaneous development of organisations, technologies, services, and multiple network relationships. Collaboration between different organisations is a prerequisite for combining the emerged innovations effectively and disseminating them rapidly.

Sustainable development is an area in which systemic innovations play a central role. Here it is essential to go beyond technical aspects and include social mobilisation and acceptance, institutional arrangements (e.g. laws and stakeholder roles), and financial and operational requirements. Emphasis on the interaction between technology and human behaviour is necessary in the development of infrastructure, transport and energy systems. For instance, the success of new technologies linked to the follow-up and management systems of energy consumption is highly dependent on user preferences and know-how. Also the sustainable development in the technology-intensive sectors of water supply and the waste treatment depends on consumer choices (Van de Klundert and Anschütz, 2001).

Sustainable urban living is a significant component in the overall sustainability (Simmie and Strambach, 2006). Along with the global trend of urbanisation, cities have become the main places of distributing and consuming goods and services. Their fast growth is anticipated to continue throughout the 21st century and have considerable effects on the environment (Mcdonalda *et al.*, 2008). The concept of sustainable city (eco-city) aims to integrate different aspects of environmental and social viability in city planning, human-nature interaction, infrastructure systems and energy production, housing, and the provision of services. Public-private cooperation is an integral part of city innovations. Digital technologies promote sustainability via effective governmental operations and public services, advanced industries, and a smart information infrastructure (Caragliu and Nijkamp, 2011).

Sustainability-oriented and smart systems need a growing number of advisory and consultancy services, and new KIBS have emerged in this context. Environmental consultancy is often mentioned as one of the most important new KIBS sectors (Miles, 2005). At a more general level, integrative service practices – illustrated by the following example – are an answer to the need for a more holistic perspective:

> Interserve is a large international company whose origins are in construction and engineering design. Nowadays its offering covers not only all stages and assets linked to buildings and infrastructure, ranging from sustainability and energy issues to estates planning and facilities management, but also key frontline services such as back-to-work and care at home services. An example that illustrates the 'human side' of its portfolio is its bespoke change management programme – a nine step model which starts from strategic engagement and ends with building commitment amongst the employees.
>
> (HLG, 2014)

The importance of systemic views is also growing within individual businesses and business networks. In the manufacturing sector, the development of the industrial internet drives this development. It includes 'machine to machine' – communication via the internet, enabling the telemetry of various meters, remote control of devices and mobile control of work. Currently, the data generated is mainly used for the improvement of internal efficiency of companies, but the information that companies get from each other also favours the emergence of new, more deeply networked business models. The idea of business ecosystems, which still today is more a metaphor than a description of reality, may become a genuine framework for the coevolving of producers, subcontractors and clients. They may examine their own business models as a part of a whole, or participate in joint efforts which develop the whole ecosystem in a co-operatively agreed manner (Westerlund *et al.*, 2014).

On the other hand, there are problems that slow down the development. Organisational structures and operating models in the majority of companies are not adjusted to innovation processes that deviate from an R&D project: it is often unclear whose responsibility these kinds of development tasks are. There are also concerns about the protection of innovations created in this way. Incentives to sell integrated service solutions are often underdeveloped and users may be unaware of their application potential (Westerlund *et al.*, 2014). Thus, measures should be taken to support both the supply and the demand side.

All in all, system innovations require new kinds of policy measures and governance structures. Promoting cross-sectorial collaboration – typical of system innovations – through regulation and related initiatives would be important, because such collaboration is difficult. There is also demand for the development of reasonable indicators for system innovations. A good starting point for this development is the emerging practices of impact assessment: these practices are closely linked with learning and continuous improvement instead of the earlier focus on past activities. Regarding KIBS, in particular, collecting together dispersed good practices of integrative services and brokering activities would be an important step forward. There is also much to do in exploring how KIBS could be successfully involved in different institutional settings regionally and nationally (Muller and Doloreux, 2009).

The desirable alternative in this scenario is the increase of collaborative ways of working, and the support given to this development by policy stakeholders. Changes are needed, not only in processes and organisational structures, but also in attitudes and values. A successful implementation of collaborative practices requires the combination of top-down and bottom-up approaches: a 'managerialist' approach that secures efficiency and effectiveness in innovation and implementation, and an empowerment approach that emphasises grassroots initiatives of citizens, users, employees and individual entrepreneurs. Particularly important are initiatives that develop interactive practices reaching beyond the traditional R&D model of innovation, foster open innovation and support the development of networks.

The avoidable alternative is a situation in which willingness to develop systemic innovations does not increase, because systemic problems are more difficult to tackle than developments needs in individual products and services. This unwillingness neglects the consequences of the tight coupling of different parts of the system and the feedback mechanisms included. The solutions made one by one in unconnected innovation processes often result in the emergence of a new problem in another part of the system while one problem is solved.

Service-driven scenario

This scenario focuses on the need to develop the business models of KIBS towards increasing client-orientation, with the focus on the service value. During the last decade, client- and value-based strategies have gained ground generally and are linked to the striving for innovative practices. These strategies highlight that profitable growth is the result of succeeding in value offerings. If a company concentrates on how to match or beat competitors, it easily restricts itself to the conventional context, which also the competitors know and in which all seek a competitive advantage by means of minor improvements. A value-based strategy considerably extends the creative scope of companies and provides them with a wide range of options even irrespective of the general situation in their industry. (Hoover *et al.*, 2001)

In KIBS, the ability to solve the clients' problems has always been an important characteristic. However, this ability was earlier linked to the idea of knowledge asymmetry favouring the professional over the client. Nowadays the clients increasingly take the core expertise of the professional (e.g. legal or technical expertise) for granted, the real issue being responsiveness to their specific needs. In order to contribute to the client's value, KIBS need a comprehensive understanding of the overall value creation process of their clients, and of the role of their service in this process. A successful service also requires a thorough understanding of the client's strategy and goals, and know-how in the elicitation of the client's input and in the management of collaboration (Muller and Doloreux, 2009).

The approach of service-dominant logic has provided deeper insights into the reasons behind the central role of clients (Vargo and Lusch, 2011). The question is not only about the visible demand in the markets, but the clients carry out important tasks in value generation. Value is not an inherent property of technology, goods or services. It cannot be first produced and then sold, but its emergence requires that an individual product or service is linked to other products and services. Thus, value is unfolded in the use context. This perspective changes the way in which the relationship with the client should be understood and arranged: the client is not a target but co-creator of value. Dialogue, transparency and risk sharing are core elements in 'user-facing' parts of business, supported by internal development of corporate responsibility, workplace quality etc.

From the viewpoint of innovation, these views include the same message as the network- and systems-driven scenario. The focus should be transferred from inside-out to outside-in perspective: continuous collaboration with clients throughout the innovation process. Innovation can be related, more than to products, to solutions and experiences and based on the collaboration with a network of actors with different competences (Prahalad and Ramaswamy, 2004). Actually, this collaboration is increasingly going on in the form of idea acquisition, testing and piloting. A bigger change would be the emergence of alternative process models for innovation that would supplement or replace the highly formalised stage-gate process which nowadays is the ideal and norm, and usually the only model recognised in policy programs. Collaboration with clients favours more experimental approaches that merge the planning and implementation to reduce time-to-market. In addition to the acceleration of innovation, more radical results are possible because experimental approaches often aim to change the market logics instead of seeking a share of the existing markets (Vargo and Lusch, 2011).

In KIBS, the immediate clients are, in general, not the ultimate recipients of services. Thus, a value-orientation also means a more intensive collaboration with end-users. Two groups of end-users have to be taken into account: the employees of the client organisation and persons or entities external to the organisation (other companies, citizens and society in general). End-users are in a unique position to make judgments on the extent to which services are likely to satisfy requirements. They might also offer insights into how a service can be performed in a more efficient and cost-effective way. However, listening to the end-users is not a simple task because they are often far from the decision makers who purchase specific services. Thus, their engagement has to be explicitly planned and managed. This is particularly important to take into account in the public context in which the number of stakeholders and citizens is usually very large.

Attention to end-users is one expression of the value perspective reaching the whole value chain. Another expression is the brokering activity of KIBS among suppliers and customers. The former refers to the promotion of joint ventures in the provision of services – a practice increasingly demanded if SMEs want to develop their service capability. The latter refers to joint operations of two or more customer organisations in the procurement of services; it is an important direction in the development of procurement practices (see the next section). In the best case, the two types of collaborations meet and broader networks are formed as the following case shows:

> Pitti Immagine is an Italian KIBS devoted to promoting the fashion industry worldwide; recently its scope has been expanded to food and fragrance

as well. The company is focused on the modern trade fair as a platform of renewal and development. By offering information and knowledge, it aims to make the trade fair an event that creates stimulating relationships involving the exhibitors, their collections and the buyers and public. Starting from a deep collaboration with the exhibitors and visitors, Pitti has developed innovative services involving many external partners: IT system providers, catering, fitters, logistic operators, communication agencies, and designers. It acts as a resource integrator, which activates different partners based on the trade fair typology and the selected exhibitors. It has also created a new virtual concept: all the material of 'the traditional fair' is reproduced for the virtual fair accessible two weeks later.

(HLG, 2014)

The desirable alternative in this scenario is the deepening of a genuine service mind set: appreciating the client as a value co-creator. Experiential and accelerated innovation processes also play a central role in this scenario. Their development is based on the collaboration between providers and clients, but needs support from policy instruments. A further source for the progress of collaborative innovation is the pressure coming the empowered consumers and citizens. In this end-user based prospect, the technology-driven and service-driven scenarios are mutually reinforcing (Kiely *et al.*, 2004). New consumer communities emerging in the internet exemplify joint innovations and also reflect the importance of experience as an element of value (Prahalad and Ramaswamy, 2004; Lush and Vargo, 2006). The avoidable alternative is that use value-based practices are seen as opposite to short-sighted strivings for profit, and therefore both awareness raising activities and the practical improvement of necessary competences will be neglected.

Procurement-driven scenario

Earlier, the growth of KIBS was explained as a simple shift of intra-organisational tasks to external actors. Several studies have, however, shown that service upgrading instead of pure replacement outsourcing is usually the main motive in the purchase of knowledge intensive services (Kox, 2002). Along with the growing need for external expertise, also the significance of this knowledge-accessing motive has increased. Today's outsourcing strategies go beyond the 'make versus buy' decision to encompass technology assessing, risk sharing, joint development, and marketing activities. Outsourcing has shifted from an operational tool to activities with strategic importance, closer to the heart of the firm (Contractor *et al.*, 2010).

On the other hand, there are big problems in the practical procurement: it is not in line with the above-described strategic aims but overlooks service providers as a source of know-how and innovation. Particularly in the public sector, there is little incentive to promote innovative practices, because of the lack of transparency and reluctance to stray from rigid definitions of service, and the principle of 'lowest price.' Public procurers operating under the common European guidelines have to be competent in technical issues of tenders; consequently, they often concentrate on formal and normative procedures rather than on the content of the service to buy. This also makes it difficult to include the end-users' and/or customers' needs and suggestions within the tender specifications in an effective and appropriate way.

The concept of 'informed client function' is central in the context of procurement. In the public sector, it is developed but highly prescribed. In the private context, failure to develop and maintain this function is common and leads to unmet expectations; it also exposes the parties involved to many kinds of risks. Procurers may not be aware of their organisation's needs and they typically cannot express them in terms that will enable suppliers to offer the most appropriate services in the most efficient and cost-effective way. The lack of agreement on the appropriate arrangement for delivering services prevents the achievement of the desired end-user experience and best value for money. Suppliers may even offer services of inferior quality or performance.

The need for informed client function applies irrespective of the type or size of the procurer organisation. An efficient and cost-effective procurement is as important in small organisations as it is in their larger counterparts. In the latter, the informed client function should, however, be more 'professional' to achieve the benefits of economies of scale and internal efficiency. Some procurers might be defining their needs and service requirements for the first time. In such cases, it is possible that they specify a higher level of service than is required, with the respective higher costs. A way to avoid this problem is to discuss with prospective service providers before a formal tender/bid process. This might reveal over-specification which can be corrected without sacrificing the quality, performance or targets of the service. It requires, however, that the client is capable of engaging with service providers as an equal: understand the issues involved and the most appropriate option.

In this scenario, the activity of procurers drives the development, providing new opportunities for business service companies and supporting innovation in them. There is an emerging procurement model, which includes a transparent process, performance-related specifications, multiple selection criteria and long-term relationships. The model is based around the following activities:

1 establishment of needs – stakeholder interests and end-user needs
2 development of a procurement strategy – the broad plan for service provision
3 request for information – the pre-qualification of service providers
4 request for proposals – effectively, the tender/bid stage
5 evaluation and commercial close – formalizing the arrangement
6 mobilisation – start-up for service delivery
7 performance review – monitoring of service delivery; corrective actions

A benefit of a transparent process is that service providers can be more assured of honest dealing on the part of procurers. They understand what they are expected to provide, and where the risks lie. Even more effective processes are those where service providers are encouraged to offer innovative solutions within the overall framework. Central to this desire are performance-related specifications as opposed to those of a prescriptive nature. Determining the fulfilment of targets is much easier when performance indicators have been agreed at the outset and used, over time, to fine-tune service provision. At best, performance specifications focus on deliverables that end-users define in consultation with stakeholders in the client organisation.

The quality of the relationship between procurer and service provider is important. Building productive relationships takes time and is most likely to come from a longer-term perspective on service delivery than one motivated by short-termism. Often, this implies closer working between procurer and provider, but a looser relationship might also be possible. It is necessary to determine the most appropriate relationship for any given combination of service provision against the needs of the client. Formal reviews of service delivery provide the opportunity to consider changes that might be necessary to raise performance, realise targets and agree incentives. The greater the openness in this regard, the more likely it is that procurers and providers will be able to explore novel ways of delivering end-user satisfaction with a best value for money.

A defined procurement process also includes decision-making concerning the options for service delivery: single service provider, bundled (or multiple) services provider, total service provider, and agency. Single service or bundled services are not mutually exclusive; combining them is often the 'best-fit' solution. The total service provider is offering a single-point of responsibility based on the integration of a number of services ('one-stop shop'). Despite its intention, there is no one model of total service provision. In dynamic markets, one procurer's interpretation of its needs, expressed as total service provision, is likely to be different from another. Lastly, the use of agency means that personnel are hired from a supplier as and when required. Agencies can answer a temporary need for personnel, but this alternative ought not to be considered as a permanent solution.

The selection of the most appropriate option has to take into account the resources and costs involved in managing the relationship with the service providers. Each option brings with it some risks, including the ability to attract competitive bids. It is also important to note that procurers (units and individuals) act for and on behalf of the client organisation and have to balance many and often competing interests to ensure that the organisation achieves its objectives. It can be a complex situation that is subject to change as the organisation responds to the business environment.

The most desirable alternative in this scenario is the generalisation of the above-described procurement model both in the private and public sectors. Due to the common rules, this requires active policy support at the European level. The avoidable alternative is that the current practices persist, and in the regulatory framework the focus continues to be on the form instead of the content.

Policy options

While the main driver is different, the four scenarios include common emphases. In particular, the increasing need for openness and collaboration is characteristic of all scenarios and includes the topics of open innovation, best practice sharing and connections across value chains. Other policy highlights are the importance of service-based business models, the need for SME support, and the renewals required in service contents and in the procurement of knowledge intensive services.

The importance of a broad view on innovation came out in all scenarios: the emergence of novelties should not be considered one sidedly as a formal process inside R&D functions. Innovation is increasingly based on entrepreneurial interactions between different subjects: be they single individuals (employees, independent inventors, end users, etc.) or organisations (suppliers, clients, public authorities, etc.). These new, open innovation models leverage on the networks' power of aggregating resources and competencies far beyond the scope of a single firm. Supporting open innovation systems would allow SMEs to take part – and benefit – of innovation activities otherwise out of their reach.

On the other hand, there is not one generally applicable model for innovation. Open practices include a continuum from the use of purposive inflows and outflows of specific knowledge to loose development networks. In addition, openness is not the only issue to be considered in innovation activities; for instance, acceleration of the innovation process within a firm may be the most urgent task. Also structured R&D still plays an important role in many innovation efforts.

Today, open access often means openness in the global scale. Here, a balance between open practices and the protection of intellectual property (IP) has to be strived for and be taken into account in the respective policies. IP issues are particularly sensible for SMEs, which need collaboration with large corporations, and challenging for service companies that only rarely can use patents. Informal protection means – secrecy, restricted access to certain databases, loyalty building among employees, and keeping important experts in-house are plausible alternatives but not systematically applied (Kuusisto and Päällysaho, 2008).

Along with the blurring sectorial boundaries, industries are increasingly interconnected and the distinction between products and services is losing prominence. Consequently, also managerial best practices are less context-specific, which enlarges their application scope. For instance, showcasing how changes in workplace practices can reduce the carbon footprint is widely applicable. Also the number of plausible alternatives is growing, which highlights the dissemination of knowledge about the related experiences. For instance, outsourcing can be implemented in many ways and includes a variety of solutions: embedded working, site sharing, knowledge transfer arrangements, risk sharing etc.

Diffusion and sharing of best practices among firms of different sizes, among regions and countries, and among sectors (both private and public) fosters innovation and competitiveness. A useful approach could be 'learning circles' that facilitate knowledge transfer from large to small companies. In order to motivate companies to disseminate their experiences, the simultaneous development of IP protection means is crucial. On the other hand, it is important to highlight that copying a service practice developed elsewhere is rarely successful – the learning inputs included must be skillfully modified to be suitable to a new context.

Collaboration within and between value chains has been highlighted as a promoter of innovation in terms of generation, development and execution of new ideas. Encouraging and supporting common initiatives of the procurers and providers is an important policy option in this context. The lack of orchestration capabilities is a typical problem and requires both awareness raising and concrete support: the importance of the integrator function has to be emphasised and the core capabilities created as a part of development programs. More concrete discussion of orchestrators' tasks would promote the visibility of this role. The following key tasks have been identified: balancing the interests of divergent parties, taking care of knowledge mobility and competence leveraging, legitimizing the common activities, and visioning about the benefits of collaboration (Ritala *et al.*, 2009).

The broadened view on innovation has brought to the discussion the concept of 'business model innovation', which means that innovation is considered – not only as novel products, services, and processes – but also as a change in the components of the corporate strategy. Identifying new business models and understanding their structures and logics, with the aim of sharing this knowledge, is an important policy option. An essential aspect is to understand clients, end-users and various stakeholders as co-creators of value. This perspective highlights the significance of service orientation in all kinds of companies (not only in service sectors).

Service innovations have often significant impacts on consumers and societal welfare, both at the individual and at the community level. Many services even produce eco-socio-cultural consequences on the well-being of entire nations.

Public procurement, and the role of SMEs in it, is a key issue in the EU. SMEs as suppliers could be an important source of innovation for the procuring organisations, but these companies often do not have competencies and resources to participate in the procurement tenders. While the procurers hereby miss ideas for innovation, the providers loose significant market opportunities. There are several options to support the active role of SMEs. Development programs that help SMEs to create the competences needed for dialogue and interaction with public authorities are one alternative. Also the arenas for best practices sharing can be used for this purpose. On the procurers' side, fostering the adoption of more transparent and systematic tendering practices is a central approach and also concerns the private sector. It helps clarifying the contents of the service being procured – difficulties in this respect are one central problem today. It can also reveal the excessive number of tender authorities and compel the public bodies to define in more detail at which level different types of procurement decisions are made and are most reasonable.

Skills shortage and labour mismatch are typical today and reflect the rapid development of new service contents and means of delivery. The advent of new media and the availability of big data require skills and competencies which differ from those in traditional KIBS and which are also hard to find on the labour markets. Highlighting the changing contents of knowledge intensive services is important in order to keep the focus of policies up-to-date and anticipatory.

Mapping the existent skills shortages in KIBS could promote the lining up of educational and training programs with jobs requirements.

The general sufficiency and quality of KIBS should be included in discussions on the important success factors of the economy. Polarisation of the sector and the regional imbalance in service provision are issues in which policy measures are needed to prevent negative development. Ways of stimulating both efficient supply and skillful demand for KIBS should be included in regional strategies and innovation programs. In these strategies and programs, the benefits of local proximity and the possibilities for the utilisation of extra-regional knowledge should be seen as mutually reinforcing. The specific emphasis and concrete applications in each region depend on the local pool of expertise, the industrial structure, and the central institutional factors that support regional competitiveness.

Attention should be paid to the smallest KIBS in particular, as they often have difficulties in developing their knowledge base while concentrating on daily routines. There are also needs in training the small clients in the use of KIBS. The clients would need, not only concrete support in qualified procurement of services, but also information – clearly more than they have today – on the importance of external expert services for successful business.

Awareness of the role of KIBS should be visible in innovation policy. Inclusion of the KIBS sector in innovation systems projects, conducted at the regional, national and cross-border levels, could increase the contribution of this sector to the economy, and it could also promote the framing of the research questions in this field.

Concluding remarks and future research

During the last two decades, research into service innovations has rapidly accumulated. However, deviating from the technology studies, a linkage between foresight and innovation has not emerged in the area of services – foresight exercises in services have been rare (Miles, 2005). The exercise described in this chapter aims to narrow this gap in the area of KIBS, which several studies have recognised as central actors in innovation: as sources, facilitators and disseminators.

The HLG work, on which this chapter is based, was policy-oriented. It aimed to provide meaningful futures information and shared visions, which policy makers can use to anticipate the consequences of their developmental strategies and choices (cf. Havas *et al.*, 2010). While this approach is useful from the viewpoint of practical implications, it does not as such benefit the theory in the best possible way. In order to strengthen the theoretical contribution, we have supplemented the scenarios and identified policy options with a conceptual analysis that covers the central arguments of KIBS studies from their emergence in the mid-1990s to the most recent findings.

The work raises ideas for several types of further studies. The first group includes studies that would analyse deeper the developments included in each scenario. The role of KIBS as the promoters of digitalisation needs new insights, as do service-based business models and the multi-actor collaboration that crosses sectorial boundaries. Research into the linkages between service innovations and systemic innovations is only beginning and should take rapid steps forward, because many sustainability issues need the improvement of service systems and their integration into the transfer of the institutional landscape. The internal development of KIBS and the procurement of knowledge intensive services also need further research. Empirical evidence on the conditions that influence the positive contribution of these services is still scarce. The impact of the geographical scope of knowledge acquisition is an important issue in this context and includes the topic of the internationalisation of KIBS, among others.

The second group concerns studies with more direct policy relevance. A better overview on the current stage of knowledge transfer mechanisms is needed in order to support the sharing of best practices. The development of policy programs and indicators based on the interactive and non-liner models of innovation require additional research to gain ground. Innovative research projects are also needed for the construction of a deeper understanding of how innovation in and through KIBS could enhance sustainability and foster the well-being of the present and future generations. These projects should be evaluated not only in terms of economic returns, but also from the point of view of equity, social justice, ecology, consumer freedom, etc. Concepts such as the 'triple bottom line', focused on people, planet and profit, could be adopted to assess initiatives, encouraging firms to consider also ecological and social outcomes of innovations to be pursued.

Our final point concerns the concept 'knowledge-intensive'. After the 'discovery' of the importance of KIBS, the relativity of this concept has become apparent. When defining specific sectors as KIBS, Miles *et al.*, (1995) anticipated this situation; they stated that the sectors they pointed out were more knowledge intensive than others at that time, but in the future other sectors may also feature as an arena for knowledge intensive services. Today, many sectors that are not specialised in expert offerings resemble KIBS. For instance, manufacturers providing industrial services do not offer only maintenance and repair of equipment, but also integrated solutions that actually include business consultancy. Another example is security services, which earlier were considered low skilled 'manual and personal services' but nowadays provide remote control and respective consultancy concerning the security of buildings and areas. Thus, a central challenge is to broaden KIBS studies to cover knowledge intensive activities irrespective of the sector – a perspective that has already been applied to some extent (OECD, 2006).

References

Antonelli, C. (1999) *The microdynamics of technological change*, London and New York: Routledge.

Bessant, J. and Rush, H. (1995) 'Building bridges for innovation: the role of consultants in technology transfer,' *Research Policy* 24: 97–114.

Bilbao-Osorio, B. and Rodríguez-Pose, A. (2004) 'From R&D to innovation and economic growth in the EU,' *Growth and Change* 35(4): 434–55.

Boone, T., Ganeshan, R. and Hicks R.L. (2008) 'Learning and knowledge depreciation in professional services,' *Management Science* 54(7): 1231–36.

Boyd, D. and Crawford, K. (2012) 'Critical questions for big data: provocations for a cultural, technological and scholarly phenomenon,' *Information, Communication & Society* 15(5): 662–79.

Burt, G., Wright, G., Bradfield, B., Cairns, G. and van der Heijden, K. (2006) 'The role of scenario planning in exploring the environment in view of the limitations of PEST and its derivatives,' *International Studies of Management and Organization* 36(3): 50–76.

Caloghirou Y., Kastelli, I. and Tsakanikas, A. (2004) 'Internal capabilities and external knowledge sources: complements or substitutes for innovative performance?' *Technovation* 24: 29–39.

Caragliu A., Bo, C.D. and Nijkamp, P. (2011) 'Smart cities in Europe,' *Journal of Urban Technology* 18(2): 65–82.

Chesbrough, H.W. and Crowther, A.K. (2006) 'Beyond the high-tech: early adopters of open innovation in other industries,' *R&D Management* 36(3): 229–36.

Contractor, F., Kumar, V., Kundu, S.K. and Pedersen, T. (2010) 'Reconceptualizing the firm in a world of outsourcing and offshoring: the organizational and geographical relocation of high-value company functions,' *Journal of Management Studies* 47(8): 1417–33.

Corrocher N. and Cusmano, L. (2014) 'The "KIBS engine" of regional innovation systems: empirical evidence from European regions,' *Regional Studies* 48(7): 1212–26.

den Hertog, P. (2000) 'Knowledge-intensive business services as co-producers of innovation,' *International Journal of Innovation Management* 4(4): 491–528.

den Hertog P. (2002) 'Co-producers of innovation: on the role of knowledge-intensive business services in innovation', in Gadrey, J. and Gallouj, F. (eds) *Productivity, innovation and knowledge in services: new economic and socio-economic approaches*, Cheltenham and Northampton: Edward Elgar, 223–55.

den Hertog, P. and Bilderbeek, R. (2000) 'The new knowledge-infrastructure: the role of technology-based knowledge-intensive business services in national innovation systems,' in Boden, M. and Miles, I. (eds) *Services and the knowledge-based economy*, London and New York: Continuum, 222–46.

Escribano, A., Fosfuri, A. and Tribó, J.A. (2009) 'Managing external knowledge flows: the moderating role of absorptive capacity,' *Research Policy* 38: 96–105.

Eurostat, *Innovation in high-tech sectors (CIS 2008, CIS 2010, CIS 2012)*, http://ec.europa.eu/eurostat/web/products-datasets/-/htec_cis6.

Fosstenløkken, S.M., Løwendahl, B.R. and Revang, Ø. (2003) 'Knowledge development through client interaction: a comparative study,' *Organization Studies* 24(6): 859–79.

Freel, M. (2006) 'Patterns of technological innovation in knowledge-intensive business services,' *Industry and Innovation* 13(3): 335–58.

Gallouj, F. (2002) 'Knowledge-intensive business services: processing knowledge and producing innovation,' in Gadrey, J. and Gallouj, F. (eds) *Productivity, innovation and knowledge in services. New economic and socio-economic approaches*, Cheltenham and Northampton: Edward Elgar, 256–84.

Geels, F.W. (2005) 'Processes and patterns in transitions and system innovations: refining the co-evolutionary multi-level perspective,' *Technological Forecasting and Social Change* 72(6): 681–96.

Godet, M. (2000) 'The art of scenarios and strategic planning: tools and pitfalls,' *Technological Forecasting and Social Change* 65(1): 3–22.

Greenwood, R., Li, S.X., Prakash, D.L. and Deephouse, D.L. (2005) 'Reputation, diversification and organizational explanations of performance in professional service firms,' *Organization Science* 16(6): 661–73.

Haller, S., Karnouskos, S. and Schroth, C. (2009) 'The internet of things in an enterprise context,' in Dominique, J., Fensel, D. and Traverso, P. (eds) *First Future Internet Symposium – FIS 2008, LNCS 5468*, Springer Verlag, 14–28.

Hargadon, A. B., and Bechky, B. A. (2006) 'When collections of creatives become creative collectives: a field study of problem-solving at work,' *Organization Science* 17(4): 484–500.

Havas, A., Schartinger, D. and Weber, M. (2010) 'The impact of foresight on innovation policy-making: recent experiences and future perspectives,' *Research Evaluation* 19(2): 91–104.

Hipp, C. and Grupp, H. (2005) 'Innovation in the service sector: the demand for service-specific innovation measurement concepts and typologies,' *Research Policy* 34(4): 517–35.

HLG (2014) *High Level Group on business services*, European Commission, http://ec.europa.eu/growth/industry/policy/renaissance/high–level–group/index_en.htm

Hoover, W.E., Eloranta, E., Holmström, J. and Huttunen, K. (2001) *Managing the demand-supply chain – value innovations for customer satisfaction*, New York: John Wiley & Sons, Inc.

Howells, J. and Roberts, J. (2000) 'Global knowledge systems in a service economy,' in Andersen, B., Howells, J., Hull, B., Miles, I. and Roberts, J. (eds) *Knowledge and innovation in the new service economy*, Cheltenham and Northampton: Edward Elgar, 248–66.

Janssen, M., Charalabidis, Y. and Zuiderwijk, A. (2012) 'Benefits, adoption barriers and myths of open data and open government,' *Information Systems Management* 29: 258–68.

Kiely, J., Beamisch, N. and Armistead, C. (2004) 'Scenarios for future service encounters,' *Service Industries Journal* 24(3): 131–49.

Kox, H. (2002) *Growth challenges for the Dutch business services industry. International comparison and policy issues*, CPB Netherlands Bureau for Economic Policy Analysis.

Kuusi, O. (1999) *Expertise in the future use of generic technologies*, Government Institute for Economic Research – VATT, Research Reports 59, Helsinki.

Kuusisto, J. and Päällysaho, S. (2008) 'Intellectual property protection as a key driver of service innovation: an analysis of innovative KIBS businesses in Finland and the UK,' *International Journal of Services Technology and Management* 9(3/4): 268–84.

Leiponen, A. (2001) *Knowledge services in the innovation system*, ETLA, The Research Institute of the Finnish Economy, Series B 185, Helsinki.

Leminen, S., Rajahonka, M., Westerlund, M. and Siuruainen, R. (2014) 'Ecosystem business models for the internet of things,' Proceedings of the XXIV RESER conference, 11–13th September 2014, Helsinki, Finland.

Lusch R. F. and Vargo S. L. (2006) 'Service-dominant logic: reactions, reflections and refinements,' *Marketing Theory* 6(3): 281–88.

Macpherson (1997) 'The role of producer service outsourcing in the innovation performance of New York State manufacturing firms,' *Annals of the Association of American Geographers* 87(1): 52–71.

Mcdonalda, R., Kareiva, P. and Forman, R. (2008) 'The implications of current and future urbanization for global protected areas and biodiversity conservation,' *Biological Conservation* 141: 1695–1703.

Miles, I. (2000) 'Services innovation: coming of age in the knowledge-based economy,' *International Journal of Innovation Management* 4: 371–89.

Miles, I. (2005) 'Knowledge intensive business services: prospects and policies,' *Foresight* 7: 39–63.

Miles, I., Kastrinos, N., Flanagan, K., Bilderbeek, R., Hertog, B., Huntink, W. and Bouman, M. (1995) *Knowledge-intensive business services: users, carriers and sources of innovation*, EIMS Publication No. 15, Luxembourg.

Miller, D.J., Fern M.J. and Cardinal, L.B. (2007) 'The use of knowledge for technological innovation within diversified firms,' *Academy of Management Journal* 50(2): 307–25.

Miozzo, M. and Miles, I. (2003) 'Introduction,' in Miozzo, M. and Miles, I. (eds) *Internationalization, technology and services*, Cheltenham and Northampton: Edward Elgar, 1–11.
Muller, E. and Doloreux, D. (2009) 'What we should know about knowledge–intensive business services,' *Technology in Society* 31: 64–72.
Muller, E. and Zenker, A. (2001) 'Business services as actors of knowledge transformation: the role of KIBS in regional and national innovation systems,' *Research Policy* 30: 1501–16.
OECD (1999) *Business services: trends and issues*, Paris: OECD.
OECD (2006) *Innovation and knowledge-intensive service activities*, Paris: OECD.
Pedersen, J., Kocsis, D., Tripathi, A., Tarrell, A., Weerakoon, A., Tahmasbi, N., Xiong, J., Deng, W., Oh, O. and Vreede, G. (2013) 'Conceptual foundations of crowdsourcing: a review of IS research,' in Proceedings of the 46th International Conference on System Sciences, 2013, Hawaii, Maui, USA, pp. 579–88.
Popper, R. (2008) 'How are foresight methods selected?' *Foresight* 10(6): 62–89.
Porter, M.E. (2000) 'Location, competition, and economic development: local clusters in a global economy,' *Economic Development Quarterly* 14(1): 15–34.
Prahalad, C.K., Ramaswamy, V. (2004) *The future of competition: co-creating unique value with customers*, Boston: Harvard Business School Press.
Preissl, B. (2000) 'Service Innovation: What makes it different? Empirical evidence from Germany,' in Metcalfe, J.S. and Miles, I. (eds) *Innovation systems in the service economy. measurement and case study analysis*, Boston, Dordrecht and London: Kluwer Academic Publishers, 125–48.
Ritala, P., Armila, L. and Blomqvist, K. (2009) 'Innovation orchestration capability – defining the organizational and individual level determinants,' *International Journal of Innovation Management* 13(4): 569–91.
Santos-Vijande, M.L., González-Mieres, C. and López-Sánchez, J.A. (2013) 'An assessment of innovativeness in KIBS: implications on KIBS' co-creation culture, innovation capability, and performance,' *Journal of Business & Industrial Marketing* 28(2): 86–102.
Shearmur, R. and Doloreux, D. (2014) 'Knowledge-intensive business services (KIBS) use and user innovation: high-order services, geographic hierarchies and internet use in Quebec's manufacturing sector,' *Regional Studies*, doi:10.1080/00343404.2013.870988
Simmie, J. and Strambach, S. (2006) 'The contribution of KIBS to innovation in cities: an evolutionary and institutional perspective,' *Journal of Knowledge Management* 10(5): 26–40.
Strambach S. (2001) 'Innovation processes and the role of knowledge-intensive business services (KIBS),' in Koschatzky, K., Kulicke, M. and A Zenker, A. (eds) *Innovation networks: concepts and challenges in the European perspective*. Technology, Heidelberg: Physica-Verlag, 53–68.
Strambach, S. (2008) 'Knowledge-intensive business services (KIBS) as drivers of multilevel knowledge dynamics,' *International Journal of Services technology and Management* 10 (2/3/4): 152–74.
Suddaby, R. and Greenwood, R. (2001) 'Colonizing knowledge: commodification as a dynamic of jurisdictional expansion in professional service firms,' *Human Relations* 54(7): 933–53.
Toivonen, M. (2007) 'Innovation policy in services – the development of knowledge-intensive business services (KIBS) in Finland,' *INNOVATION: Management, Policy and Practice* 9(3–4): 249–261.
Toivonen, M., Brax, S. and Tuominen, T. (2008) 'Client-oriented multicompetence – the core asset in KIBS,' *International Journal of Services Technology and Management* 10(2–4): 175–89.

Tomlinson, M. (2001) 'A new role for business services in economic growth,' in Archibugi, D. and Lundvall, B Å. (eds) *The globalizing learning economy*, Oxford: Oxford University Press, 97–107.

Van de Klundert, A. and Anschutz, J. (2001) 'Integrated sustainable waste management – the concept; tools for decision-makers – experiences from the urban waste expertise programme (1005–2001),' WASTE, Gouda, The Netherlands.

Vargo, S. and Lusch, R. (2011) 'It's all B2B…and beyond: toward a systems perspective of the market,' *Industrial Marketing Management* 40: 181–87.

Wagner, S. Hoisl, K. and Thoma, G. (2014) 'Overcoming localization of knowledge – the role of professional service firms,' *Strategic Management Journal* 35: 1671–88.

Westerlund, M., Leminen, S. and Rajahonka, M. (2014) 'Designing business models for the internet of things,' *Technology Innovation Management Review* July: 5–14.

Windrum P. and Tomlinson M. (1999) 'Knowledge-intensive services and international competitiveness: a four country comparison,' *Technology Analysis & Strategic Management* 11(3): 391–408.

Wolpert, J.D. (2002) 'Breaking out of the innovation box,' *Harvard Business Review* 80: 77–83.

Wood, P. (2002) 'How may consultancies be innovative,' in Wood, P. (ed) *Consultancy and innovation – the business service revolution in Europe*, London and New York: Routledge, 72–89.

Wright, G. and Goodwin, P. (2009) 'Decision making and planning under low levels of predictability: enhancing the scenario method,' *International Journal of Forecasting* 25: 813–25.

Zhang, Y. and Li, H. (2010) 'Innovation search of new ventures in a technology cluster: the role of ties with service intermediaries,' *Strategic Management Journal* 31: 88–109.

Index